Data Analysis

Data Analysis

Edited by
Gérard Govaert

First published in France in 2003 by Hermes Science/Lavoisier entitled: *Analyse des données*
© LAVOISIER, 2003
First published in Great Britain and the United States in 2009 by ISTE Ltd and John Wiley & Sons, Inc.

ISTE Ltd
27-37 St George's Road
London SW19 4EU
UK

www.iste.co.uk

John Wiley & Sons, Inc.
111 River Street
Hoboken, NJ 07030
USA

www.wiley.com

© ISTE Ltd, 2009

Library of Congress Cataloging-in-Publication Data

Analyse des données. English.
 Data analysis / edited by Gérard Govaert.
     p. cm.
 Includes bibliographical references and index.
 ISBN 978-1-84821-098-1
 1. Mathematical statistics. I. Govaert, Gérard. II. Title.
 QA276.D325413 2009
 519.5--dc22
                                                                                          2009016228

British Library Cataloguing-in-Publication Data
A CIP record for this book is available from the British Library
ISBN: 978-1-84821-098-1

Printed and bound in Great Britain by CPI/Antony Rowe, Chippenham and Eastbourne.

Contents

Preface

Statistical analysis has traditionally been separated into two phases: an exploratory phase, drawing on a set of descriptive and graphical techniques, and a decisional phase, based on probabilistic models. Some of the tools employed as part of the exploratory phase belong to *descriptive statistics*, whose elementary exploratory methods consider only a very limited number of variables. Other tools belong to *data analysis*, the subject matter of this book. This topic comprises more elaborate exploratory methods to handle multidimensional data, and is often seen as stepping beyond a purely exploratory context.

The first part of this book is concerned with methods for obtaining the pertinent dimensions from a collection of data. The variables so obtained provide a synthetic description, often leading to a graphical representation of the data. A considerable number of methods have been developed, adapted to different data types and different analytical goals. Chapters 1 and 2 discuss two reference methods, namely Principal Components Analysis (PCA) and Correspondence Analysis (CA), which we illustrate with examples from statistical process control and sensory analysis. Chapter 3 looks at a family of methods known as Projection Pursuit (less well known, but with a promising future), that can be seen as an extension of PCA and CA, which makes it possible to specify the structures that are being sought. Multidimensional positioning methods, discussed in Chapter 4, seek to represent proximity matrix data in low-dimensional Euclidean space. Chapter 5 is devoted to functional data analysis where a function such as a temperature or rainfall graph, rather than a simple numerical vector, is used to characterize individuals.

The second part is concerned with methods of clustering, which seek to organize data into homogenous classes. These methods provide an alternative means, often complementary to those discussed in the first part, of synthesizing and analyzing data. In view of the clear link between clustering and discriminant analysis – in pattern recognition the former is termed unsupervised and the latter supervised learning – Chapter 6 gives a general introduction to discriminant analysis. Chapter 7 then provides an overall picture of clustering. The statistical interpretation of clustering in terms of mixtures of probability distributions is discussed in Chapter 8 and Chapter 9 looks at how this approach can be applied to spatial data.

I would like to express my heartfelt thanks to all the authors who were involved in this publication. Without their expertise, their professionalism, their invaluable contributions and the wealth of their experience, it would not have been possible.

Gérard GOVAERT

Chapter 1

Principal Component Analysis: Application to Statistical Process Control

1.1. Introduction

Principal component analysis (PCA) is an exploratory statistical method for graphical description of the information present in large datasets. In most applications, PCA consists of studying p variables measured on n individuals. When n and p are large, the aim is to synthesize the huge quantity of information into an easy and understandable form.

Unidimensional or bidimensional studies can be performed on variables using graphical tools (histograms, box plots) or numerical summaries (mean, variance, correlation). However, these simple preliminary studies in a multidimensional context are insufficient since they do not take into account the eventual relationships between variables, which is often the most important point.

Principal component analysis is often considered as the basic method of factor analysis, which aims to find linear combinations of the p variables called components used to visualize the observations in a simple way. Because it transforms a large number of correlated variables into a few uncorrelated principal components, PCA is a dimension reduction method. However, PCA can also be used as a multivariate outlier detection method, especially by studying the last principal components. This property is useful in multidimensional quality control.

Chapter written by Gilbert SAPORTA and Ndèye NIANG.

1.2. Data table and related subspaces

1.2.1. *Data and their characteristics*

Data are generally represented in a rectangular table with n rows for the individuals and p columns corresponding to the variables. Choosing individuals and variables to analyze is a crucial phase which has an important influence on PCA results. This choice has to take into account the aim of the study; in particular, the variables have to describe the phenomenon being analyzed.

Usually PCA deals with numerical variables. However, ordinal variables such as ranks can also be processed by PCA. Later in this chapter, we present the concept of supplementary variables which afterwards integrates nominal variables.

1.2.1.1. *Data table*

Let $\mathbf{X}$ be the (n, p) matrix of observations:

$$\mathbf{X} = \begin{pmatrix} x_1^1 & \cdots & x_1^p \\ \vdots & \vdots & \vdots \\ x_i^1 & x_i^j & x_i^p \\ \vdots & \vdots & \vdots \\ x_n^1 & \cdots & x_n^p \end{pmatrix}.$$

where x_i^j is the value of individual i for variable j (denoted $\mathbf{x}^j$) which is identified with a vector of n components $(x_1^j, \ldots, x_n^j)'$. In a similar way, an individual i is identified to a vector $\boldsymbol{x}_i$ of p components with $\boldsymbol{x}_i = (x_i^1, \ldots, x_i^p)'$.

Table 1.1 is an example of such a data matrix. Computations have been carried out using SPAD 5 software, version 5 [1], kindly provided by J.-P. Gauchi.

The data file contains 57 brands of mineral water described by 11 variables defined in Table 1.2. The data come from the bottle labels. Numerical variables are homogenous; they are all active variables (see section 1.4.3). A variable of a different kind such as price would be considered as a supplementary variable. On the other hand, qualitative variables such as country, type and whether still or sparkling (PG) are necessarily supplementary variables.

---

1. DECISIA (former CISIA-CERESTA), Building Hoche, 13 rue Auger, 93697 Pantin cedex.

Name	Country	Type	PG	CA	MG	NA	K	SUL	NO3	HCO3	CL
Evian	F	M	P	78	24	5	1	10	3.8	357	4.5
Montagne des Pyrénées	F	S	P	48	11	34	1	16	4	183	50
Cristaline-St-Cyr	F	S	P	71	5.5	11.2	3.2	5	1	250	20
Fiée des Lois	F	S	P	89	31	17	2	47	0	360	28
Volcania	F	S	P	4.1	1.7	2.7	0.9	1.1	0.8	25.8	0.9
Saint Diéry	F	M	G	85	80	385	65	25	1.9	1350	285
Luchon	F	M	P	26.5	1	0.8	0.2	8.2	1.8	78.1	2.3
Volvic	F	M	P	9.9	6.1	9.4	5.7	6.9	6.3	65.3	8.4
Alpes/Moulettes	F	S	P	63	10.2	1.4	0.4	51.3	2	173.2	1
Orée du bois	F	M	P	234	70	43	9	635	1	292	62
Arvie	F	M	G	170	92	650	130	31	0	2195	387
Alpes/Roche des Ecrins	F	S	P	63	10.2	1.4	0.4	51.3	2	173.2	10
Ondine	F	S	P	46.1	4.3	6.3	3.5	9	0	163.5	3.5
Thonon	F	M	P	108	14	3	1	13	12	350	9
Aix les Bains	F	M	P	84	23	2	1	27	0.2	341	3
Contrex	F	M	P	486	84	9.1	3.2	1187	2.7	403	8.6
La Bondoire Saint Hippolite	F	S	P	86	3	17	1	7	19	256	21
Dax	F	M	P	125	30.1	126	19.4	365	0	164.7	156
Quézac	F	M	G	241	95	255	49.7	143	1	1685.4	38
Salvetat	F	M	G	253	11	7	3	25	1	820	4
Stamna	GRC	M	P	48.1	9.2	12.6	0.4	9.6	0	173.3	21.3
Iolh	GR	M	P	54.1	31.5	8.2	0.8	15	6.2	267.5	13.5
Avra	GR	M	P	110.8	9.9	8.4	0.7	39.7	35.6	308.8	8
Rouvas	GRC	M	P	25.7	10.7	8	0.4	9.6	3.1	117.2	12.4
Alisea	IT	M	P	12.3	2.6	2.5	0.6	10.1	2.5	41.6	0.9
San Benedetto	IT	M	P	46	28	6.8	1	5.8	6.6	287	2.4
San Pellegrino	IT	M	G	208	55.9	43.6	2.7	549.2	0.45	219.6	74.3
Levissima	IT	M	P	19.8	1.8	1.7	1.8	14.2	1.5	56.5	0.3
Vera	IT	M	P	36	13	2	0.6	18	3.6	154	2.1
San Antonio	IT	M	P	32.5	6.1	4.9	0.7	1.6	4.3	135.5	1
La Française	F	M	P	354	83	653	22	1055	0	225	982
Saint Benoit	F	S	G	46.1	4.3	6.3	3.5	9	0	163.5	3.5
Plancoët	F	M	P	36	19	36	6	43	0	195	38
Saint Alix	F	S	P	8	10	33	4	20	0.5	84	37
Puits Saint Georges/Casino	F	M	G	46	33	430	18.5	10	8	1373	39
St-Georges/Corse	F	S	P	5.2	2.43	14.05	1.15	6	0	30.5	25
Hildon bleue	B	M	P	97	1.7	7.7	1	4	26.4	236	16
Hildon blanche	B	M	G	97	1.7	7.7	1	4	5.5	236	16
Mont Roucous	F	M	P	1.2	0.2	2.8	0.4	3.3	2.3	4.9	3.2
Ogeu	F	S	P	48	11	31	1	16	4	183	44
Highland Spring	B	M	P	35	8.5	6	0.6	6	1	136	7.5
Parot	F	M	G	99	88.1	968	103	18	1	3380.51	88
Vernière	F	M	G	190	72	154	49	158	0	1170	18
Terres de Flein	F	S	P	116	4.2	8	2.5	24.5	1	333	15
Courmayeur	IT	M	P	517	67	1	2	1371	2	168	1
Pyrénées	F	M	G	48	12	31	1	18	4	183	35
Puits Saint Georges/Monoprix	F	M	G	46	34	434	18.5	10	8	1373	39
Prince Noir	F	M	P	528	78	9	3	1342	0	329	9
Montcalm	F	S	P	3	0.6	1.5	0.4	8.7	0.9	5.2	0.6
Chantereine	F	S	P	119	28	7	2	52	0	430	7
18 Carats	F	S	G	118	30	18	7	85	0.5	403	39
Spring Water	B	S	G	117	19	13	2	16	20	405	28
Vals	F	M	G	45.2	21.3	453	32.8	38.9	1	1403	27.2
Vernand	F	M	G	33.5	17.6	192	28.7	14	1	734	6.4
Sidi Harazem	MO	S	P	70	40	120	8	20	4	335	220
Sidi Ali	MO	M	P	12.02	8.7	25.5	2.8	41.7	0.1	103.7	14.2
Montclar	F	S	P	41	3	2	0	2	3	134	3

Table 1.1. *Data table*

1.2.1.2. *Summaries*

1.2.1.2.1. Centroid

Let $\overline{x}$ be the vector of arithmetic means of each of the p variables, defining the centroid:

$$\overline{x} = (\overline{x}^1, \ldots, \overline{x}^p)'$$

Name	Complete water name as labeled on the bottle
Country	Identified by the official car registration letters; sometimes it is necessary to add a letter, for example Crete: GRC (Greece Crete)
Type	M for mineral water, S for spring water
PG	P for still water, G for sparkling water
CA	Calcium ions (mg/litre)
MG	Magnesium ions (mg/litre)
NA	Sodium ions (mg/litre)
K	Potassium ions (mg/litre)
SUL	Sulfate ions (mg/litre)
NO3	Nitrate ions (mg/litre)
HCO3	Carbonate ions (mg/litre)
CL	Chloride ions (mg/litre)

Table 1.2. *Variable description*

where $\overline{x}^j = \sum_{i=1}^n p_i x_i^j$.

If the data are collected following a random sampling, the n individuals all have the same importance in the computations of the sample characteristics. The same weight $p_i = 1/n$ is therefore allocated to each observation.

However, it can be useful for some applications to use weight p_i varying from one individual to another as grouped data or a reweighted sample. These weights, which are positive numbers summing to 1, can be viewed as frequencies and are stored in a diagonal matrix of size n:

$$\mathbf{D}_p = \begin{pmatrix} p_1 & & \\ & \ddots & \\ & & p_n \end{pmatrix}.$$

We then have the matrix expression $\overline{\boldsymbol{x}} = \mathbf{X}'\mathbf{D}_p\mathbf{1}_n$ where $\mathbf{1}_n$ represents the vector of $\mathbb{R}^n$ with all its components equal to 1. The centered data matrix associated with $\mathbf{X}$ is then $\mathbf{Y}$ with $y_i^j = x_i^j - \overline{x}^j$ and $\mathbf{Y} = \mathbf{X} - \mathbf{1}_n\overline{\boldsymbol{x}}' = (\mathbf{I}_n - \mathbf{1}_n\mathbf{1}_n'\mathbf{D}_p)\mathbf{X}$, where $\mathbf{I}_n$ is the unity matrix of dimension n.

1.2.1.2.2. Covariance matrix and correlation matrix

Let $s_j^2 = \sum_{i=1}^n p_i(x_i^j - \overline{x}^j)^2$ and $s_{k\ell} = \sum_{i=1}^n p_i(x_i^k - \overline{x}^k)(x_i^\ell - \overline{x}^\ell)$, the variance of variable j and the covariance between variables k and ℓ, respectively. They are stored in the covariance matrix $\mathbf{S} = \mathbf{X}'\mathbf{D}_p\mathbf{X} - \overline{\boldsymbol{x}}\overline{\boldsymbol{x}}' = \mathbf{Y}'\mathbf{D}_p\mathbf{Y}$.

We define the linear correlation coefficient between variables k and ℓ by:

$$r_{k\ell} = \frac{s_{k\ell}}{s_k s_\ell}.$$

If $\mathbf{Z}$ is the standardized data table associated with $\mathbf{X}$, $z_i^j = (x_i^j - \overline{x}^j)/s_j$, we have $\mathbf{Z} = \mathbf{Y}\mathbf{D}_{1/s}$ where $\mathbf{D}_{1/s}$ the diagonal matrix of the inverse of standard deviations:

$$\mathbf{D}_{1/s} = \begin{pmatrix} 1/s_1 & & \\ & \ddots & \\ & & 1/s_p \end{pmatrix}.$$

$\mathbf{R}$ is the correlation matrix containing the linear correlation coefficients between all pairs of variables; we have $\mathbf{R} = \mathbf{D}_{1/s}\mathbf{S}\mathbf{D}_{1/s}$. $\mathbf{R}$ is the covariance matrix of standardized variables. It summarizes linear dependency structure between the p variables.

Tables 1.3 and 1.4 list the numerical summaries associated with the dataset example.

Variable	Mean	Standard deviation	Minimum	Maximum
CA	102.46	118.92	1.20	528.00
MG	25.86	28.05	0.20	95.00
NA	93.85	195.51	0.80	968.00
K	11.09	24.22	0.00	130.00
SUL	135.66	326.31	1.10	1371.00
NO3	3.83	6.61	0.00	35.60
HCO3	442.17	602.94	4.90	3380.51
CL	52.47	141.99	0.30	982.00

Table 1.3. *Simple statistics for continuous variables*

	CA	MG	NA	K	SUL	NO3	HCO3	CL
CA	1.00							
MG	0.70	1.00						
NA	0.12	0.61	1.00					
K	0.13	0.66	0.84	1.00				
SUL	0.91	0.61	0.06	−0.03	1.00			
NO3	−0.06	−0.21	−0.12	−0.17	−0.16	1.00		
HCO3	0.13	0.62	0.86	0.88	−0.07	−0.06	1.00	
CL	0.28	0.48	0.59	0.40	0.32	−0.12	0.19	1.00

Table 1.4. *Correlation matrix*

1.2.2. *The space of statistical units*

The Pearson geometrical approach is based on a data cloud associated with the observations: each unit defined by p coordinates is then considered as an element of a vector space of p dimensions, referred to as the space of statistical units. The centroid $\overline{x}$ defined in section 1.2.1.2 is then the barycenter of the data cloud.

PCA consists of visualizing the most reliable data cloud possible within a low dimensional space. The analysis is based on distances between points representing the individuals. The method by which these distances are computed influences the results to a large extent. It is therefore essential to determine it before any analysis.

1.2.2.1. 'The metric'

In the usual 3D physical space, computing a distance is simple using Pythagoras' formula. However, in statistics, the problem is more complicated: how can distances between individuals described by variables having measurement units as different as euros, kg, km, etc. be calculated? The Pythagoras formula is as arbitrary as any other. The following general formulation therefore has to be used: $\mathbf{M}$ is a positive definite symmetric matrix of size p and the distance between two individual $\boldsymbol{x}_i$ and $\boldsymbol{x}_j$ is defined by the quadratic form:

$$d^2(\boldsymbol{x}_i, \boldsymbol{x}_j) = (\boldsymbol{x}_i - \boldsymbol{x}_j)'\mathbf{M}(\boldsymbol{x}_i - \boldsymbol{x}_j) = d^2(i, j).$$

In theory, the choice of $\mathbf{M}$ depends on the user who is the only one to precisely determine the adequate metric. In practice, however, the usual metrics in PCA are $\mathbf{M} = \mathbf{I}_p$ if the variances are not too different and are expressed in the same measurement unit; otherwise, the metric $\mathbf{M} = \mathbf{D}_{1/s^2}$, the diagonal matrix of the variance inverses, is preferred. This latter metric is the most used (default option in many PCA programs) because as well as suppressing the measurement units, it gives the same importance in the computation of distances to each variable, whatever its variance. Using this metric is equivalent to standardizing the variables, setting them dimensionless and setting them all the same variance of 1. In the example, the variable standard deviations are very different (Figure 1.3). The variables will then be standardized.

REMARK 1.1.– Every symmetric positive matrix $\mathbf{M}$ can be written as $\mathbf{M} = \mathbf{T}\mathbf{T}'$. We therefore have: $\boldsymbol{x}_i'\mathbf{M}\boldsymbol{x}_j = \boldsymbol{x}_i'\mathbf{T}'\mathbf{T}\boldsymbol{x}_j = (\mathbf{T}\boldsymbol{x}_i)'(\mathbf{T}\boldsymbol{x}_j)$. It is then possible to use $\mathbf{X}$ and the metric $\mathbf{M}$ rather than the identity metric $\mathbf{I}_p$ and the transformed data matrix $\mathbf{X}\mathbf{T}'$. PCA usually consists of standardizing the variables and using the identity metric $\mathbf{I}_p$, referred to as standardized PCA.

1.2.2.2. Inertia

Inertia is a fundamental notion of PCA. The total inertia of a data cloud is the weighted mean of square distances between points and the centroid. It represents the dispersion of the data cloud around the barycenter. Note that

$$I_g = \sum_{i=1}^{n} p_i(\boldsymbol{x}_i - \overline{\boldsymbol{x}})'\mathbf{M}(\boldsymbol{x}_i - \overline{\boldsymbol{x}}) = \sum_{i=1}^{n} p_i\|\boldsymbol{x}_i - \overline{\boldsymbol{x}}\|^2.$$

It can be shown that the inertia around a particular point, defined by:

$$I_a = \sum_{i=1}^{n} p_i(\boldsymbol{x}_i - \mathbf{a})'\mathbf{M}(\boldsymbol{x}_i - \mathbf{a}),$$

may be written according to Huyghens formula:

$$I_a = I_g + (\overline{x} - a)'M(\overline{x} - a) = I_g + ||\overline{x} - a||^2.$$

It can also be shown that twice the total inertia is equal to the average of all of the pairs of square distances between the n individuals. However, the most used equation is:

$$I = \text{tr}(MS) = \text{tr}(SM).$$

If $M = I_p$, the inertia is then equal to the sum of the p variances. In the case of the metric $M = D_{1/s^2}$, the inertia equals the trace of the correlation matrix i.e. p, the number of variables. The inertia then does not depend on the variables values but only on their number.

In the following chapter, this last case will be considered. For PCA with a general metric M, see the books of Saporta [SAP 06] or Lebart *et al.* [LEB 06].

1.2.3. *Variables space*

Each variable is defined by n coordinates; it is then considered as a vector of a space of n dimensions referred to as variable space. To compute the 'distances' between variables, we use D_p, the diagonal weight matrix, which has (in case of zero-mean variables) the following properties:

– the scalar product between two variables x^k and x^ℓ is

$$(x^k)'D_p x^\ell = \sum_{i=1}^{n} p_i x_i^k x_i^\ell$$

which is the covariance $v_{k\ell}$;

– the square norm of a variable is then equal to its variance

$$||x^j||_{D_p}^2 = s_j^2$$

and the standard deviation represents the variable 'length';

– by denoting the angle between two variables as $\theta_{k\ell}$, we have

$$\cos\theta_{k\ell} = \frac{< x^k, x^\ell >}{||x^k|| \cdot ||x^\ell||} = \frac{v_{k\ell}}{s_k s_\ell} = r_{k\ell},$$

which is the linear correlation coefficient.

In the variables space, we are interested in angles rather than distances and the variables will be represented as vectors rather than points.

1.3. Principal component analysis

1.3.1. *The method*

Recall that the purpose of PCA is to find synthetic representations of large numerical datasets, in particular by using 2D plots. If the initial spaces of statistical units and variables representation have too many dimensions, it is impossible to visualize the data cloud. We therefore look for spaces with few dimensions best fitting the data cloud, that is, those which save the best initial cloud configuration.

The method consists of projecting the data cloud in order to minimize the shrinkage of the distances which are inherent to the projection. This is equivalent to choosing the projection space F which maximizes the criterion:

$$\sum_{i=1}^{n}\sum_{j=1}^{n} p_i p_j d^2(i,j).$$

The subspace we look for is such that the average of the square distances between projections is maximal (the projection reduces distances); in other words, the inertia of projections cloud has to be maximal. It is shown [SAP 06] that the search of the subspace F can be sequential: first we look for the 1D subspace with maximal inertia then we look for the 1D subspace orthogonal to this with maximal inertia, and so on.

1.3.2. *Principal factors and principal components*

We begin by looking for a 1D subspace i.e. a straight line defined by a unit vector $\mathbf{u} = (u_1, \ldots, u_p)'$. As explained in the previous section, the vector has to be defined such that the points projected onto its direction have maximal inertia. The projection, or coordinate c_i, of an individual i onto Δ is defined by: $c_i = \sum_{j=1}^{p} \mathbf{x}_i^j u_j$ (Figure 1.1).

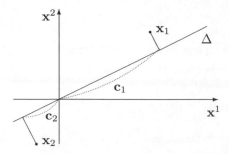

Figure 1.1. *Projection onto direction* Δ

The list of the individual coordinates c_i on Δ forms a new artificial variable $\mathbf{c} = (c_1, \ldots, c_n)' = \sum_{j=1}^{p} \mathbf{x}^j u_j = \mathbf{Xu}$; it is a linear combination of the original variables. The inertia (or variance) of points projected onto Δ is then:

$$\text{Var}(\mathbf{c}) = \sum_{i}^{n} p_i c_i^2 = \mathbf{c}' \mathbf{D}_p \mathbf{c} = \mathbf{u}' \mathbf{X}' \mathbf{D}_p \mathbf{Xu} = \mathbf{u}' \mathbf{Su}.$$

Recall that the usual case of standardized PCA is considered; the covariance matrix of standardized data then corresponds to the correlation matrix $\mathbf{R}$. The criterion of maximizing the inertia of projected points onto Δ is then written as:

$$\max_{\mathbf{u}} \mathbf{u}' \mathbf{Su} = \max_{\mathbf{u}} \mathbf{u}' \mathbf{Ru}$$

under the constraint $\mathbf{u}'\mathbf{u} = 1$.

The solution of this quadratic maximization problem is $\mathbf{u}_1$, the eigenvector of $\mathbf{R}$ associated with the largest eigenvalue λ_1. We then search the vector $\mathbf{u}_2$ orthogonal to $\mathbf{u}_1$ such that the inertia of points projected onto this direction is maximal. Similarly, it is shown that $\mathbf{u}_2$ is the eigenvector of $\mathbf{R}$ associated with the second largest eigenvalue λ_2. More generally, the subspace of q dimensions which we are looking for is spanned by the first q eigenvectors of the matrix $\mathbf{R}$ associated with the largest eigenvalues.

Vectors $\mathbf{u}_j$ are called *principal factors*. They contain the coefficients to be applied to the original variables in the linear combination $\mathbf{c} = \mathbf{Xu}$.

Principal components are artificial variables defined by principal factors: $\mathbf{c}^j = \mathbf{Xu}_j$; they contain the coordinates of the orthogonal projections of individuals onto the axes defined by the $\mathbf{u}_j$.

In practice, PCA will consist of diagonalizing the $\mathbf{R}$ matrix to obtain the $\mathbf{u}_j$ and computing the principal components $\mathbf{c}^j = \mathbf{Xu}_j$.

See Tables 1.5 and 1.6 for the example results.

Number	Eigenvalues	%	% cumulated
1	3.8168	47.71	47.71
2	2.0681	25.85	73.56
3	0.9728	12.16	85.72
4	0.7962	9.95	95.67
5	0.1792	2.24	97.91
6	0.0924	1.16	99.07
7	0.0741	0.93	100.00
8	0.0004	0.00	100.00

Table 1.5. *Eigenvalues λ_1, λ_2, etc.*

Variables	Eigenvectors				
	1	2	3	4	5
CA	-0.28	-0.54	-0.17	-0.20	0.01
MG	-0.47	-0.18	-0.04	-0.17	-0.46
NA	-0.44	0.29	-0.03	0.17	0.62
K	-0.43	0.32	0.01	-0.12	-0.45
SUL	-0.23	-0.60	-0.03	-0.03	0.33
NO3	0.12	0.06	-0.97	0.15	-0.06
HCO3	-0.40	0.35	-0.13	-0.35	0.24
CL	-0.32	-0.07	0.07	0.86	-0.15

Table 1.6. *Eigenvectors*

1.3.3. *Principal factors and principal components properties*

1.3.3.1. *Principal component variance*

The variance of a principal component is equal to the eigenvalue λ: $\mathrm{Var}(\mathbf{c}^j) = \lambda_j$. The variance of $\mathbf{u}$ is defined by $\mathbf{Su} = \mathbf{Ru}$, $\mathbf{u'u} = 1$ and:

$$\mathrm{Var}(\mathbf{c}) = \mathbf{c'D}_p\mathbf{c} = \mathbf{u'X'D}_p\mathbf{Xu} = \mathbf{u'Su} = \mathbf{u'Ru} = \mathbf{u'}(\lambda\mathbf{u}) = \lambda\mathbf{u'u} = \lambda.$$

The principal components are therefore linear combinations of original variables with maximal variance.

1.3.3.2. *A maximal association property*

The variable $\mathbf{c}^1$ has the greatest link to $\mathbf{x}^j$ in the sense of the square correlations sum: $\sum_{j=1}^{p} r^2(\mathbf{c}, \mathbf{x}^j)$ is maximal. It is shown [SAP 06] that:

$$\sum_{j=1}^{p} r^2(\mathbf{c}, \mathbf{x}^j) = \frac{\mathbf{c'D}_p\mathbf{ZZ'D}_p\mathbf{c}}{\mathbf{c'D}_p\mathbf{c}}$$

where $\mathbf{Z}$ is the standardized data table. The maximum of this ratio is reached when $\mathbf{c}$ is the eigenvector of $\mathbf{ZZ'D}_p$ associated with its largest eigenvalue: $\mathbf{ZZ'D}_p\mathbf{c} = \lambda\mathbf{c}$.

The principal component $\mathbf{c}$ is then a linear combination of the columns of $\mathbf{Z}$: $\mathbf{c} = \mathbf{Zu}$ and then $\mathbf{ZZ'D}_p\mathbf{c} = \lambda\mathbf{c}$ becomes $\mathbf{ZZ'D}_p\mathbf{Zu} = \lambda\mathbf{Zu}$. Since we have $\mathbf{Z'D}_p\mathbf{Z} = \mathbf{R}$ and $\mathbf{ZRu} = \lambda\mathbf{Zu}$ and, if the rank of $\mathbf{Z}$ is p, we obtain $\mathbf{Ru} = \lambda\mathbf{u}$.

1.3.3.3. *Reconstitution formula*

Post-multiplying both members of $\mathbf{Xu}_j = \mathbf{c}^j$ by $\mathbf{u}'_j$ and summing over j, we have

$$\mathbf{X}\sum_{j=1}^{p} \mathbf{u}_j\mathbf{u}'_j = \sum_{j=1}^{p} \mathbf{c}^j\mathbf{u}'_j.$$

It can easily be shown that $\sum_{j=1}^{p} \mathbf{u}_j\mathbf{u}'_j = \mathbf{I}_p$ since the $\mathbf{u}_j$ are orthonormal. We then find $\mathbf{X} = \sum_{j=1}^{p} \mathbf{c}^j\mathbf{u}'_j$. The centered data table may be reconstituted using factors and principal components. If we only use the first q terms corresponding to the first q largest eigenvalues, we have the best approximation of $\mathbf{X}$ by a matrix of rank q in the least-squares sense (Eckart–Young theorem).

To summarize, it can be said that PCA consists of transforming original correlated variables $\mathbf{x}^j$ into new variables, the principal components $\mathbf{c}^j$, which are uncorrelated linear combinations of the $\mathbf{x}^j$ with maximal variance and with the greatest link to the $\mathbf{x}^j$. PCA is therefore a linear factorial method.

Non-linear extensions of PCA exist: we look for variable transformations, for example, by splines [DEL 88] available in some software (prinqual procedure in SAS).

1.4. Interpretation of PCA results

PCA provides graphical representations allowing the visualization of relations between variables and the eventual existence of groups of individuals and groups of variables. PCA results are 2D figures and tables. Their interpretation is the most delicate phase of the analysis and has to be carried out according to a precise scheme to be explained later.

Before beginning the interpretation itself, it is useful to start with a brief preliminary reading of the results in order to roughly verify the dataset contents. It is possible that by examining the first principal plane, we observe some individuals completely outside the rest of the population. This implies either (1) the presence of erroneous data such as typing errors or measurement error which have to be corrected, or (2) individuals totally different from others which must be removed from the analysis to better observe the remaining individuals (they can be reintroduced afterwards as supplementary elements).

After this preliminary study, PCA results can then be examined more closely; we begin with the interpretation phase which consists of several stages.

REMARK 1.2.– Although simultaneous representations of individuals and variables called 'biplot' [GOW 96] exist, we recommend representing the set separately in order to avoid confusion.

1.4.1. *Quality of representations onto principal planes*

PCA allows us to obtain graphical representation of individuals in a space of fewer dimensions than p, but this representation is only a deformed vision of the reality. One of the most crucial points in interpreting the results of PCA consists of appreciating this deformation (or, in other words, the loss of information due to the dimension reduction) and in determining the number of axes to retain.

The criterion usually employed to measure PCA quality is the percentage of total inertia explained. It is defined:

$$\frac{\lambda_1 + \lambda_2 + \ldots + \lambda_k}{\lambda_1 + \lambda_2 + \ldots + \lambda_k + \ldots + \lambda_p} = \frac{\lambda_1 + \lambda_2 + \ldots + \lambda_k}{I_g}.$$

This is a global measure which has to be completed with other considerations. First, the number of variables must be taken into account: a 10% inertia does not have the same interest for a 20-variable table or for a 100-variable table.

Second, at the individual level, it is necessary to look at the reliability of the representation of each individual, independently of the global inertia percentage. It is possible to have a first principal plane with a large total inertia and to find that two individuals, far from each other in the full space, have very close projections (Figure 1.2).

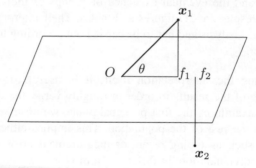

Figure 1.2. *Close projections of distant points*

The most widely used measure of an individual representation quality is the cosine of the angle between the principal plane and the vector x_i. If this cosine is large, x_i is close to the plane and we can then examine the position of its projection onto the plane with respect to other points; if the cosine is small, we will be wary of any conclusion.

1.4.2. *Axis selection*

Axis selection is an essential point of PCA but has no rigorous solution. There are theoretical criteria based on statistical tests or eigenvalue confidence intervals but the latter are useful only for non-standardized PCA and the p-dimensional Gaussian case. In the most frequent practical case of correlation matrices, only empirical criteria are applicable. The best known is the Kaiser rule: for standardized data, principal components corresponding to eigenvalues larger than 1 are retained; this means only components which 'bring' more than original variables are of interest.

It is also usual to employ a scree test, which consists of detecting the existence of a significant decay on the eigenvalues diagram. This is not always easy in practice, however.

In the example we use the Kaiser rule combined with the eigenvalues diagram (see Table 1.7). A break is detected after the second eigenvalue and we retain two axes corresponding to 73.56% explained inertia. The third axis is easily interpreted, but as it is identified with the variable NO3 (correlation –0.96) and is not correlated with other variables, it is not of great interest.

Number	Eigenvalues	%	% cumulated	
1	3.8168	47.71	47.71	**
2	2.0681	25.85	73.56	**
3	0.9728	12.16	85.72	**********************
4	0.7962	9.95	95.67	******************
5	0.1792	2.24	97.91	****
6	0.0924	1.16	99.07	**
7	0.0741	0.93	100.00	**
8	0.0004	0.00	100.00	*

Table 1.7. *Eigenvalues scree plot*

1.4.3. *Internal interpretation*

PCA results are obtained from variables and individuals called active elements, in contrast to supplementary elements which do not participate directly in the analysis. Active variables and individuals are used to compute principal axes; supplementary variables and individuals are then projected onto these axes.

Active variables (numerical) are those with interesting intercorrelations: they are the main variables of the study. Supplementary variables provide useful information for characterizing the individuals but are not directly used in the analysis. We are only interested by the correlations of supplementary variables and active variables via principal components, and not by the correlations between supplementary variables. Internal interpretation consists of analyzing the results using active variables and individuals. Supplementary elements study is carried out in the external interpretation phase.

1.4.3.1. *Variables*

PCA yields principal components, which are new artificial variables defined by linear combinations of original variables. We must be able to interpret these principal components (according to original variables). This is done simply through the computations of linear correlation coefficients $r(\mathbf{c}, \mathbf{x}^j)$ between principal components and original variables. The largest coefficients (in absolute value) close to 1 are those of interest (see Table 1.8). In standard PCA, we use standardized data and the computation of $r(\mathbf{c}, \mathbf{x}^j)$ is particularly simple. It may be shown that $r(\mathbf{c}, \mathbf{x}^j) = \sqrt{\lambda}\mathbf{u}_j$.

Variables	Coordinates				
	1	2	3	4	5
CA	−0.55	−0.78	−0.17	−0.18	0.01
MG	−0.91	−0.25	−0.04	−0.15	−0.20
NA	−0.86	0.41	−0.03	0.15	0.26
K	−0.84	0.46	0.01	−0.11	−0.19
SUL	−0.45	−0.87	−0.03	−0.03	0.14
NO3	0.23	0.09	0.96	0.13	−0.03
HCO3	−0.78	0.50	−0.13	−0.31	0.10
CL	−0.62	−0.10	0.07	0.77	−0.06

Table 1.8. *Variable-factor correlations or variables coordinates*

Usually the correlation between the variables for a couple of principal components is synthesized on a graph called the 'correlation display' on which each variable x^j is positioned by abscissa $r(c^1, x^j)$ and ordinate $r(c^2, x^j)$. Analyzing the correlation display allows detection of possible groups of similar variables or opposite groups of variables having different behavior, giving a sense of principal axes.

In the example (Figure 1.3), axis 1 is negatively correlated to all the variables (except NO3 which is not significant). Observations with the largest negative coordinate on the horizontal axis correspond to water with the most important mineral concentrations. Along the vertical axis, waters with high calcium and sulfate concentration are in opposition with waters with high potassium and carbonate concentration.

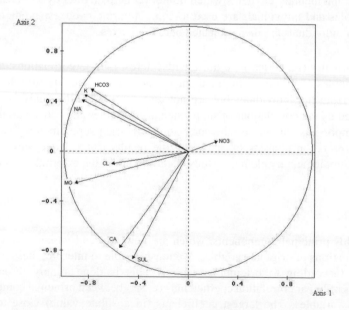

Figure 1.3. *Variable representation onto plane 1, 2*

REMARK 1.3.– When all original variables are positively correlated with each other, the first principal component defines a 'size effect'. It is known that a symmetric matrix with all terms positive has a first eigenvector with all its components having the same sign, which can be chosen to be positive. Then the first principal component is positively correlated to all original variables and individuals are ranked along axis 1 according to the increase of all variables (on average). The second principal component distinguishes individuals with similar 'size' and is referred to as the 'shape factor'.

1.4.3.2. *Observations*

Interpreting observations consists of examining their coordinates and especially their resulting graphical representations, referred to as principal planes (Figure 1.4). The aim is to see how the observations are scattered, which observations are similar and which observations differ from the others. In case of non-anonymous observations, they can then help to interpret principal axes; for example, we will look for opposite individuals along an axis.

Conversely, using results of variables analysis allows interpretation of observation. When, for example, the first component is highly correlated with an original variable, it means that individuals with large positive coordinates along axis 1 are characterized by this variable of value much larger than average (the origin of the axes represents the centroid of data cloud).

In observation study, it is also very useful to look at the individual contributions to each axis for help in interpreting axes. Contribution is defined by $\frac{p_i(c_i^k)^2}{\lambda_k}$, where c_i^k represents the value for individual i of the kth component $\mathbf{c}^k$ and $\lambda_k = \sum_{i=1}^n p_i(c_i^k)^2$.

The important contributions are those that exceed observation weight. However, it is necessary to be careful when an individual has an excessive contribution which can produce instability. Removing it can highly modify the analysis results. The analysis should be made without this point, which can be added as a supplementary element. Observations such as ARVIE and PAROT (Figure 1.4) are examples of such points.

It should be noted that, for equal weight, contributions do not provide more information than coordinates. Table 1.9 lists coordinates, contributions and square cosines of angles with principal axes which allow the evaluation of the quality of the representation (see section 1.4.1).

1.4.4. *External interpretation: supplementary variables and individuals*

Recall that supplementary elements do not participate in the computations of principal axes, but are very useful afterwards in consolidating and enriching the interpretation.

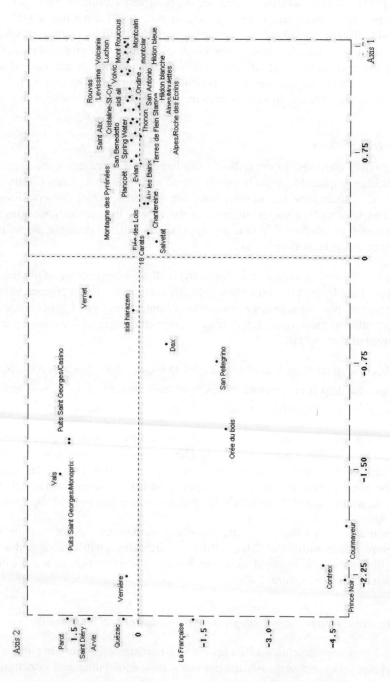

Figure 1.4. *Observations representation on plane 1, 2*

Principal Component Analysis   17

Individuals			Coordinates					Contributions					Square cosines				
Identifier	P. Rel.	DISTO	1	2	3	4	5	1	2	3	4	5	1	2	3	4	5
Evian	1.75	0.71	0.72	0.06	0.06	-0.21	-0.18	0.2	0.0	0.0	0.1	0.3	0.73	0.01	0.01	0.06	0.04
Montagne des Pyrénées	1.75	1.08	0.95	0.19	0.15	0.33	0.01	0.4	0.0	0.0	0.2	0.0	0.84	0.04	0.02	0.10	0.00
Cristaline–St–Cyr	1.75	1.38	0.98	0.16	0.54	0.00	0.07	0.4	0.0	0.5	0.0	0.0	0.69	0.02	0.21	0.00	0.00
Fiée des Lois	1.75	0.80	0.38	-0.11	0.60	-0.21	-0.22	0.1	0.0	0.7	0.1	0.5	0.18	0.02	0.45	0.05	0.06
Volcania	1.75	2.81	1.45	0.33	0.71	0.15	0.07	1.0	0.1	0.9	0.1	0.0	0.75	0.04	0.18	0.01	0.00
Saint Diéry	1.75	16.07	-3.54	1.48	0.13	0.56	-0.95	5.8	1.9	0.0	0.7	8.9	0.78	0.14	0.00	0.02	0.06
Luchon	1.75	2.36	1.40	0.25	0.52	0.12	0.11	0.9	0.1	0.5	0.0	0.1	0.83	0.03	0.12	0.01	0.00
Volvic	1.75	2.12	1.32	0.41	-0.12	0.24	-0.11	0.8	0.1	0.0	0.1	0.1	0.82	0.08	0.01	0.03	0.01
Alpes/Moulettes	1.75	1.31	1.07	0.01	0.40	-0.06	0.04	0.5	0.0	0.3	0.0	0.0	0.87	0.00	0.12	0.00	0.00
Orée du bois	1.75	6.37	-1.22	-2.02	0.16	-0.49	-0.37	0.7	3.5	0.0	0.5	1.4	0.23	0.64	0.00	0.04	0.02
Arvie	1.75	52.52	-6.51	2.67	0.10	0.35	-1.25	19.5	6.1	0.0	0.3	15.4	0.81	0.14	0.00	0.00	0.03
Alpes/Roche des Ecrins	1.75	1.31	1.07	0.01	0.40	-0.06	0.04	0.5	0.0	0.3	0.0	0.0	0.87	0.00	0.12	0.00	0.00
Ondine	1.75	1.93	1.14	0.22	0.74	-0.03	0.06	0.6	0.0	1.0	0.0	0.0	0.67	0.03	0.28	0.00	0.00
Thonon	1.75	2.35	0.96	0.05	-1.17	0.01	-0.10	0.4	0.0	2.5	0.0	0.1	0.39	0.00	0.58	0.00	0.00
Aix les Bains	1.75	0.99	0.66	-0.03	0.59	-0.30	-0.12	0.2	0.0	0.6	0.2	0.1	0.44	0.00	0.35	0.09	0.02
Contrex	1.75	25.50	-2.18	-4.29	-0.57	-1.40	0.07	2.2	15.6	0.6	4.3	0.0	0.19	0.72	0.01	0.08	0.00
La Bondoire Saint Hippol	1.75	6.57	1.33	0.26	-2.13	0.41	0.00	0.8	0.1	8.2	0.4	0.0	0.27	0.01	0.69	0.03	0.00
Dax	1.75	1.78	-0.62	-0.64	0.61	0.61	-0.07	0.2	0.3	0.7	0.8	0.0	0.22	0.23	0.21	0.21	0.00
Quézac	1.75	15.10	-3.37	0.36	-0.18	-1.56	-0.78	5.2	0.1	0.1	5.4	6.0	0.75	0.01	0.00	0.16	0.04
Salvetat	1.75	3.00	0.11	-0.41	0.14	-0.77	0.25	0.0	0.1	0.0	1.3	0.6	0.00	0.06	0.01	0.20	0.02
Stamna	1.75	1.66	1.04	0.15	0.73	0.06	0.04	0.5	0.0	1.0	0.0	0.0	0.66	0.01	0.32	0.00	0.00
Iolh	1.75	1.00	0.73	0.09	-0.24	-0.05	-0.35	0.2	0.0	0.1	0.0	1.2	0.53	0.01	0.06	0.00	0.12
Avra	1.75	24.03	1.45	0.22	-4.63	0.58	-0.22	1.0	0.0	38.7	0.7	0.5	0.09	0.00	0.89	0.01	0.00
Rouvas	1.75	1.63	1.20	0.23	0.32	0.14	-0.04	0.7	0.0	0.2	0.0	0.0	0.88	0.03	0.06	0.01	0.00
Alisea	1.75	2.43	1.44	0.30	0.45	0.16	0.06	0.9	0.1	0.4	0.1	0.0	0.85	0.04	0.08	0.01	0.00
San Benedetto	1.75	1.13	0.83	0.18	-0.29	-0.09	-0.30	0.3	0.0	0.2	0.0	0.9	0.61	0.03	0.08	0.01	0.08
San Pellegrino	1.75	4.15	0.74	-1.79	0.33	-0.21	-0.15	0.3	2.7	0.2	0.1	0.2	0.13	0.77	0.03	0.01	0.01
Levissima	1.75	2.40	1.38	0.27	0.58	0.11	0.07	0.9	0.1	0.6	0.0	0.0	0.80	0.03	0.14	0.01	0.00
Vera	1.75	1.42	1.15	0.18	0.21	0.03	-0.07	0.6	0.0	0.1	0.0	0.0	0.93	0.02	0.03	0.00	0.00
San Antonio	1.75	1.80	1.30	0.27	0.13	0.10	0.02	0.8	0.1	0.0	0.0	0.0	0.94	0.04	0.01	0.01	0.00
La Française	1.75	68.27	-5.64	-2.83	0.45	5.28	0.55	14.6	6.8	0.4	61.3	3.0	0.47	0.12	0.00	0.41	0.00
Saint Benoit	1.75	1.93	1.14	0.22	0.74	-0.03	0.06	0.6	0.0	1.0	0.0	0.0	0.67	0.03	0.28	0.00	0.00
Plancoët	1.75	1.10	0.68	0.19	0.73	0.11	-0.12	0.2	0.0	1.0	0.0	0.2	0.42	0.03	0.49	0.01	0.01
Saint Alix	1.75	1.88	1.04	0.33	0.74	0.28	-0.02	0.5	0.1	1.0	0.2	0.0	0.58	0.06	0.29	0.04	0.00
Puits Saint Georges/Casi	1.75	6.28	-1.29	1.62	-0.79	-0.20	1.03	0.8	2.2	1.1	0.1	10.3	0.27	0.42	0.10	0.01	0.17
St–Georges/Corse	1.75	2.70	1.33	0.32	0.84	0.28	0.08	0.8	0.1	1.3	0.2	0.1	0.66	0.04	0.26	0.03	0.00
Hildon bleue	1.75	13.11	1.51	0.27	-3.22	0.54	-0.08	1.0	0.1	18.8	0.6	0.1	0.17	0.01	0.79	0.02	0.00
Hildon blanche	1.75	1.52	1.13	0.07	-0.15	0.07	0.12	0.6	0.0	0.0	0.0	0.1	0.84	0.00	0.02	0.00	0.01
Mont Roucous	1.75	2.84	1.53	0.35	0.50	0.23	0.08	1.1	0.1	0.5	0.1	0.1	0.82	0.04	0.09	0.02	0.00
Ogeu	1.75	1.09	0.97	0.19	0.15	0.29	0.01	0.4	0.0	0.0	0.2	0.0	0.87	0.03	0.02	0.08	0.00
Highland spring	1.75	1.79	1.17	0.21	0.61	0.04	0.02	0.6	0.0	0.7	0.0	0.0	0.77	0.02	0.21	0.00	0.00
Parot	1.75	63.44	-6.61	3.99	-0.40	-1.58	1.09	20.1	13.5	0.3	5.5	11.6	0.69	0.25	0.00	0.04	0.02
Vernière	1.75	7.65	-2.27	0.26	0.19	-1.27	-0.88	2.4	0.1	0.1	3.6	7.6	0.67	0.01	0.00	0.21	0.10
Terres de Flein	1.75	1.33	0.86	-0.03	0.45	-0.14	0.16	0.3	0.0	0.4	0.0	0.2	0.55	0.00	0.16	0.02	0.02
Courmayeur	1.75	29.42	-1.90	-4.83	-0.46	-1.30	0.46	1.7	19.8	0.4	3.7	2.1	0.12	0.79	0.01	0.06	0.01
Pyrénées	1.75	1.06	0.98	0.19	0.14	0.23	0.00	0.4	0.0	0.0	0.1	0.0	0.90	0.03	0.02	0.05	0.00
Puits Saint Georges/Mono	1.75	6.37	-1.32	1.62	-0.80	-0.20	1.02	0.8	2.2	1.1	0.1	10.2	0.27	0.41	0.10	0.01	0.16
Prince Noir	1.75	30.69	-2.29	-4.80	-0.23	-1.46	0.33	2.4	19.6	0.1	4.7	1.1	0.17	0.75	0.00	0.07	0.00
Montcalm	1.75	2.94	1.49	0.31	0.71	0.17	0.09	1.0	0.1	0.9	0.1	0.1	0.76	0.03	0.17	0.01	0.00
Chantereine	1.75	0.87	0.38	-0.20	0.54	-0.42	-0.15	0.1	0.0	0.5	0.4	0.2	0.17	0.05	0.33	0.20	0.02
18 Carats	1.75	0.51	0.17	-0.22	0.48	-0.23	-0.25	0.0	0.0	0.4	0.1	0.6	0.06	0.09	0.45	0.10	0.13
Spring Water	1.75	6.53	0.88	0.10	-2.37	0.23	-0.24	0.4	0.0	10.1	0.1	0.6	0.12	0.00	0.86	0.01	0.01
Vals	1.75	7.28	-1.54	1.82	0.24	-0.43	1.15	1.1	2.8	0.1	0.4	12.9	0.33	0.46	0.01	0.03	0.18
Vernand	1.75	1.87	-0.29	1.13	0.44	-0.33	0.18	0.0	1.1	0.3	0.2	0.3	0.04	0.68	0.10	0.06	0.02
sidi harazem	1.75	1.91	-0.38	0.13	0.12	1.10	-0.44	0.1	0.0	0.0	2.7	1.9	0.08	0.01	0.01	0.64	0.10
sidi ali	1.75	1.98	1.11	0.27	0.78	0.12	0.06	0.6	0.1	1.1	0.0	0.0	0.62	0.04	0.30	0.01	0.00
montclar	1.75	1.93	1.32	0.23	0.31	0.08	0.09	0.8	0.0	0.2	0.0	0.1	0.91	0.03	0.05	0.00	0.00

Table 1.9. *Coordinates, contributions and square cosine of individuals*

The case of numerical supplementary variables has to be distinguished from the categorical variables. The former are positioned on the correlation display after having simply computed the correlation coefficient between each supplementary variable and the principal components. The interpretation is made in the same way as for active variables through the detection of significant correlations.

For supplementary categorical variables, we generally represent each category by its barycenter in the observation space. Some software (especially SPAD, for example) provide help with interpretation by giving test values which measure the distance of the point representing a category of the origin.

More precisely, the test value measures this distance in number of standard deviations of a normal distribution. They allow an extremal position of a subgroup of observations to be displayed. A category will be considered as significant for an axis if its associated test value is larger in absolute value than 2 (with a 5% risk) (see Table 1.10).

In the example, the barycenter of sparkling waters (and consequently, that of still waters) is more than 3 standard deviations from the origin $(-3, 5)$. Sparkling waters are significantly very far from the origin.

It is easy to plot supplementary individuals onto the principal axes. Since we have the formulae allowing principal components computations, we simply have to compute linear combinations of these supplementary points characteristics.

Categories		Test values					Coordinates				
Label	Frequency	1	2	3	4	5	1	2	3	4	5
1. Country											
France	40	–1.9	0.7	2.1	–0.5	0.7	–0.33	0.09	0.18	–0.04	0.03
Britain	4	1.2	0.2	–2.7	0.5	–0.2	1.17	0.16	–1.28	0.22	–0.05
Greece	2	0.8	0.2	–3.5	0.4	–1.0	1.09	0.15	–2.44	0.26	–0.29
Greece-Crete	2	0.8	0.2	0.8	0.2	0.0	1.12	0.19	0.52	0.10	0.00
Italy	7	0.7	–1.5	0.4	–0.5	0.1	0.49	–0.77	0.13	–0.17	0.01
Morocco	2	0.3	0.2	0.6	1.0	–0.6	0.36	0.20	0.45	0.61	–0.19
2 . Type											
Mineral	38	–2.5	–0.5	–1.2	–0.7	0.4	–0.46	–0.07	–0.11	–0.06	0.01
Spring	19	2.5	0.5	1.2	0.7	–0.4	0.92	0.13	0.22	0.11	–0.03
3 . PG											
Sparkling	16	–3.5	2.7	–0.5	–1.8	0.3	–1.44	0.82	–0.11	–0.34	0.03
Still	41	3.5	–2.7	0.5	1.8	–0.3	0.56	–0.32	0.04	0.13	–0.01

Table 1.10. *Coordinates and test values of the categories*

1.5. Application to statistical process control

1.5.1. *Introduction*

Online statistical process control is essentially based on control charts for measurements, drawing the evolution of a product or process characteristics. A control chart is a tool which allows a shift of a location (mean) or a dispersion (standard deviation, range) parameter regarding fixed standard or nominal values to be detected through successive small samples $(x_i, i = 1, 2, \ldots, n)$.

Several types of control charts exist [MON 85, NIA 94], all based on the assumption that the distribution of x_i is $\mathcal{N}(\mu_0, \sigma_0)$. Standard or nominal values μ_0 and σ_0 are assumed known or fixed. If this is not the case, they are replaced by unbiased estimations.

Here, we are only interested in classical Shewhart control charts for the detection of process mean shifts. In Shewhart control charts, at each instant i we use only $\overline{x}_i$, the mean value of observations available, which is compared to lower (LCL) and upper (UCL) control limits:

$$LCL = \mu_0 - 3\sigma_0/\sqrt{n} \quad \text{and} \quad UCL = \mu_0 + 3\sigma_0/\sqrt{n}.$$

This control chart can be seen as a graphical representation of a succession of statistical tests $H_0 : \mu = \mu_0$ against $H_1 : \mu \neq \mu_0$ for a set of samples; the standard deviation σ_0 is assumed known. The critical region corresponds to the control chart from the control area. This equivalence to hypothesis tests will facilitate extension to several variables.

In most of the cases there are not one but several characteristics to simultaneously control. The usual practice consists of using as many charts as characteristics. This method has the major drawback that it does not take into account the correlations between variables representing these p characteristics. That then leads to undesired situations of false alarms (Figure 1.5). The univariate charts may signal an out-of-control situation while the multivariate process is under control (region B and C) or, more severe, a non-detection of a multivariate process shift (region A) may occur.

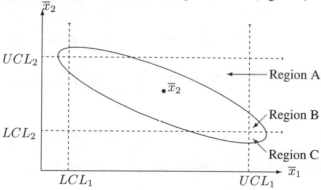

Figure 1.5. *Multivariate control chart*

A global approach through multivariate charts is therefore the only adequate approach (see [NIA 02]). Principal component analysis, which provides artificial but uncorrelated variables, is, in some sense, a first solution to the problem of correlated characteristics. We will see later that once a shift has been detected, adequate univariate charts may help to determine which variables are responsible for this shift, or assignable causes.

1.5.2. *Control charts and PCA*

1.5.2.1. *PCA and outliers*

Multivariate control charts are based on a transformation of a $\mathbb{R}^p$ vector in a scalar through a quadratic function. They can be seen as methods for multidimensional outlier detection. These methods consist of finding a sub-order in $\mathbb{R}^p$ generally based on a multivariate distance measure (a detailed study can be found in [BAR 84]). This measure is then used in a statistical test, to decide if an observation is an outlier when the standardized statistic has an abnormally large or small value, under a model hypothesis (in quality control, normality assumption is often made).

In the multidimensional case, an outlier may be the result of an extreme value for only one of the p principal components or the result of small systematic errors in several directions. This latter type of outlier corresponds to the problem of orientation (correlation) and not of location (mean) or dispersion (variance). Using principal component analysis, not as a dimension reduction method but rather as an outlier detection method, facilitates the search for extreme directions.

To best summarize the data structure, not only the first components should be retained but also the last components considered as the residuals of the analysis. Jolliffe [JOL 86] has shown that the first components allow the detection of outliers which inflate the variances and covariances. These outliers are also extreme on original variables, so they can be directly detected. The first components do not yield supplementary information.

On the other hand, outliers not visible on original variables (those that perturb the correlation between variables) will be detected on the last principal components. Several methods of outliers detection based on PCA have been proposed by many authors, specially Hawkins [HAW 74], Gnanadesikan [GNA 77], Jolliffe [JOL 86] and Barnett and Lewis [BAR 84].

Proposed techniques consist of applying formal statistical tests to principal components individually or conjointly. These tests are based on residual statistics computed using the last q principal components. The most widely used residual statistics are:

$$R_{1i}^2 = \sum_{k=p-q+1}^{p} (c_i^k)^2 = \sum_{k=p-q+1}^{p} \lambda_k (y_i^k)^2,$$

where $y_k = \frac{c^k}{\sqrt{\lambda_k}}$, R_{1i}^2 is a weighted sum of the standardized principal components which give more importance to principal components with large variances and

$$R_{2i}^2 = \sum_{k=p-q+1}^{p} (c_i^k)^2 / \lambda_k = \sum_{k=p-q+1}^{p} (y_i^k)^2.$$

The distributions of these statistics are easily obtained. If the observations are normally distributed with known mean μ and variance Σ, y_k have an exact Gaussian distribution $\mathcal{N}(0, 1)$. If there are no outliers in the data, the residual statistics R_{1i}^2 and R_{2i}^2 have χ_q^2 distribution. When μ and Σ are unknown, it is also possible (using their estimations) to obtain approximate distributions of these residual statistics and then to perform statistical tests.

1.5.2.2. Control charts associated with principal components

In quality control, PCA is used as a method for detecting shifts considered as outliers. The last principal components may be as interesting as the first components, since the type or the direction of the shifts are *a priori* unknown.

Recall that principal components are defined as linear combinations of original variables, which best summarize the data structure. They take into account the correlations between variables and then, even taken individually, they help to detect shifts (unlike original variables).

Note that principal components charts should not be used instead of, but in conjunction with, multivariate charts. The problem of false alarms and non-detection of out-of-control (noted on Figure 1.5) is attenuated but not completely suppressed for uncorrelated principal components. The 3-sigma control limits for standardized principal components are then $\pm 3/\sqrt{n}$. The presence of outliers can be tested with a control chart on the R_{2i}^2, whose upper control limit corresponds to the fractile $1 - \alpha$ of χ_q^2.

For residual statistics, the presence of outliers can be tested with a control chart defined by:

$$UCL = \chi_{q,1-\alpha}^2, \quad LCL = 0 \quad \text{and} \quad Stat = R_{2i}^2.$$

EXAMPLE.– We have simulated 30 samples of 5 observations from a multinormal distribution $\mathcal{N}_3(0, R)$; the three variables are assumed to have zero mean and variances equal to 0.5, 0.3 and 0.1, respectively. The correlation matrix is:

$$R = \begin{pmatrix} 1 & & \\ 0.9 & 1 & \\ 0.1 & 0.3 & 1 \end{pmatrix}.$$

We have then simulated a mean shift for the last five samples which consists of increasing the first variable mean and diminishing the second variable mean by half of their standard deviation. This situation is detected by the adequate multidimensional control chart [NIA 94] as well as the last principal component control chart. In Figure 1.6, note that the last five control points are clearly detected while the phenomenon is not visible on the first two principal components.

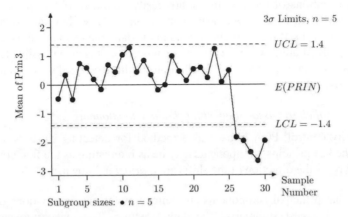

Figure 1.6. *Control chart for the 3rd principal component (Prin 3)*

When the number of characteristics is small, it is possible to find a simple interpretation for the principal components based on a small number of variables among the p original variables. Control charts based on principal components not only allow the detection of shifts but also help with the detection of assignable causes.

On the other hand, if the number of variables is very large, the proposed methods require many control charts for the first and the last components, which may be unpractical. We may only use the first q components, as in the dimension reduction approach of PCA, but then it is necessary (1) to test the quality of the representation of the p original variables by the q components and (2) to use methods based on residuals for outliers or shifts detection.

Furthermore, even if we find q principal components summarizing at best the information present in the p original variables, these q components depend on a large number of variables or on all original variables. To simplify the principal components interpretation, other methods of *projection pursuit* have been proposed [NIA 94]. The work done by Caussinus *et al.* [CAU 03] (see Chapter 3) is useful in improving these methods.

1.6. Conclusion

PCA is a very efficient method for representing correlated data. It is widely used in market study, opinion surveys and in the industrial sector more and more.

We have presented principal components analysis essentially as a linear method for dimension reduction, in which we are generally interested in the first principal components. Through its application to statistical process control, we have seen that

PCA can also be used as a multidimensional outlier detection technique, based on the last components.

Non-linear extensions of PCA exist and will be used more frequently [DEL 88, SCH 99].

1.7. Bibliography

[BAR 84] BARNETT V., LEWIS T., *Outliers in Statistical Data*, Wiley, New York, 1984.

[CAU 03] CAUSSINUS H., FEKRI M., HAKAM S., RUIZ-GAZEN A., "A monitoring display of multivariate outliers", *Computational Statistics and Data Analysis*, vol. 44, num. 1–2, p. 237–252, 2003.

[DEL 88] DE LEEUW J., VAN RIJCKEVORSEL J. L. A., *Component and Correspondence Analysis: Dimension Reduction by Functional Approximation*, Wiley, New York, 1988.

[GNA 77] GNANADESIKAN R., *Methods for Statistical Data Analysis of Multivariate Observations*, Wiley, New York, 1977.

[GOW 96] GOWER J. C., HAND D. J., *Biplots*, Chapman & Hall, London, 1996.

[HAW 74] HAWKINS D. M., "The detection of errors in multivariate data using principal component", *Journal of the American Statistical Association*, vol. 69, num. 346, p. 340–344, 1974.

[JOL 86] JOLLIFFE I. T., *Principal Component Analysis*, Springer-Verlag, New York, 1986.

[LEB 06] LEBART L., MORINEAU A., PIRON M., *Statistique Exploratoire Multidimension-nelle*, Dunod, Paris, 4th edition, 2006.

[MON 85] MONTGOMERY D. C., *Introduction to Statistical Quality Control*, Wiley, New York, 1985.

[NIA 94] NIANG N. N., Méthodes multidimensionnelles pour la maîtrise statistique des procédés, PhD thesis, Paris Dauphine University, France, 1994.

[NIA 02] NIANG N. N., "Multidimensional methods for statistical process control: some contributions of robust statistics", LAURO C., ANTOCH J., ESPOSITO V., SAPORTA G., Eds., *Multivariate Total Quality Control*, Heidelberg, Physica-Verlag, p. 136–162, 2002.

[SAP 06] SAPORTA G., *Probabilités, Analyse de Données et Statistique*, Technip, Paris, 2006.

[SCH 99] SCHÖLKOPF B., SMOLA A., MULLER K., "Kernel principal component analysis", SCHÖLKOPF B., BURGES C., SMOLA A., Eds., *Advances in Kernel Methods – Support Vector Learning*, MIT Press, p. 327–352, 1999.

Chapter 2

Correspondence Analysis: Extensions and Applications to the Statistical Analysis of Sensory Data

2.1. Correspondence analysis

2.1.1. *Data, example, notations*

In correspondence analysis (CA), the data are composed of n individuals described by two qualitative (or categorical) variables. For example, voters were asked their social and economic category (variable SEC having, for example, 10 categories: workman, employee, etc.) while voting in the last election (variable with each category representing a candidate). It is convenient to gather these data in a table, referred to as a contingency table or cross table (see Figure 2.1), in which:

– the rows represent the categories of one variable (e.g. a row = a SEC);

– the columns represent the categories of the other variable (e.g. a column = a candidate);

– the number n_{ij} of people who belong to SEC i and have voted for the candidate j can be found at the intersection of row i and column j.

Let us note that, unlike for an *individuals* × *variables* table, the rows and the columns of a contingency table play symmetric roles. It is clear that such a table cannot be analyzed using principal component analysis (PCA).

Chapter written by Jérôme PAGÈS.

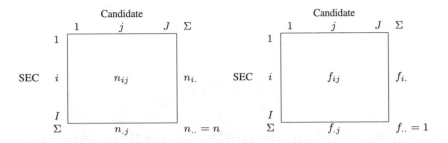

Figure 2.1. *Example of contingency table and associated probability*

The sum of the terms of the same row (number of voters per SEC) or of the same column (number of votes per candidate) is referred to as the margin and is classically denoted by replacing the index on which the summation is made by a point. Thus:

$$n_{i.} = \sum_j n_{ij} \qquad n_{.j} = \sum_i n_{ij} \qquad n = n_{..} = \sum_{ij} n_{ij}.$$

Finally, it is useful to consider the associated relative frequency (or probability) table, with f_{ij} as the general term:

$$f_{ij} = \frac{n_{ij}}{n} \qquad f_{i.} = \sum_j f_{ij} = \frac{n_{i.}}{n} \qquad f_{.j} = \sum_i f_{ij} \qquad f_{..} = \sum_{ij} f_{ij} = 1.$$

2.1.2. Questions: independence model

Such a table is constructed from the point of view of studying the relationship between two qualitative variables, e.g. is the SEC related to the vote? This general formulation, symmetric from the point of view of the two variables, is always accompanied by an 'asymmetric' question: which are the 'preferred' candidates of a given SEC? Which SEC constitutes the greatest part of the electorate of a given candidate?

In these latter two questions, the table is examined by row and by column. In this spirit, the contingency tables are often published expressed in percentages (of the sum of the rows or of the columns) according to the preferred question. In the terminology of correspondence analysis, a row (respectively, a column) transformed into percentages is referred to as a row profile (column profile). The row profile i has $n_{ij}/n_{i.}$ (or $f_{ij}/f_{i.}$) as general term. The sum of these terms equals 1 and a profile can also be regarded as a conditional probability distribution: $n_{ij}/n_{i.} = f_{ij}/f_{i.} =$ the probability of voting for the candidate j knowing that one belongs to the SEC i.

Lastly, the row profile and the column profile, calculated on the whole of the table (having $f_{i.} = n_{i.}/n$ or $f_{.j} = n_{.j}/n$ as the general term) are referred to as marginal profiles ($f_{i.}$: column margin; $f_{.j}$: row margin) or average profiles.

Studying the relationship between two qualitative variables consists of studying the difference between the observed data and the theoretical situation of independence. This theoretical situation corresponds to the table having $\tilde{n}_{ij}$ (theoretical frequency) or $\tilde{f}_{ij}$ (theoretical probability) as the general term:

$$\tilde{n}_{ij} = \frac{n_{i.}n_{.j}}{n} \qquad \tilde{f}_{ij} = f_{i.}f_{.j}.$$

This 'theoretical' table constitutes the independence model. It has the same margins as the observed table:

$$n_{i.} = \tilde{n}_{i.} \qquad n_{.j} = \tilde{n}_{.j}.$$

If the data table satisfies the relation of independence ($n_{ij} = \tilde{n}_{ij}$), then (on the one hand) all the row profiles and (on the other hand) all the column profiles are equal to the corresponding average profile: $f_{ij}/f_{i.} = f_{.j}$ and $f_{ij}/f_{.j} = f_{i.}$ $\forall i, j$. The variability between profiles therefore directly expresses the relationship between the two variables. For example, we can comment that these two candidates have very different electorates.

The objective of CA is to study the relationship between the two qualitative variables defining the table, precisely by describing the variability of the profiles, rows and columns.

2.1.3. Intensity, significance and nature of a relationship between two qualitative variables

The difference between the observed data and the independence model is classically measured by:

$$\chi^2 = \sum_{ij} \frac{(n_{ij} - \tilde{n}_{ij})^2}{\tilde{n}_{ij}} = n \sum_{ij} \frac{(f_{ij} - \tilde{f}_{ij})^2}{\tilde{f}_{ij}} = n\Phi^2.$$

This indicator is 0 if the observed data exactly satisfy the independence model. It is all the greater as the difference between the observed probability table (f_{ij}) and the theoretical table increases, measured by the Φ^2 criterion, and for greater n.

The Φ^2 criterion measures the intensity of the relationship between the two qualitative variables, an intensity that depends only on the probabilities. (The relationship is intense if the candidates have very different electorates, independent of the size n of the studied population).

The χ^2 criterion measures the significance of the relationship when the n studied individuals constitute a sample extracted from a population. This point of view is missed in CA, but the reference to χ^2 is essential because this criterion, unlike Φ^2, is well known. The nature of the relationship between the two variables can be described by decomposing Φ^2:

– by case: which are the most remarkable associations between a row category i and a column category j?

– by row: which SEC has a particular voter profile?

– by column: which candidates have a particular electorate?

Concurrently to this *a priori* decomposition of the relationship, correspondence analysis proposes an *a posteriori* decomposition, i.e. depending on the data themselves. In a similar way to that of PCA, which highlights dimensions expressing (as much as possible) the variability of the individuals, i.e. their discrepancy from the average, CA highlights dimensions expressing (as much as possible) the discrepancy from the independence model.

CA can be presented in multiple ways, partly due to the richness of this method. Even if it is not the most traditional presentation, it is convenient to exploit the analogies with PCA; many properties of PCA can be easily transposed.

2.1.4. *Transformation of the data*

The discrepancy of the independence model is taken into account by considering table **X** having:

$$x_{ij} = \frac{f_{ij}}{f_{i.}f_{.j}} - 1$$

as a general term.

When the data satisfy an independence model, the x_{ij} are zero. When the categories i and j are associated more (respectively, less) than in the independence model, x_{ij} is positive (respectively, negative).

Let us note that the analyzed table does not depend on the number n of the individuals. Multiplying the data table by a constant does not alter the results of CA. CA is a descriptive method, which yields only the intensity and the nature of the relationship between two variables.

2.1.5. *Two clouds*

Each row is a point in a space, denoted $\mathbb{R}^J$, whose dimension is associated with a column of the table **X**. The entire set of points constitutes the cloud $\mathcal{N}_I$ (see Figure 2.2). The weight $f_{i.}$ is assigned to the row i. The center of gravity of $\mathcal{N}_I$ therefore lies in the origin of the axes ($\sum_i f_{i.}x_{ij} = 0 \ \forall j$). It corresponds to the average row profile (calculated from the n individuals).

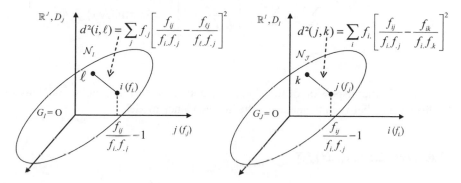

Figure 2.2. *The two clouds in CA*

The cloud of the columns, denoted $\mathcal{N}_J$, is built in the space $\mathbb{R}^I$ in a symmetric way: the weight $f_{.j}$ is assigned to the column j; the cloud $\mathcal{N}_J$ is centered and the origin corresponds to the average column profile.

The weights of the columns induce the metric in $\mathbb{R}^I$. The distance between the rows i and l is therefore equal to:

$$d^2(i, \ell) = \sum_j f_{.j} \left[\frac{f_{ij}}{f_{i.}f_{.j}} - \frac{f_{\ell j}}{f_{\ell .}f_{.j}} \right]^2.$$

It quantifies to which extent the two rows i and ℓ deviate from the independence model in the same way. Symmetrically, the weights of the rows induce the metric in $\mathbb{R}^J$. The distance between the columns j and h is:

$$d^2(j, h) = \sum_j f_{i.} \left[\frac{f_{ij}}{f_{i.}f_{.j}} - \frac{f_{ih}}{f_{i.}f_{.h}} \right]^2.$$

This use of the weights of the points of a cloud as a metric in the other space is natural. We attach the same importance to a row as an element of a cloud, on the one hand, and as a dimension on the other hand. It induces a set of relations between the clouds $\mathcal{N}_I$ and $\mathcal{N}_j$ referred to as 'duality' . The first of these relations expresses that the total inertia of each of these two clouds is the same:

$$
\begin{aligned}
\text{Inertia}(\mathcal{N}_I) \quad &= \quad \sum_i f_{i.} \sum_j f_{.j} \left[\frac{f_{ij}}{f_{i.}f_{.j}} - 1 \right]^2 \\
&= \quad \sum_j f_{.j} \sum_i f_{i.} \left[\frac{f_{ij}}{f_{i.}f_{.j}} - 1 \right]^2 = \text{Inertia}(\mathcal{N}_J) = \Phi^2 = \frac{1}{n}\chi^2.
\end{aligned}
$$

Analyzing the dispersion of the clouds $\mathcal{N}_I$ and $\mathcal{N}_J$ therefore consists of studying the discrepancy between the data and the independence model.

2.1.6. *Factorial analysis of* X

The factorial analysis of **X** consists of projecting $\mathcal{N}_I$ and $\mathcal{N}_J$ on their principal axes (see Figure 2.3). Regarding the cloud of individuals in PCA (Chapter 1), we seek a sequence of axes having a maximum projected inertia, each being orthogonal to those already found. For the analysis of the rows cloud, we are led to the singular values decomposition of $\mathbf{X'D}_I\mathbf{XD}_J$, where $\mathbf{D}_I$ is the diagonal matrix of the row weights $(f_{i.})$ and $\mathbf{D}_J$ is that of the column weights $(f_{.j})$ [ESC 08, p. 111]. As in PCA, the eigenvectors are the unit vectors of the directions (in $\mathbb{R}^I$) maximizing the $\mathcal{N}_I$ inertia; the eigenvalues are the inertias of the $\mathcal{N}_I$ projection on these directions.

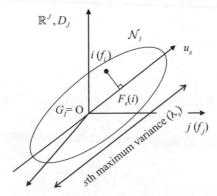

Figure 2.3. *Projection of the cloud $\mathcal{N}_I$ on the rank s axis. $\mathbf{u}_s$ is the sth maximum inertia direction of $\mathcal{N}_I$. Since the origin O is confounded with G_I, center of gravity of $\mathcal{N}_I$, this inertia can be interpreted as a variance (item i being weighted with f_i)*

Formally, denoting the rank s eigenvalue of $\mathbf{X'D}_I\mathbf{XD}_J$ as λ_s and the unit eigenvector associated with λ_s as $\mathbf{u}_s$, we have:

$$\mathbf{X'D}_I\mathbf{XD}_J\mathbf{u}_s = \lambda_s\mathbf{u}_s \quad \text{and} \quad \|\mathbf{u}_s\|^2 = \mathbf{u}'_s\mathbf{D}_J\mathbf{u}_s = 1.$$

The projection $F_s(i)$ of the row i on $\mathbf{u}_s$ is $F_s(i) = [\mathbf{XD}_J\mathbf{u}_s]_i$ and we have $\lambda_s = \sum_i f_{i.}[F_s(i)]^2$. Recalling the fact that the total inertia is a measurement of the relationship between the two qualitative variables defining the contingency table, CA is an analysis of this relationship by means of an *a posteriori* decomposition of this measurement. We seek associations and oppositions of categories that express most of this relationship.

As in PCA, duality relations relate the analysis of the clouds $\mathcal{N}_I$ and $\mathcal{N}_J$. Denoting the coordinate of the column j on the rank s axis as $G_s(j)$:

– inertias of clouds $\mathcal{N}_I$ and $\mathcal{N}_J$ in projection on their principal rank s axis are equal:

$$\lambda_s = \sum_i f_{i.}\left[F_s(i)\right]^2 = \sum_j f_{.j}\left[G_s(j)\right]^2 ;$$

– the coordinates of the rows and the columns on the rank s axes are related by the transition or barycentric (or quasi-barycentric) formulae:

$$F_s(i) = \frac{1}{\sqrt{\lambda_s}}\sum_j \frac{f_{ij}}{f_{i.}}G_s(j) = \frac{1}{\sqrt{\lambda_s}}\sum_j \left[\frac{f_{ij}}{f_{i.}} - f_{.j}\right]G_s(j),$$

$$G_s(j) = \frac{1}{\sqrt{\lambda_s}}\sum_i \frac{f_{ij}}{f_{.j}}F_s(i) = \frac{1}{\sqrt{\lambda_s}}\sum_i \left[\frac{f_{ij}}{f_{.j}} - f_{i.}\right]F_s(i).$$

These relations use the row profiles ($\{f_{ij}/f_{i.}; j = 1, J\}$) and the column profiles ($\{f_{ij}/f_{.j}; i = 1, I\}$). By connecting the coordinates of the rows ($F_s(i)$) and those of the columns ($G_s(j)$), superimposing projections of clouds $\mathcal{N}_I$ and $\mathcal{N}_J$ on axes having the same rank is suggested. They then provide an intuitive interpretation rule of the relative positions of the rows and columns.

In particular, they are known with the first formulation. Regardless of a coefficient ($\sqrt{\lambda_s}$), the row i lies at the barycenter of the columns, each column j bearing the weight $f_{ij}/f_{i.}$. This clearly shows 'the attraction' of a row by the columns with which it is most associated. The coefficients $f_{ij}/f_{i.}$ are positive and allow interpretation in terms of the barycenter (or quasi-barycenter, because of the coefficient $\sqrt{\lambda_s}$).

The second formulation, less known, immediately results from the first since the origin is the center of gravity of the clouds:

$$\sum_i f_{i.}F_s(i) = 0 \quad \text{and} \quad \sum_j f_{.j}G_s(j) = 0.$$

It explicitly reveals the independence model in the coefficients expressing the coordinate of a row as a linear combination of those of the columns. A row is attracted by the columns with which it is more associated (compared to the independence model) $(f_{ij}/f_{i.} > f_{.j})$ and 'pushed away' by those with which it is less associated (compared to the independence model).

2.1.7. *Aid to interpretation*

As in PCA, the results of CA always include the following indicators.

2.1.7.1. *Quality of representation of a cloud by an axis*

For the axis of rank s, this is measured by the ratio:

$$\frac{\text{Cloud inertia projected on the axis } s}{\text{Global cloud inertia}} = \frac{\lambda_s}{\sum_t \lambda_t}.$$

Usually multiplied by 100 and referred to as the percentage of inertia, this percentage is used as in PCA. On the other hand, projected inertia λ_s is interpreted in a specific way in CA. It is always lower than 1, this value being reached in the case of an exclusive association between a group of rows and a group of columns (see Figure 2.4).

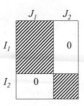

Figure 2.4. *Structure of the data leading to an eigenvalue equal to 1 in CA. The rows and columns can each be divided into two groups: (I_1, I_2) and (J_1, J_2). The categories of I_1 (respectively, I_2) are associated only with those of J_1 (respectively, J_2)*

The value 1 is of course seldom reached in practice, but it constitutes an important reference. The quantity λ_s, as a share of the Φ^2, measures the intensity of the share of the relationship highlighted by the rank s axis; this share is maximum in the case of an exclusive association between a block of rows and a block of columns. A value λ_s close to 1 indicates that the structure highlighted by the axis s is easy to illustrate from the raw data.

2.1.7.2. *Contribution of a point to the inertia of an axis or a plan*

This is measured by the ratio: (inertia of the point)/(inertia of the cloud), i.e. for the row i and the sth axis: $f_{i.}[F_s(i)]^2/\lambda_s$. This ratio is significant in CA because the weights of the rows or of the columns imposed by the method are generally different from one another. The displays therefore do not necessarily highlight the most contributive elements. (When the projected elements have the same weight, the displays are sufficient to roughly appreciate the contributions; this is the case for the variables in standardized PCA, for example.)

The contributions of several rows (or columns) to the inertia of the same axis can be summed. We can therefore calculate the contribution of a set of rows (or columns). The contribution of a point to the inertia of a plan is obtained according to the same principle, i.e. for the row i and plan (s, t):

$$\frac{f_{i.}[F_s(i)^2 + F_t(i)^2]}{\lambda_s + \lambda_t}.$$

2.1.7.3. *Quality of representation of a point by an axis or a plan*

The same ratio (for the whole cloud) is used, i.e. (projected inertia)/(total inertia) which, for a point, can be interpreted as a square cosine. Thus, for the row i and rank s axis:

$$\frac{\text{Inertia of the point } i \text{ projected on the axis } s}{\text{Global inertia of the point } i} = \frac{f_{i.}[F_s(i)]^2}{f_{i.}d^2(0, i)} = \cos^2 \theta_i^s$$

where θ_i^s is the angle between $\overrightarrow{Oi}$ and the axis of rank s (see Figure 2.5). Qualities of representation of the same point on several axes can be summed. The quality of representation of the row i by the plan P is therefore equal to the square cosine of the angle between i and its projection on P (see Figure 2.5):

$$(cos\theta_i^P)^2 = (\cos \theta_i^s)^2 + (\cos \theta_i^t)^2.$$

A low quality of representation of the row i on the plan P does not mean that the plan P does not bring any information about i but that the discrepancy from the independence model expressed by i (i.e. the difference between its profile and the average profile) can only be completely described by taking into account at least one other dimension.

2.1.8. *Some properties*

2.1.8.1. *Maximum number of principal axes*

The cloud $\mathcal{N}_I$ lies in a space having J dimensions. Moreover, each point satisfies the relation $\sum_j f_{.j}x_{ij} = 0$. The cloud $\mathcal{N}_I$ therefore belongs to a subspace having $J - 1$ dimensions, orthogonal (for the metric $\mathbf{D}_J$) with the 'first bisector' (containing the vectors with identical coordinates). According to this point of view, $\mathcal{N}_I$ can be

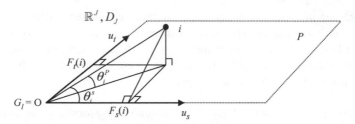

Figure 2.5. *Quality of representation of a point. The plan P is generated by the axes of rank s and t. The quality of representation of i by the rank s axis (respectively, the plan P) is measured by* $\cos^2 \theta_i^s$ *(respectively,* $\cos^2 \theta_i^P$ *)*

perfectly represented by a number of axes less than or equal to $J - 1$. However, this cloud contains I points having their center of gravity at the origin. According to this point of view, $I - 1$ axes (or less) are enough to perfectly represent it. Finally, the maximum number of axes having a non-null inertia is: $\min\{(J - 1), (I - 1)\}$. The same reasoning, beginning from cloud $\mathcal{N}_J$, leads to the same result.

2.1.8.2. *Consequence for the maximum value of* Φ^2

In CA, the eigenvalues are always lower than 1. The number of dimensions necessary to perfectly represent the clouds (that of the rows and that of the columns) therefore coincides with the maximum value of the Φ^2 for a table of given dimensions. This value is reached when it is possible, using permutations of the rows and of the columns, to display the data like a diagonal table (if $I = J$) or a diagonal block table with $\inf(I - 1, J - 1)$ blocks (if $I \neq J$). In this case, the set with the most members is functionally related to the other (each of its categories is associated with only one category of the other set).

2.1.8.3. *Distributional equivalence*

Let us assume we have two proportional rows i and ℓ. In this case, carrying out CA on the initial contingency table or on the table in which the two rows i and ℓ were summed leads to the same result. This property, seemingly technical, has a concrete impact. To build the SEC × candidates table quoted at the beginning of this chapter, it is therefore necessary to first choose the level of detail of the variable SEC. We can suspect, for example, that distinguishing the various levels of qualification of the professional workmen brings nothing in the sense that the classes thus defined have the same profile of votes. In that case, the fact that it amounts to the same thing to consider these categories separately or gathered brings a certain serenity to the user. According to Lebart *et al.* [LEB 75, p. 231], distributional equivalence guards against 'the arbitrary choice of the nomenclatures'.

The proof of this property is straightforward. If two rows i and ℓ are proportional in the contingency table, they are identical in the transformed table. In cloud $\mathcal{N}_I$, the two items i and ℓ are confounded and we can consider them separately with the weights $f_{i.}$ and $f_{\ell.}$ or as only one point of weight $f_{i.} + f_{\ell.}$.

2.1.9. Relationships to the traditional presentation

2.1.9.1. Transformation of the data

In the traditional presentation (Figure 2.6b), the study of the rows is carried out by considering their profile $\{f_{ij}/f_{i.}; j = 1, J\}$. The cloud $\mathcal{N}_I$ thus built has the marginal profile $G_I = \{f_{.j}; j = 1, J\}$ as center of gravity. The study of the $\mathcal{N}_I$ dispersion around its center of gravity G_I amounts to studying the differences between $f_{ij}/f_{i.}$ and $f_{.j}$ rather than comparing $\frac{f_{ij}}{f_{i.}f_{.j}}$ with 1. The data tables transformed into profiles according to whether we study the rows $(f_{ij}/f_{i.})$ or the columns $(f_{ij}/f_{.j})$ are different.

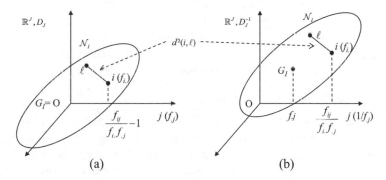

Figure 2.6. *Equivalence between two presentations of the cloud of the rows: (a) presentation adopted here and (b) traditional presentation*

2.1.9.2. Weights and metrics

In the traditional presentation, the weights of the rows and columns are, as adopted here, the marginal frequencies $f_{i.}$ and $f_{.j}$. However, the metrics, which are diagonal, are indirectly derived from these weights. In $\mathbb{R}^J$ (respectively, $\mathbb{R}^I$), each dimension is given the coefficient $1/f_{.j}$ (respectively, $1/f_{i.}$). Taking into account the different transformation of the data, the same distances are obtained in both cases as follows:

$$d^2(i, \ell) = \sum_j f_{.j} \left[\frac{f_{ij}}{f_{i.}f_{.j}} - \frac{f_{\ell j}}{f_{\ell.}f_{.j}} \right]^2 = \sum_j \frac{1}{f_{.j}} \left[\frac{f_{ij}}{f_{i.}} - \frac{f_{\ell j}}{f_{\ell.}} \right]^2.$$

ASSESSMENT.– From the point of view of the user, the interest of the traditional presentation is to be based on the concept of profile, natural in terms of interpretation. The richness of the analysis increases by considering the two approaches.

2.1.10. *Example: recognition of three fundamental tastes*

2.1.10.1. *Data*

Thirty tasters were asked to identify sweet, acidic or bitter solutions. The data were gathered in Table 2.1, having as the general term (i, j) the number of times that the taste i was identified as j. In practice, the analysis of such a small table does not require CA. The direct observation of the table shows that the perceived taste is in good agreement with the stimulus: the sweet solution is perfectly recognized as sweet and there is a slight confusion between sour and bitter, the bitterness being slightly less recognized than the sourness. Nevertheless, for this reason, this table is useful for illustrating the outputs provided by CA.

		Perceived taste			
		sweet	sour	bitter	Σ
	sweet	10	0	0	10
Taste	sour	0	8	2	10
stimulus	bitter	0	4	6	10
	Σ	10	12	8	30

Table 2.1. *Recognition of tastes: raw data (e.g. of the 10 people with the stimulus acid, 8 identified it as sour and 2 as bitter)*

2.1.10.2. *Transformation of the data*

The discrepancies of the independence model analyzed by CA are evident in Table 2.2c. The strong values on the first diagonal express the good recognition of tastes, which are excellent for sweet. The negative values relative to sweet $(-1$, which is the minimum) expresses the absence of confusion between this taste and the others. Lastly, the zero or the low values (0 and -0.25) at the intersection of sour and bitter tastes expresses a confusion between these two tastes. The transformation of the data correctly translates our direct analysis of the table.

	sweet	sour	bitter	Σ
sweet	0.333	0.000	0.000	0.333
sour	0.000	0.267	0.067	0.333
bitter	0.000	0.133	0.200	0.333
Σ	0.333	0.400	0.267	1.000

(a) observed frequencies f_{ij}

	sweet	sour	bitter	Σ
sweet	0.111	0.133	0.089	0.333
sour	0.111	0.133	0.089	0.333
bitter	0.111	0.133	0.089	0.333
Σ	0.333	0.400	0.267	1.000

(b) independence model $f_{i.}f_{.j}$

	sweet	sour	bitter
sweet	2.00	-1.00	-1.00
sour	-1.00	1.00	-0.25
bitter	-1.00	0.00	1.25

(c) discrepancy from independence

$$\frac{f_{ij}}{f_{i.}f_{.j}} - 1$$

	sweet	sour	bitter	Σ
sweet	13.333	4	2.667	20.0
sour	3.333	4	0.167	7.5
bitter	3.333	0	4.167	7.5
Σ	20.000	8	7.000	35.0

(d) decomposition of χ^2

$$n\frac{(f_{ij}-f_{i.}f_{.j})^2}{f_{i.}f_{.j}}$$

Table 2.2. *Data and independence model*

Let us note that, regarding the discrepancy with the independence model, the recognition of the bitterness is slightly more remarkable than that of the sourness, whereas direct inspection of the table suggests a (slightly) better recognition for sour. Even if surprising, this contradiction is apparent. The calculation of the discrepancy with the independence model takes into account the two margins of the data table, which shows that perceptions identified as sour are more frequent than those identified as bitter (12 against 8). Is the better recognition of sourness due to the fact that sour comes more readily to mind than bitter?

2.1.10.3. *Correspondence analysis*

The first axis of Figure 2.7 sets sweet taste, real or perceived, against the other two. The eigenvalue of 1 expresses the exclusiveness in the association between *stimulus sweet* and *perceived as sweet*. The equality of the coordinates of *sour* and *bitter* (*stimulus or perceived taste*), implicated by the eigenvalue of 1, shows that these two tastes are identical in their opposition to sweet. The barycentric properties are particularly clear in this extreme case. The second axis exclusively sets tastes *sour* and *bitter* against one another, as stimulus or as perceived taste. If the association for each taste between stimulus and perceived taste is clear, the rather low eigenvalue of 0.1667 indicates that we are far from the exclusiveness seen in the first axis and that there is some confusion between these two tastes. In detail, we note that *perceived as bitter* is more extreme than *perceived as sour*, which corresponds well to the least specificity of *perceived as sour*, associated in two-thirds of the cases to the stimulus *sour*, whereas *perceived as bitter* is associated in three-quarters of the cases to the stimulus *bitter*. We find here one of the two (quasi) barycentric properties which, in this very simple case, can easily be clarified (Figure 2.8); see the following section.

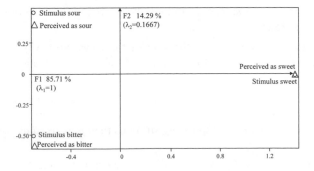

Figure 2.7. *CA of Table 2.1: first (and unique) plane*

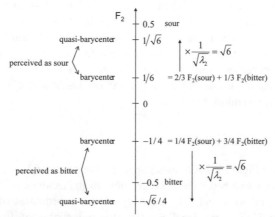

Figure 2.8. *(Quasi) Barycentric property applied to axis 2. Since the stimuli, acid and bitter, are fixed, the coordinate of each perception is calculated in two steps*

CONCLUSION.– In this small example, CA provides a grid for the analysis of the table at any point. Moreover, since it takes into account the margins through the reference to the independence model, CA provides a more in-depth visualization of the associations between rows and columns than that obtained by direct inspection of the table (see identification of the sourness and the bitterness).

2.1.10.4. *Barycentric property and simultaneous representation*

The coordinate of a column (*perceived as bitter* and *perceived as sour*) is calculated in two steps. *Perceived bitter* is initially located at the barycenter of the stimuli weighted with the coefficients of its profile (0, 2/8, 6/8) (Figure 2.8); then this coordinate is dilated by a coefficient equal to the inverse of the root of inertia associated with the axis (here $\sqrt{6}$). The relationship between rows (stimuli) and columns (perceptions) is expressed in the cloud of the barycenters at the same time from the point of view:

– of its nature (a column is on the side of the rows with which it is the most associated; here the stimulus and the perception of the same taste are on the same side);

– of its intensity (the cloud of the barycenters becomes wider as the columns are exclusively associated with the rows located on the same side; here *perceived as bitter* is more associated with the corresponding stimulus than *perceived as sour*).

In the representation of CA, this cloud is dilated so that it has the same inertia as the cloud of the stimuli, i.e. all the more strongly as the intensity of the relationship is low since the coefficient is the inverse of (the root of) the eigenvalue, which measures the share of the relationship (in the meaning of Φ^2) expressed by the axis. This results in the fact that the analysis of the graphs only informs us about the nature of the relationship (association between categories). To appreciate the intensity of this relationship, it is necessary to refer to the eigenvalues and, of course, to the raw data.

A complementary illustration from this property, if it is needed, can be obtained from Table 2.1. Since the margins are fixed, we decrease the confusion between *sour* and *bitter*. For example, by replacing the sub-table {8,2,4,6} in Table 2.1 by {9,1,3,7}, we obtain the same second factor regardless of a homothety (the first one is, of course, unchanged). λ_2 is altered from 0.1667 to 0.3757. The nature of the relationship did not change (same graphic); only its intensity increased.

2.2. Multiple correspondence analysis

2.2.1. *Data, notations and example*

A set of n individuals is described by Q qualitative variables. The traditional example is a survey in which n people answered Q closed questions. For each closed question, respondents answer by choosing one (and only one) category within a set list. For example, for the question 'Do you like Brahms?' this list could be {*yes*; *no*; *does not know Brahms*}. The category 'missing data' belongs to all the lists (generally implicitly); thus, each respondent always has one and only one category per question. The whole of the data is gathered in the table known as 'raw data table' (see Figure 2.9a and Table 2.3(a)). From the raw data table, we derive two other tables to which factorial analysis can be directly applied (see Figure 2.9).

No.	Sourness Threshold	Bitterness Threshold
1	high (1)	high (1)
2	high (1)	high (1)
3	high (1)	high (1)
4	middle (2)	high (1)
5	middle (2)	low (2)
6	middle (2)	low (2)
7	low (3)	low (2)
8	low (3)	low (2)
9	low (3)	low (2)

(a)

	Sourness threshold	
Bitterness threshold	high(1)	low(2)
high (1)	3	0
middle (2)	2	1
low (3)	0	3

(b)

Table 2.3. *(a) Raw data and (b) associated contingency table*

The *complete disjunctive table* (CDT, also known as indicator matrix) crosses the individuals and the categories. Its columns are the indicators of the categories. It presents a remarkable property that has important effects on the analysis: $\sum_{j \in q} y_{ij} = 1 \ \forall q$. The sum of the columns belonging to the same variable is constant and all its terms are equal to 1. The first consequence is, of course, that the margin column on

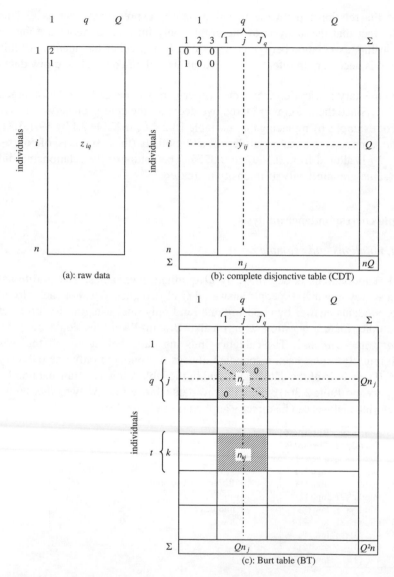

Figure 2.9. *Three presentations of the data in MCA: (a) raw data, where z_{iq} is the label (or number) of the category of the variable q possessed by i, n is the number of individuals and Q is the number of variables; (b) complete disjonctive table (CDT), where $y_{ij} = 1$ if i possesses the category j (of q) and 0 if not, n_j is the number of individuals possessing the category j, J_q is the number of categories of the variable q and $J_1 = 3$; and (c) Burt table (BT), where n_{kj} is the number of individuals possessing both the category j (of q) and the category k (of t). Note that dark gray represents the contingency table crossing the variables t and q and light gray represents the diagonal table comprising frequencies of the categories of the variable q*

the whole of the table is constant. The *Burt table* (BT) crosses the categories with themselves. It juxtaposes the whole of the contingency tables, crossing the variables two by two. The sub-tables lying on the first diagonal cross each variable with itself; they are diagonal and this first diagonal contains the frequencies of the categories. The Burt table is symmetrical and its margins are proportional to the frequencies of the categories.

2.2.2. Aims

Since the data table crosses individuals and variables, the aims associated with it are basically the same as those of the PCA, i.e.:

– description of the variability of the individuals, highlighting a typology *sensu lato* i.e. clusters or tendencies;

– description of the relationships between the variables (these being qualitative, this description refers to CA and consists of highlighting associations and oppositions between categories);

– construction of *a posteriori* synthetic variables i.e. depending on the data (these synthetic variables being quantitative (they are the principal components in PCA), the assignment of a numerical value to the individuals is made by assigning a coefficient to each category and by summing, for each individual, the coefficients of the possessed categories).

In the latter case, we expect that these synthetic variables summarize the variability of the individuals as best as possible i.e. have a maximum variance.

2.2.3. MCA and CA

Multiple correspondence analysis (MCA) is the factorial method suited to the study of *individuals* × *qualitative variables* tables. It can be presented in multiple ways. In France, where CA is widely used, it is useful to present it as an extension of CA to the study of the relationships between more than two qualitative variables. We follow this use, which leads to the major properties of MCA in a convenient way. The results of MCA can be obtained by applying CA software to the CDT or to the BT. This remarkable equivalence, as for the usual CA in the case of two variables, was established and described in [LEB 77, p. 126]. We return the reader to this work, or a more recent release of this work [LEB 06], for the proofs of these equivalences on which we base our presentation of MCA, mainly as a CA on CDT and, for some properties, as a CA on BT.

42   Data Analysis

2.2.4. *Spaces, clouds and metrics*

2.2.4.1. *MCA as CA of the CDT*

After transformation, the analyzed table has as a general term:

$$x_{ij} = \frac{ny_{ij}}{n_j} - 1.$$

This transformation, designed for a contingency table, does not seem very natural. It will be indirectly justified from the properties of the clouds of points, i.e. taking into account weights and metrics.

The cloud of the individuals $\mathcal{N}_I$ lies in the space $\mathbb{R}^J$ provided by the diagonal metric (denoted $\mathbf{D}_J$) with n_j/nQ as general term. The individual i has the weight $1/n$. The origin, centre of gravity of the cloud $\mathcal{N}_I$, corresponds to a fictitious individual who could choose, for each question, all the categories of answer (each with a weight equal to its frequency within the n studied individuals). Similarly, the cloud of categories $\mathcal{N}_J$ lies in the space $\mathbb{R}^n$ provided by the diagonal metric (denoted $\mathbf{D}_I$) with $1/n$ as general term. The category j is given the weight n_j/nQ. The origin, center of gravity of the cloud $\mathcal{N}_J$, corresponds to a fictitious category possessed by all the individuals.

2.2.4.2. *MCA as CA of BT*

After transformation, the analyzed table has as general term:

$$x_{kj} = \frac{nn_{kj}}{n_j n_k} - 1.$$

This term is identical to its counterpart in CA, applied to the contingency sub-table crossing the two variables, including the categories k and j. It is therefore directly justified. The cloud of the categories induced by this table (denoted $\mathcal{N}_j^*$ to distinguish it from the previous cloud) lies in the same space, provided by the same metric, as the cloud $\mathcal{N}_I$ of the individuals defined from the CDT (see Figure 2.10). These two clouds are closely related: the category k (of $\mathcal{N}_j^*$) is located at the center of gravity of the individuals (of $\mathcal{N}_I$) who possess it. Indeed,

$$\frac{1}{n_k} \sum_i y_{ik} \left[\frac{ny_{ij}}{n_j} - 1 \right] = \frac{nn_{kj}}{n_k n_j} - 1.$$

This induces concrete aspects to the equivalence between CA of the CDT and CA of BT. The following are equivalent:

– to analyze the cloud of individuals ($\mathcal{N}_I$) or that of the whole of the centers of gravity associated with categories ($\mathcal{N}_j^*$);

– to regard a category as an indicator variable ($\mathcal{N}_J$) or as a center of gravity ($\mathcal{N}_j^*$).

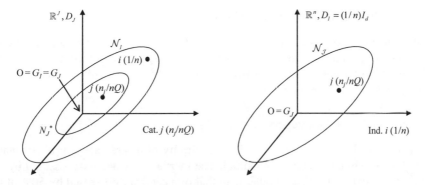

Figure 2.10. *Clouds in MCA. The individual i (respectively, category j) has the same weight $1/n$ (respectively, n_j/nQ), as an element of the cloud $\mathcal{N}_I$ (respectively, $\mathcal{N}_J$) and as an axis, via the metric $\mathbf{D}_I$ (respectively, $\mathbf{D}_J$), in space $\mathbb{R}^n$ (respectively, $\mathbb{R}^J$). The clouds $\mathcal{N}_I$ of the individuals (associated with the CDT) and $\mathcal{N}_J^*$ of the categories (associated with BT) lie in the same space and have the same principal axes*

2.2.5. *Properties of the clouds in CA of the CDT*

2.2.5.1. *Individuals*

In the distance between i and the origin 0, we can reveal the contribution of each variable q. Denoting the category (of the variable q) possessed by the individual i as k,

$$d^2(i,0) = \sum_{q \in Q} \sum_{j \in q} \frac{n_j}{nQ} \left[\frac{ny_{ij}}{n_j} - 1 \right]^2 = \frac{1}{nQ} \sum_q \frac{(n - n_k)n}{n_k}.$$

This contribution is low if n_k (frequency of the category k of q possessed by i) is large. A category possessed by many individuals has little effect on the character of each. Conversely, a rare category strongly moves its owners far from the origin. To fix the ideas, from a category possessed by 50% of the individuals to another possessed by 5%, the contribution of the corresponding variable, for the individuals possessing this category, is multiplied by approximately 36.

The distance between the individuals i and ℓ is written:

$$d^2(i,\ell) = \frac{1}{Q} \sum_j \frac{n}{n_j} (y_{ij} - y_{\ell j})^2.$$

This increases with the number of variables for which i and ℓ do not possess the same category, and this is particularly true if these categories are rare.

2.2.5.2. *Categories*

From CA of the CDT the following properties concerning the categories j and k arise.

2.2.5.2.1. Distance from the origin: $d^2(0, j) = n/n_j - 1$

A distant category is even further away if it is not very frequent, a phenomenon exacerbated for rare categories. To quantify, a category possessed by 5% of the individuals is at a greater distance from the origin by a factor of 4 than a category possessed by 50% of the individuals.

2.2.5.2.2. Inertia of the category j: $\frac{1}{Q}(1 - \frac{n_j}{n})$

In the calculation of its inertia, the eccentricity of a rare category is attenuated by its low weight but remains influential. The inertia of a category possessed by 5% of the individuals is 1.8 times higher than that of a category possessed by 50% of the individuals. It is therefore advisable to avoid the rare categories as far as possible, even if the total number of individuals is very high. In this way, when an *a priori* gathering of categories is not available, software SPAD (système pour l'analyse des données) [SPA 01] proposes a random ventilation (in the other categories) of the categories whose frequency is lower than a threshold fixed by the user (the default value is 2%).

2.2.5.2.3. Distance between two categories: $d^2(j, k) = \frac{n}{n_k n_j}$ (frequency of individuals possessing one and only one of the categories j and k)

Two categories are even closer to one another if they are possessed by the same individuals.

2.2.5.2.4. Inertia of the whole set of categories belonging to the same variable q: $\frac{J_q - 1}{Q}$

This inertia increases with the number J_q of categories of the variable q. This encourages the use, as far as possible, of variables having the same number of categories. However, this influence of the number of categories of a variable is less serious than it first appears. It can be shown that the J_q categories of the variable q generate a subspace having $J_q - 1$ dimensions and that their inertia is distributed in an isotropic way in this subspace (this inertia is $1/Q$, whatever the direction). Compared to a variable having few categories, a variable having many categories can therefore intervene on a greater number of axes but does not have *a priori* reasons to influence the first axis.

2.2.5.3. *Total inertia*

Considering the previous section, it is easy to show that the total inertia of the cloud $\mathcal{N}_J$ of the categories (and, by duality, that of the cloud $\mathcal{N}_I$ of the individuals) is equal to $J/Q - 1$, J/Q being the average number of categories per variable. In CA of the CDT, total inertia therefore depends only on the 'format' of the data and not on the data themselves. This property is similar to that of the standardized PCA (the total inertia equals the number of variables) and differs largely from that of the simple CA, in which the total inertia is essential as a measurement of relationship.

2.2.6. *Transition formulae*

2.2.6.1. *From CA of the CDT*

Applying the transition formulae from CA (see section 2.1.6), we obtain the two fundamental relations of MCA:

$$F_s(i) = \frac{1}{\sqrt{\lambda_s}} \sum_j \frac{y_{ij}}{Q} G_s(j) \qquad G_s(j) = \frac{1}{\sqrt{\lambda_s}} \sum_i \frac{y_{ij}}{n_j} F_s(i).$$

By discarding a coefficient on each axis, an individual lies in the barycenter of the categories it possesses and a category lies in the barycenter of the individuals who possess it. This double (quasi, because of the coefficient) barycentric property is particularly simple due to the fact that y_{ij} is equal to 0 or 1. It is quite sufficient to interpret the graphs, which makes MCA the factorial method whose graphs are the easiest to interpret.

2.2.6.2. *From CA of BT*

From the BT, we obtain the relation:

$$F_s(j) = \frac{1}{\lambda_s} \sum_k \frac{n_{kj}}{Q n_j} F_s(k).$$

The coefficient is λ_s (not its square root) because the eigenvalues resulting from CA of BT are the square of those of CA applied to the CDT. From this, two rules of interpretation identical to those from the simple CA result. A category appears on the direction of the categories with which it is the most associated; two categories appear close one another if they are associated with the other categories in the same manner. This latter property gives an account of proximities, on a factorial display, between two categories only having few (or no) common individuals (which is disconcerting for the novice user).

2.2.7. *Aid for interpretation*

The outputs of MCA software always comprise, in addition to the displays, the usual indicators of quality of representation (global and by point) and of contribution. They are used in the same way as for CA. Some comments are required.

2.2.7.1. *Eigenvalues*

From CA, $\lambda_s \leq 1$. The value 1 is reached when there is a subset of categories, including at least one category per variable, exclusively associated with a subset of individuals (they are possessed only by them and they do not possess other ones). We are close to this situation in surveys when missing data are primarily due to a subset of individuals.

2.2.7.2. *Contribution of categories and variables*

The absolute contribution of the category j to the rank s axis is $(n_j/nQ)[G_s(j)]^2$. According to the barycentric properties, $G_s(j)$ is proportional to the coordinate of the barycenter (denoted g_j) of the individuals who have j, i.e.:

$$G_s(j) = F_s(g_j)/\sqrt{\lambda_s}.$$

Thus, the absolute contribution of the variable q to the rank s axis, sum of the absolute contributions of its categories, is (where the squared correlation ratio between the factor F_s and the variable q is denoted $\eta^2(F_s, q)$):

$$\sum_{j \in q} \frac{n_j}{nQ} G_s(j)^2 = \frac{1}{Q} \frac{\sum_j \frac{n_j}{n} F_s(g_j)^2}{\lambda_s} = \frac{1}{Q} \eta^2(F_s, q).$$

This contribution makes it possible to quickly select the variables related to a given factor. It also requires to be calculated for the supplementary variables. This result also shows that the factors of MCA are the quantitative variables most related to the whole of the studied qualitative variables (in the sense of the average of the squared correlation ratios). This property is similar to that of the principal components from PCA (which maximize the sum of the squared correlation coefficients; see section 1.3.3.2 in Chapter 1).

2.2.7.3. *Quality of representation of the categories*

The whole set of the J_q categories of the same variable q generates a subspace having $(J_q - 1)$ dimensions. A variable with 5 categories therefore requires at least 4 axes to be perfectly represented. Hence:

– the percentages of inertia expressed by the axes are generally low in MCA, particularly if the number of categories per variable is large;

– when we recode a quantitative variable in a qualitative variable, by segmenting its range of variation in intervals which are defined as many categories, it is not useful to carry out too fine a segmenting, even if the number of individuals seems to allow it (8 categories seems a reasonable empirical maximum).

2.2.8. *Example: relationship between two taste thresholds*

The interpretation rules in MCA will be illustrated through an application to a dataset, taking as a starting point a real example but of a very small size (see Table 2.3) convenient for highlighting several properties. Moreover, the case of two qualitative variables presents a theoretical interest since it concerns both the simple CA and MCA.

The formal relations between these two methodologies are established in Lebart *et al.* [LEB 06, p. 211].

Nine people underwent two tests aiming to determine their perception threshold of two fundamental tastes: sourness and bitterness. The higher the threshold, the less sensitive is the person. These tests are fast and do not allow the thresholds to be determined with precision; hence the use of qualitative variables. The test concerning the bitterness is less precise, yielding two variables having different numbers of categories.

This example refers to a classical question of the sensory analysis. From a physiological point of view, the detection of each taste is carried out using specific receptors; the perception thresholds of the different tastes should not be linked. In practice, however, relationships are observed (not very intense but significant); roughly, the same people have the lowest (or highest) thresholds for the whole range of tastes.

The display from MCA (Figure 2.11) clearly shows the main properties of the method:

– the origin of the axes is the barycenter of the individuals: as for PCA, MCA highlights the principal directions of dispersion of the individuals;

– the origin of the axes is the barycenter of the categories of each variable: MCA opposes the categories within each variable (and cannot oppose the categories of a variable to categories of other variables since that would not have any meaning);

– axis by axis, discarding the coefficient $1/\sqrt{\lambda_s}$, an individual is at the barycenter of its categories (e.g. 7 has a low threshold for the two tastes). Since $1/\sqrt{\lambda_s} \geq 1$, compared to the exact barycenters the cloud of the individuals is dilated which brings them closer to the extreme categories (therefore 7 is closer to the low threshold of the sourness than to that of the bitterness).

The latter point above needs to be commented upon since an individual has or does not have a category; a category is all the more characteristic of an axis if it is far along this axis. Therefore 7 is characterized better by its sensitivity to sourness than by its sensitivity to bitterness. The dilation coefficient is not the same one according to the axes, the technical argument (applicable to any factorial analysis) represents the planes associated with axes having comparable inertias.

The first axis expresses a very clear structure ($\lambda_1 = 0.918 \simeq 1$) which opposes the low thresholds to the high thresholds: we summarize it by 'general sensory sensitivity'. The clearness of this structure is quite visible in the contingency table (see Table 2.3b) whose restriction on these four categories is diagonal. This axis ranks the individuals by increasing 'sensitivity'.

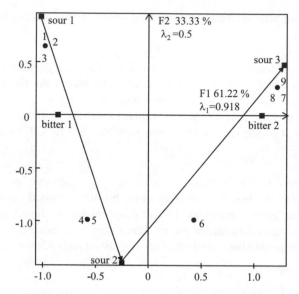

Figure 2.11. *MCA of Table 2.3a: first factorial display*

The barycentric property indicates how MCA built this quantitative synthetic variable from qualitative variables: with each category a coefficient is associated, proportional to its coordinate and the value of the synthetic variable for an individual, the sum of the coefficients of its categories. Thus, the possession of the category *bitter 2* (low threshold) or *sour 3* (low threshold) accounts positively and that of *bitter 1* (high threshold) or *sour 1* (high threshold) accounts negatively. In detail, *sour 3* accounts more than *bitter 2*: it is exclusively associated with *bitter 2* whereas the reverse is not true, which is taken into account in the barycentric property. The second factor is independent of the bitterness. It opposes, for sourness, the average sensitivity to the extreme sensitivities.

On the display (1,2), all of the points, rows and columns show a parabolic form. This shape of cloud, frequent in correspondence analysis, is referred to as the Guttman effect [BEN 73] or horseshoe effect [GRE 84]. It corresponds to the following structure of the data, called a 'scalogramme': if the rows and the columns of the data table are ranked according to their coordinate along the first axis, the highest values lie around the first diagonal. This can be observed in this example (see Table 2.4). Thus presented, the matrix reveals a scalogramme structure. The strong values (1 in the case of a CDT) are gathered around the first diagonal.

REMARK 2.1.– The low number of individuals: this example shows that MCA works with a low number of individuals in the sense it produces displays that accurately illustrate the data. In practice, apart from methodological tests, using MCA with so

	sour 1	bitter 1	sour 2	bitter 2	sour 3
1	1	1	0	0	0
2	1	1	0	0	0
3	1	1	0	0	0
4	0	1	1	0	0
5	0	1	1	0	0
6	0	1	1	1	0
7	0	0	0	1	1
8	0	0	0	1	1
9	0	0	0	1	1

Table 2.4. *CDT associated with the data of Table 2.3; the rows and the columns are ordered according to the first axis to reveal the scalogramme structure*

few individuals is dangerous because the results are very unstable. To be certain, we can alter one value in the data table and perform the analysis again.

REMARK 2.2.– Case of two variables; see Figure 2.12. We illustrate here the theoretical results stated in [LEB 06, p. 211]. The study of the relationship between the two perception thresholds *via* CA of the contingency Table 2.3b leads only to one axis, taking into account the dimensions of this table. The representation of the categories obtained by this CA is homothetic to the first axis of MCA. In $\mathbb{R}^n$, the variable *sourness* generates a 2D subspace whose reconstitution by MCA requires a second axis, included in this subspace (thus associated with the eigenvalue 1/2; see the last paragraph of section 2.2.5.2) and orthogonal to F_1.

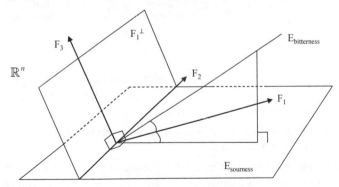

Figure 2.12. *MCA of Table 2.3: geometrical representation in $\mathbb{R}^n$ where F_1, F_2, F_3 are factorial axes, E_{sourness} is a 2D subspace generated by the 3 categories of the variable sourness, $E_{\text{bitterness}}$ is a 1D subspace generated by the 2 categories of the variable bitterness, F_1 is the bisector of the angle formed by $E_{\text{bitterness}}$ and its projection on E_{sourness} and F_2 is the direction of E_{sourness} orthogonal to F_1 (i.e. included in $F_1^{\perp}$, subspace orthogonal to F_1)*

The third factor of MCA (F_3) is related to the first by the following relations (where the coefficient of correlation is denoted $r[.,.]$ and the restriction of the vector of the coordinates of the categories along the sth axis to the categories of the variable

q is denoted $G_s(V_q)$):

$$\lambda_3 = 1 - \lambda_1 \qquad r[G_3(V_1), G_1(V_1)] = 1 \qquad r[G_3(V_2), G_1(V_2)] = -1.$$

Because of these relations, we can choose not to comment on F_3. Direct interpretation is nevertheless always possible although difficult in the example because of the very small value of λ_3 (0.0817). Let us note only that it is due, for 92%, to the individuals 4, 5 and 6. Without these, the relationship between the two variables would be perfect.

2.3. An example of application at the crossroads of CA and MCA

2.3.1. *Data*

96 students of Agrocampus each tasted 6 orange juices on the market using a tasting questionnaire. The questionairre comprised 7 sensory descriptors, each scored using a scale with 5 levels of intensity (see Table 2.5).

(a) Orange juices	(b) Descriptors	(c) Levels
1 Pampryl (ambient)	1 Odor intensity	1 Very low or zero
2 Tropicana (ambient), Florida	2 Odor typicity	2 Low
3 Fruivita (fresh), Florida	3 Pulpy	3 Mean
4 Joker (ambient)	4 Taste intensity	4 High
5 Tropicana (fresh), Florida	5 Sweet	5 Very high
6 Pampryl (fresh)	6 Sour	
	7 Bitter	

Table 2.5. *Description of the data (fresh: juice requiring conservation in refrigerated conditions)*

The whole of the data collected is initially gathered in a file. The raw data (see Figure 2.13) has the following characteristics:

– The rows are the tasting questionnaires; there is one questionnaire by subject and by product i.e. $96 \times 6 = 576$ rows.

– The columns are the variables of the questionnaire, structured in two groups: (1) the experiment variables (number of the subject, number of the product and rank order) where each subject evaluates the products in an specific order defined according to experimental design and (2) the sensory evaluation (7 descriptors).

In the current practice of sensory analysis, the descriptors are regarded as quantitative variables. For example, the averages by product are calculated. In this study we regard them as qualitative variables, for example, we calculate the frequency of each level per product. In our opinion, this point of view is not used enough in sensory analysis.

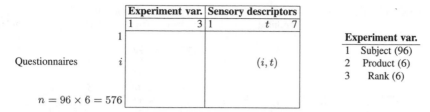

Experiment var.		Sensory descriptors		
1	3	1	t	7
		(i, t)		

Experiment var.

1	Subject (96)
2	Product (6)
3	Rank (6)

Questionnaires i

$n = 96 \times 6 = 576$

Figure 2.13. *Raw data table, where* (i, t) *is the level of the descriptor t written on the questionnaire* i

2.3.2. *Questions: construction of the analyzed table*

The objective is to obtain a synthetic visualization of the sensory evaluations of the 6 orange juices. For each descriptor t, we build the table (6×5) crossing the variable *product* and the descriptor t considered as a qualitative variable. In this table, the general term (row i, column j) contains the number of times that the level j (of the descriptor t) was given to product i. Taking into account the mode of collection of the data (each of the 96 subjects evaluated each product using each descriptor), the column margin is constant and equal to 96. All these tables are row-wise juxtaposed to constitute Table A (see Figure 2.14 and Table 2.6).

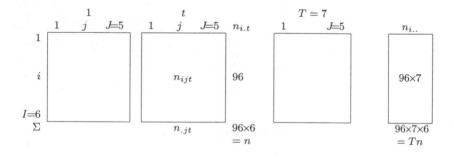

Figure 2.14. *Analyzed Table A, where* n_{ijt} *is the number of subjects having assigned the level* j *(of the descriptor t) to product* i

Row i of this table constitutes the complete sensory profile of product i, i.e. the whole set of its evaluations. The synthetic visualization is obtained by a factorial analysis of Table A.

This table is not a usual contingency table. Its treatment by CA requires to be justified. Let us note that the extension of CA to this type of table was proposed very early [BEN 73, p. 334] and has been the subject of in-depth studies [BEN 82].

Let us judge the fact of assigning a level (of a descriptor) to a product. Our data consist of $96 \times 6 \times 7 = 4032$ judgments. These 4,032 judgments can be described according to two variables: the product to which they refer (6 categories) and their type

Variable	Category	Pampryl ambient	Tropicana ambient	Fruivita fresh	Joker ambient	Tropicana fresh	Pampryl fresh	Σ	% par variable
	1	10	10	9	18	1	4	52	9
	2	34	32	31	22	22	26	167	29
Odor int.	3	19	31	29	24	36	30	169	29
	4	29	17	21	29	31	31	158	27
	5	4	6	6	3	6	5	30	5
	1	18	14	10	20	11	15	88	15
	2	31	27	26	24	22	27	157	27
Odor typ.	3	28	25	30	30	26	27	166	29
	4	16	22	26	19	28	23	134	23
	5	3	8	4	3	9	4	31	5
	1	56	46	1	51	1	5	160	28
	2	22	26	3	31	13	19	114	20
Pulpy	3	13	14	18	10	20	24	99	17
	4	5	7	47	4	43	34	140	24
	5	0	3	27	0	19	14	63	11
	1	2	5	2	3	3	2	17	3
	2	14	16	11	15	18	10	84	15
Taste int.	3	28	36	36	30	41	30	201	35
	4	42	30	36	40	32	42	222	39
	5	10	9	11	8	2	12	52	9
	1	8	12	26	13	23	6	88	15
	2	21	40	22	19	34	20	156	27
Sour	3	24	27	32	20	26	18	147	26
	4	35	13	14	38	10	42	152	26
	5	8	4	2	6	3	10	33	6
	1	13	30	47	21	34	22	167	29
	2	23	38	29	25	41	22	178	31
Bitter	3	23	20	17	25	12	23	120	21
	4	28	6	2	25	8	27	96	17
	5	9	2	1	0	1	2	15	3
	1	8	6	1	7	2	4	28	5
	2	41	11	15	31	14	32	144	25
Sweet	3	30	33	36	35	36	34	204	35
	4	15	38	35	20	37	22	167	29
	5	2	8	9	3	7	4	33	6
Σ		672	672	672	672	672	672	4032	700

Table 2.6. *Data on the sensory perception of 6 orange juices; Table A (Figure 2.14) is transposed*

i.e. the couple *level of intensity* × *descriptor* (5 × 7 = 35 types). The analyzed table crosses these two variables, *product* and *type of judgment*, for the 4,032 evaluations and, in that sense, can be seen as a contingency table and is therefore a matter for CA.

Now consider the Burt table associated with the raw data presented in Figure 2.13. Let us distinguish, in this table, the descriptors and the experiment variables, in particular the variable *product* (see Figure 2.15). The table we analyze is a subset of the Burt table; from this point of view, the CA of this subset is connected with MCA.

			Categories of the 3 experiment var.	Levels of the 7 descriptors 1 35
Categories of the 3 experiment var.	Subject			
	Product	1		Table A
		6		
	Rank order			
	Levels of the 7 descriptors	1		
		35		

Figure 2.15. *Burt table associated with the raw data table of Figure 2.13; the analyzed Table A (Figure 2.14) is the subset highlighted in gray*

REMARK 2.3.– The usual practice, in the sensory data processing, consists of carrying out a PCA on the table (*products* × *descriptors*) containing the averages of each descriptor for each product. In this PCA, the variability of the individual judgments around each average is not taken into account in the construction of the axes. This variability is taken into account in the CA presented here; on the other hand, in the construction of the axes, this CA does not take into account the order between the levels of the same descriptor.

2.3.3. *Properties of the CA of the analyzed table*

The analysis of a subset of a Burt table was considered in a general way by Cazes [CAZ 77]. We describe some important properties of a particular case which we study.

2.3.3.1. *Total inertia*

Let n_{ijt} be the general term of the analyzed table (see Figure 2.14) and $n_{..t} = \sum_{ij} n_{ijt} = n$ be the number of questionnaires distributed in each sub-table. Let us set, as usual in CA:

$$f_{ijt} = \frac{n_{ijt}}{Tn} \qquad f_{i.t} = \sum_j f_{ijt} \qquad f_{i..} = \sum_{jt} f_{ijt} \qquad f_{.jt} = \sum_i f_{ijt}.$$

The total inertia of the clouds in the CA of the Table A is:

$$\sum_i f_{i..} \sum_t \sum_j f_{.jt} \left[\frac{f_{ijt}}{f_{i..} f_{.jt}} - 1 \right]^2 = \frac{1}{T} \sum_i \Phi^2(\text{table } t) = \frac{1}{n} \frac{1}{T} \sum_i \chi^2.$$

In a similar way to the usual CA, the CA of A takes into account the whole of the relationships between, on the one hand, the variable *product* and, on the other hand, each of the 7 descriptors.

2.3.3.2. *Inertia of the whole of the levels of the same descriptor*

The share of inertia (of all the levels) of one descriptor t is proportional to Φ^2(table t). A descriptor therefore has a greater influence as it differentiates the products. The presence, in the table, of an un-discriminant descriptor is therefore not completely awkward. On the other hand, a descriptor t which is easy to use and therefore easily generates a consensus among the jury (which induces a very high Φ^2(table t)) can play a dominating role, awkward in the analysis. It is therefore useful to know the total inertia broken up by the descriptor in the full space but also projected on each axis.

2.3.3.3. *Set of columns referring to the same descriptor*

Each sub-table presents the same column margin, consequently equal (discarding the coefficient T) to the column margin of the entire Table A. This property is one of the important properties of the complete disjunctive table: it means that the barycenter of the column profiles of the categories of the same variable is confounded with the average profile, i.e. with the origin of the axes.

2.3.3.4. *Taking into account the tasting questionnaires*

The complete disjunctive table, having the questionnaires as rows and the levels of the 7 descriptors as columns, can be juxtaposed under Table A and be introduced into the analysis as supplementary elements in order to reveal the dispersion of the evaluations of each product. The cloud of the questionnaires thus obtained is that studied in a MCA of the table *questionnaires* × *descriptors* (same coordinates, same space and same metric). The rows of Table A are the barycenters of the questionnaires related to the same product. Thus, the CA of Table A, which highlights the principal axes of the cloud of the product (as barycenters), can be seen as a discriminant factorial analysis of the questionnaires to describe the variable product. This discriminant analysis, 'rough' in the sense that intra-product variance is not taken into account, is referred to as barycentric discriminant analysis [CAR 94].

2.3.4. *Results*

2.3.4.1. *Relationship between the variable product and each descriptor*

The analysis we undertake is purely descriptive. Taking into account the nature of the data (experimental, therefore likely to make an inference and sensory, therefore regarded as 'fragile'), it is important, as a preliminary, to be certain that the studied relationships cannot be reasonably attributed to random processes. The significance of the relationships between the two variables *product* and *descriptor* will be tested by considering each descriptor as qualitative (χ^2) but also as quantitative (F of the analysis of variance, with one factor, comparing the averages of the products). This process, which comprises a 1D inferential approach and then a multidimensional descriptive synthesis, is usual in the analysis of sensory data (e.g. [PAG 01]).

Table 2.7 gathers the indicators of intensity and significance of the relationship between the variable *product* and each descriptor. Let us note that, for a given descriptor $\eta^2 \leq \Phi^2$, when each value is taken by the quantitative variable, the general result is replaced by a category in the qualitative coding [SAP 75]. Regarded as quantitative variables, all the descriptors significantly discriminate at least one product. It is therefore advisable to utilize them all in the overall analysis. For two of them (*taste intensity* and *odor typicity*), this relationship is weak and insignificant from the χ^2 point of view.

Label	Descriptors as quantitative var.			Descriptors as qualitative var.		
	η^2	F	Probability	Φ^2	χ^2	Probability
Pulpy	0.515	121.18	0.000	0.551	317.117	0.000
Bitter	0.138	18.30	0.000	0.187	107.763	0.000
Sour	0.109	13.92	0.000	0.160	92.347	0.000
Sweet	0.095	12.03	0.000	0.115	65.994	0.000
Odor intensity	0.025	2.90	0.013	0.068	39.075	0.007
Taste intensity	0.023	2.71	0.019	0.035	20.313	0.438
Odor typicity	0.022	2.53	0.028	0.033	18.797	0.535
				mean=0.164		

Table 2.7. *Indicators of relationship between the variable product and each descriptor. Calculations are carried out using the raw data file (see Figure 2.13). Each descriptor is regarded both as quantitative (the intensity of the relationship is measured by the squared correlation ratio η^2 and the significance by F and its p-value from the analysis of variance with one factor) and qualitative (the intensity of the relationship is measured by Φ^2 and the significance by χ^2 and its p-value)*

The disparity within the relationship intensities is large. In the table analyzed hereafter by CA (grayed rectangle in Figure 2.15), the pulpy character therefore contributes to almost half $(0.551/(0.164 \times 7) = 0.479)$ of the total inertia. However, even for this descriptor, the characterization is far from being systematic: $\Phi^2_{pulpy} = 0.551$, a value much lower than 4, which here is the theoretical maximum taking into account the numbers of categories. This expresses a great disparity between judges despite an undeniable tendency.

2.3.4.2. *Eigenvalues*

The sequence of the eigenvalues (Table 2.8) clearly suggests preserving only the first two axes.

Rank	Eigenvalue	%	Cumulated %
1	0.1081	65.90	65.90
2	0.0340	20.70	86.60
3	0.0097	5.94	92.54
4	0.0078	4.75	97.28
5	0.0045	2.72	100.00
	Σ=0.1640		

Table 2.8. *CA of Table 2.6: eigenvalues and percentages of inertia*

These two axes (Figure 2.16) account for 86.6% of the total inertia, i.e. of the discrepancy from the independence model. This percentage is very satisfactory but unexceptional, taking into account the low number of rows (6) which ensures a perfect representation using 5 axes. The low value of the first eigenvalue (0.1081 << 1) recalls that the highlighted characterizations of the products will be necessarily qualified. To be reached, the possible maximum value of 1 required the existence of a subset of categories (including at least one by descriptor) used only for one subset of products which, in addition, would never be associated with other categories.

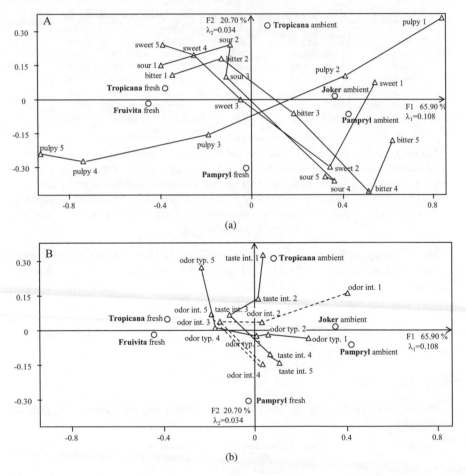

Figure 2.16. *CA of Table 2.6. Representation of the juices and the levels of the variables: (a) the most contributive variables and (b) other variables*

These data are obviously very far from this situation, which implies a perfect consensus among the tasters, a consensus never observed even in the presence of senior tasters. On the contrary, for example, the data table (Table 2.6) contains only three instances of value '0', which means that (except for these) each category was quoted at least once for each product.

2.3.4.3. *Origin of the axes*

This corresponds to the average profile. From the point of view of the categories cloud, each row having the same margin here, the origin corresponds to a uniform distribution of the products. A category almost equally distributed on the products is therefore near to the origin, whatever the plan. In the example, it is the case of *odor typicity* 3, whose frequencies all lie between 25 and 30. Recall that a point can be close to the origin on a plan, but far away from the origin on another plan in which case its profile can deviate appreciably from the average profile.

From the point of view of the products cloud, the origin corresponds to a fictitious product which would have obtained a distribution of the levels proportional to the distribution *all products confounded* for each descriptor. For example, for the descriptor *bitter*, such a product would have obtained: 29% of level 1, 31% of level 2, 21% of level 3, etc. None of the six juices are particularly close to the origin.

2.3.4.4. *Barycentric properties*

A category lies in the direction of the products with which it was generally associated. Thus, *bitter 4* was especially associated with *Joker* and with the two *Pampryl*. A product is in the direction of the categories that were generally associated with it. *Fresh Tropicana* was therefore associated, in particular, with the low levels of *sour* and *bitter*, and with the high levels of *sweet* and *pulpy*. All this can be directly checked in the data.

2.3.4.5. *Profiles*

The position of a point compared to the others depends only on its profile and not on its marginal frequency. Thus, levels 4 and 5 of pulpy, although their marginal frequencies are highly different (140 and 63), are close to one another because their profiles are very similar (they are associated almost exclusively with the fresh products).

2.3.4.6. *Representation of the categories and the juices*

To facilitate the reading of the plan, the variables are represented in two steps: (1) those that contribute the most to the plan (see Table 2.9) and (2) the others (Figure 2.16a and b).

It is useful to connect the levels of the same descriptor in their natural order. Although they are not taken into account in the construction of the axes, these orders strongly intervene in the interpretation.

Here, this reveals an organization of the variables along two directions that we assimilate by convenience to the two bisectors:

Variable	Inertia	In projection on the plan (1,2)		$\Phi^2_{1,2}$	Φ^2	$\Phi^2_{3,4,5}$
		Contribution	Quality of representation			
pulpy	0.0756	53.2	96.1	0.529	0.551	0.021
bitter	0.0209	14.7	78.4	0.147	0.187	0.040
sour	0.0189	13.3	82.7	0.133	0.160	0.028
sweet	0.0155	10.9	94.8	0.109	0.115	0.006
odor int.	0.0048	3.4	49.2	0.033	0.068	0.034
odor typ.	0.0034	2.4	72.3	0.024	0.033	0.009
taste int.	0.0029	2.1	57.9	0.020	0.035	0.015
	$\Sigma = 0.1421$	$\Sigma = 100$			mean=0.142	mean=0.164

Table 2.9. *Aids to the interpretation of the plan (1,2) concerning the variables. The contributions and qualities of representation associated with a plan are defined in sections 2.1.7.2 and 2.1.7.3. The share of Φ^2 expressed by the plan ($\Phi^2_{1,2}$) is obtained by multiplying the projected inertia by the number of variables (7). The share of Φ^2 not represented in the plan ($\Phi^2_{3,4,5}$) is obtained by difference*

– along the first one, the three descriptors *bitter*, *sour* and *sweet* evolve quite regularly (this direction opposes the juices Fruivita and Tropicana, perceived as sweet, not very sour or *bitter* compared to the other juices); note that this opposition between the juices corresponds to their origin (Florida/others).

– the second bisector is related to pulpy: it opposes the 'fresh' juices (pulpy) to the 'ambient' juices (not very pulpy).

Having established this broad outline of interpretation, several nuances can be made:

– the odor intensity takes part in the opposition fresh/ambient mostly because of the first level (the fresh juices are rarely quoted as having a very low odor intensity; see Table 2.10);

– the taste intensity and the odor typicity take part in the opposition between the two origins (the juices from Florida do not have an intense taste and have a rather typical odor).

Considering the χ^2 (see Table 2.7), these latter two variables do not significantly characterize the products. However, the relatively regular order of their levels on the plan undoubtedly suggests integrating them in the interpretation, which can be controlled in Table 2.10. Note that, for these two tables where the products are gathered by origin, the χ^2 is significant in one case and not very far from being significant in the other case.

2.3.4.7. *Non-linear relationships*

The studied data present a characteristic in comparison to the general case of the juxtaposition of contingency tables: the categories of the same variable are ordered and the analysis that we carry out is in competition with the PCA of the previously

	Odor intensity				
Type of product	very low	low	mean	high	very high
Ambient	38	88	74	75	13
Fresh	14	79	95	83	17
	Odor typicity				
Origin	very low	low	mean	high	very high
Florida	35	75	81	76	21
Other	53	82	85	58	10
	Taste intensity				
Origin	very low	low	mean	high	very high
Florida	10	45	115	98	30
Other	7	39	88	124	22

Table 2.10. *The products are gathered by type (ambient/fresh) or origin (Florida/other) and characterized by three descriptors e.g. the very low level of odor typicity was associated 14+11+10=35 times with the Florida juices (Tropicana: 14 and 11; Fruivita: 10); see Table 2.6*

mentioned average table. From the user point of view, the principal difference between these two methodologies is the possibility, *via* the proposed CA, to highlight non-linear relationships. We determine the relationship between this CA and MCA, using an interesting method applied to quantitative variables (previously recoded in classes) in order to reveal such relationships.

Generally, the most remarkable non-linear relationships are of quadratic type and appear on the factorial displays as an opposition between average levels and extreme levels (Figure 2.11). No phenomenon of this type is observed here. Discontinuities (perhaps due to the threshold effect) are less remarkable but frequent, for example:

– The case of the odor intensity has already been quoted. Only level 1 is not 'uniformly' distributed: it represents 50% of the inertia due to the sub-table corresponding to this descriptor. This distribution suggests a binary use of this descriptor.

– Perceived sourness also seems the object of a binary use in that the display strongly separates the low and mean levels (1, 2, 3) on the one hand, and the high levels (4 and 5) on the other hand. The closeness of the low and mean levels is perhaps to connect to the fact that the orange juice appears among the sourest products of frequent consumption.

– The bitterness shows an asymmetry. Levels 1 and 2 are much closer to the origin than levels 4 and 5. Generally perceived as bitter, the juices of origin other than Florida are also rather frequently perceived as slightly bitter. On the other hand, the Florida juices are seldom perceived as strongly bitter.

2.3.4.8. *Representation of supplementary variables*

Some simple physicochemical variables were measured (see Table 2.11); the weighted sum of the three sugars is sweetening power. In MCA, supplementary quantitative variables are introduced by calculating the correlation coefficient between each variable and each factor on I (set of coordinates of individuals along a factorial axis). In practice, the only constraint is to build the data file in such a way that the quantitative variables appear as columns.

Variable	PA	TA	FF	JA	TF	PF	Mean	SD
Glucose	25.32	17.33	23.65	32.42	22.70	27.16	24.76	4.57
Fructose	27.36	20.00	25.65	34.54	25.32	29.48	27.06	4.41
Saccharose	36.45	44.15	52.12	22.92	45.80	38.94	40.06	9.16
Sweeting power	89.95	82.55	102.22	90.71	94.87	96.51	92.80	6.11
Raw pH	3.59	3.89	3.85	3.60	3.82	3.68	3.74	0.12
pH after centrifugation	3.55	3.84	3.81	3.58	3.78	3.66	3.70	0.11
Overall acidity	13.98	11.14	11.51	15.75	11.80	12.21	12.73	1.62
Citric acid	0.84	0.67	0.69	0.95	0.71	0.74	0.77	0.10
Vitamin C	43.44	32.70	37.00	36.60	39.50	27.00	36.04	5.18

Raw data

Variable	PA	TA	FF	JA	TF	PF	Mean	SD
Glucose	0.12	−1.63	−0.24	1.67	−0.45	0.52	0	1
Fructose	0.07	−1.60	−0.32	1.70	−0.39	0.55	0	1
Saccharose	−0.39	0.45	1.32	−1.87	0.63	−0.12	0	1
Sweeting power	−0.47	−1.68	1.54	−0.34	0.34	0.61	0	1
Raw pH	−1.23	1.26	0.93	−1.15	0.68	−0.49	0	1
pH after centrifugation	−1.36	1.21	0.94	−1.09	0.68	−0.38	0	1
Overall acidity	0.77	−0.98	−0.75	1.86	−0.57	−0.32	0	1
Citric acid	0.75	−0.98	−0.78	1.86	−0.58	−0.27	0	1
Vitamin C	1.43	−0.65	0.19	0.11	0.67	−1.75	0	1

Centered and standardized data

Table 2.11. *PA: Pampryl ambient; TA: Tropicana ambient; FF: Fruivita fresh; JA: Joker ambient; TF: Tropicana fresh; PF: Pampryl fresh and SD: standard deviation*

In this calculation, the weights of the rows are those of CA. The cloud of the variables therefore lies in the same Euclidean space as the categories. The graph obtained by representing the variables by their coefficient of correlation with the factors on I can be interpreted as a projection. In particular, this graph can be read as the 'correlations circle' in PCA. Taking into account the physicochemical variables always introduces an interesting view of sensory data. In the example of Figure 2.17:

– pH and citric acid are negatively correlated with one another, which is expected. They are correlated with the first bisector, which opposes the juices perceived as strongly sour to those perceived as slightly or fairly sour. The interpretation of this sensory dimension is reinforced.

– The opposition between saccharose on the one hand and glucose and fructose on the other hand refers to the hydrolysis of saccharose (into glucose and fructose) favored in acid medium.

– The sweetening power is not correlated with the first bisector and thus with the perception of sweetness. This returns to the concept of gustatory balance, in particular between *sweetness* and *sourness*. It seems that, for this range of variations of sugars (the sum of the three sugars varies between 81 g/l and 101 g/l) and of the pH (between 3.59 and 3.89), the perception of sweetness mainly results from a lower sourness.

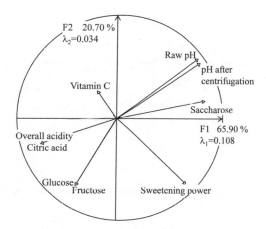

Figure 2.17. *Representation of illustrative variables*

2.3.4.9. *Representation of the tasting questionnaires*

As indicated in section 2.3.3.4, the rows of the CDT can be introduced as supplementary, mainly to visualize the variability of the perception of each juice.

The dispersion of the evaluations of the same product can be visualized by identifying the questionnaires by a sign indicating the product they refer to. Figure 2.18, which represents the tasting questionnaires concerning two juices, therefore highlights:

– in the direction of the first bisector, a moderate dispersion of the evaluations of each of the two juices and their separation (this correctly translated their distribution for the pulpy character, cf. Table 2.6);

– in the direction of the second bisector, a wide dispersion by product and a simple shift between the two clouds (Fruivita is perceived less sour and bitter).

By identifying questionnaires of the same taster, we visualize its products space. Figure 2.19 therefore shows a strong eccentricity of the evaluations of taster 19 (who judged all the products to be not very pulpy) but a configuration of the products similar to that of the whole of the jury.

2.3.4.10. *Other axes*

The decrease in value of the eigenvalues encourages comments to be limited to the first plane. Some indicators make it possible to evaluate the share of relationship not taken into account by this plan (Table 2.9).

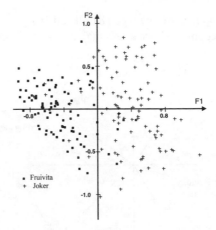

Figure 2.18. *Representations of the evaluations of the products Fruivita and Joker*

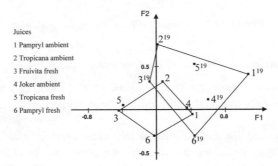

Figure 2.19. *Representation of the evaluations of the taster 19. Products 1, 2, 3 and 6 were connected for taster 19 and for the whole of the tasters, to highlight the similarity between the two configurations and their shift*

The quality of representation shows that the three variables *odor intensity, odor typicity* and *taste intensity* are the worst represented. However, these variables are also the least related to the variable product (see their Φ^2). It is therefore interesting to consult the raw inertia (and not in %) not represented ($\Phi^2_{3,4,5}$). This indicator reveals the share of the relationship between *bitter* and *product*, not represented on the plan, is equal to 0.04. This value is (slightly) higher than values which gave place to interpretations (cf. Φ^2_{12}).

The examination of the other axes (not represented here) shows that axis 4 highlights the level *bitter 5* (contribution 47%, Φ^2_4 : 0.025) associated with *ambient Pampryl* and opposed to *Joker* (contribution of these 2 juices to F_4 is 98%). The raw data (Table 2.6) shows this difference in perception of extreme bitterness between these two juices; moreover, the χ^2 calculated from the 2×2 table (*bitter 5*, other levels of *bitter* $\times$ two juices) is significant (p-value of 3%).

2.4. Conclusion: two other extensions

Since the book of Benzécri *et al.* [BEN 73], the field of CA has yielded many publications (e.g. [MOR 00]). We present in the following the principle of two extensions of CA which have a potentially important application.

2.4.1. *Internal correspondence analysis*

It is often desired to simultaneously study several contingency tables of different populations which have (at least) a common dimension. Thus, for example, we could have two tables $SEC \times candidates$ (cf. Figure 2.1) for two different elections: the SEC are the same from one table to another, but the candidates are not.

An idea is to juxtapose (row-wise) these two tables and to analyze them via CA. In this analysis, the cloud $\mathcal{N}_J$ of the candidates is composed of two sub-clouds: one per election. In the inertia of $\mathcal{N}_J$, it is possible to distinguish a between-elections inertia (due to the difference in distribution of the SEC between the two surveys which is itself due, for example, to the absence of quotas for the SEC) and a within-election inertia (expressing the relationship between SEC and candidates, analyzed in the separate CA of the two tables).

In the analysis of $\mathcal{N}_J$, between-inertia is generally awkward. In this spirit, internal correspondence analysis (ICA) enables only within-inertias to be simultaneously studied by 'eliminating' between-inertia. Geometrically, this elimination is carried out by centering each sub-cloud on the origin of the axes. Analyzed inertia is then the sum of the inertias taken into account in the separate CA. ICA was initially proposed by [BEN 83] and [ESC 84], and was generalized by [CAZ 88]. See [ESC 08] for an example.

2.4.2. *Multiple factor analysis (MFA)*

Initially described for quantitative variables, MFA also applies to the qualitative variables [ESC 08, PAG 02]. We outline here this latter case. As in MCA, a set of individuals is described by several qualitative variables. These variables are however structured in T groups. The traditional example is provided by surveys: the individuals are respondents, the variables are the questions and the groups of variables are the topics of the questionnaire. As in MCA, we initially want to describe the variability of the individuals and, in a dual way, the relationships between the variables. However, the group structure of the active variables implies these groups should be balanced in their simultaneous analysis. MFA is initially a factorial analysis in which the *a priori* influence of the groups is balanced.

Beyond this technical aspect, the group structure intervenes in the questioning, enriching it with several questions which are articulated around the comparison of the individuals cloud induced by each group (clouds with the meaning of MCA in the case of qualitative variable and with the meaning of PCA in the case of quantitative variables). The individuals cloud associated with the group t are denoted $\mathcal{N}_I^t$. Comparisons between the clouds $\mathcal{N}_I^t$ can be made:

– *Globally*: which groups induce the same shape for the individual cloud? In other words, which are the clouds $\mathcal{N}_I^t$ which globally resemble each other? This problem is similar to that of the STATIS (Structuration de Tableaux à Trois Indices de la Statistique) method [LAV 88].

– *By dimension*: are there factors common to all the groups of variables or to some of them? In other words, do the clouds $\mathcal{N}_I^t$ present correlated directions? These questions refer to the aim of canonical analysis [HOT 36].

– *By individual*: which are the individuals who, relative to the others, lie in a similar place in the clouds $\mathcal{N}_I^t$? Which individuals lie in different places? These questions return to the aim of procustean analysis [GOW 75, GRE 52].

For each of these problems, MFA proposes a solution. All these solutions are expressed within a single scope, which at the same time enriches and simplifies their interpretation.

Note that the results discussed in this chapter were calculated using SPAD (système pour l'analyse des données) [SPA 01] and the free R package FactoMineR [HUS 08].

2.5. Bibliography

[BEN 73]   BENZECRI J.-P., *L'Analyse des Données, Tome 1 : La Taxinomie, Tome 2 : l'Analyse des Correspondances*, Dunod, Paris, 1973.

[BEN 82]   BENZECRI J.-P., "Sur la généralisation du tableau de Burt et son analyse par bandes", *Les Cahiers de l'Analyse des Données*, vol. 7, p. 33–45, 1982.

[BEN 83]   BENZECRI J.-P., "Analyse de l'inertie intraclasse par l'analyse d'un tableau de correspondance", *Les Cahiers de l'Analyse des Données*, vol. 8, num. 3, p. 351–358, 1983.

[CAR 94]   CARLIER A., "Méthodes exploratoires", CELEUX G., NAKACHE J.-P., Eds., *Analyse Discriminante sur Variables Qualitatives*, p. 115–146, Polytechnica, Paris, 1994.

[CAZ 77]   CAZES P., "Etude de quelques propriétés extrèmales des facteurs issus d'un sous-tableau d'un tableau de Burt", *Les Cahiers de l'Analyse des Données*, vol. 2, num. 2, p. 143–160, 1977.

[CAZ 88]   CAZES P., CHESSEL D., DOLEDEC S., "L'analyse des correspondances interne d'un tableau partitionné: son usage en hydrobiologie", *Revue de Statistique Appliquée*, vol. 46, num. 1, p. 39–54, 1988.

[ESC 84] ESCOFIER B., "Analyse factorielle en référence à un modèle. Application au traitement des tableaux d'échanges", *Revue de statistique appliquée*, vol. 32, num. 4, p. 25–36, 1984.

[ESC 08] ESCOFIER B., PAGES J., *Analyses factorielles simples et multiples. Objectifs, méthodes et interprétation*, Dunod, Paris, 4th edition, 2008.

[GOW 75] GOWER J.-C., "Generalized procrustes analysis", *Psychometrika*, vol. 40, p. 33–51, 1975.

[GRE 52] GREEN B.-F., "The orthogonal approximation of an oblique structure in factor analysis", *Psychometrika*, vol. 17, p. 429–440, 1952.

[GRE 84] GREEN M.-J., *Theory and Applications of Correspondence Analysis*, London Academic Press, London, 1984.

[HOT 36] HOTTELLING H., "Relations between two sets of variables", *Biometrika*, vol. 8, p. 277–321, 1936.

[HUS 08] HUSSON F., JOSSE J., LE S., MAZET J., FactoMineR: Factor Analysis and Data Mining with R, 2008, R package version 1.09.

[LAV 88] LAVIT C., *Analyse Conjointe de Tableaux Quantitatifs*, Dunod, Paris, 1988.

[LEB 75] LEBART L., FENELON J.-P., *Statistique et Informatique Appliquée*, Dunod, Paris, 3rd edition, 1975.

[LEB 77] LEBART L., MORINEAU A., TABARD N., *Technique de la Description Statistique*, Dunod, Paris, 1977.

[LEB 06] LEBART L., MORINEAU A., PIRON M., *Statistique Exploratoire Multidimensionnelle*, Dunod, Paris, 4th edition, 2006.

[MOR 00] MOREAU J., DOUDIN P.-A., CAZES P., *L'analyse des Correspondances et Techniques Connexes*, Springer, New York, 2000.

[PAG 01] PAGES J., HUSSON F., "Inter-laboratory comparison of sensory profiles : methodology and results", *Food Quality and Preferences*, vol. 12, p. 297–309, 2001.

[PAG 02] PAGES J., "Analyse factorielle multiple appliquée aux variables qualitatives et aux données mixtes", *Revue de Statistique Appliquée*, vol. 50, num. 4, p. 5–37, 2002.

[SAP 75] SAPORTA G., Liaisons entre plusieurs ensembles de variables et codages de donnés qualitatives, Thesis, University of Paris VI, 1975.

[SPA 01] SPAD, *Système portable pour l'analyse des données*, Software distributed by COHERIS, Courbevoie, 2001.

Chapter 3

Exploratory Projection Pursuit

3.1. Introduction

This chapter is somewhat different from most of the other chapters in this book. Principal component analysis (PCA), correspondence analysis and all the various clustering techniques are now of routine use. They are available in all data analysis software programs and thousands or even millions of examples have been processed. Projection pursuit techniques are much more confidential; they are only available in specific software programs and few examples have been processed apart from those proposed by the originators of the various techniques.

However, less extensive use is also due to another reason. In fact, not only are projection pursuit techniques fairly young, but their purpose is ambitious – sometimes ambiguous – leading to a large variety of (often heuristic) techniques. In consequence, their use most often needs the help of a confirmed statistician. Moreover, it must be recognized that classical 'generalist' techniques are sometimes sufficient to meet the same goal (e.g. PCA in Chapter 1 for the detection of outliers). Nonetheless, the projection pursuit techniques are often very useful to replace, or at least to complement, the most popular analyses by refining their exploratory capability; they therefore should belong to the current toolboxes of the data analyst. The aim of this chapter is to help the reader share this opinion by providing a survey of the questions to be solved, of the theoretical approaches which have been proposed and of the practical tools which are now available.

Chapter written by Henri CAUSSINUS and Anne RUIZ-GAZEN.

3.2. General principles

3.2.1. *Background*

As in many questions of data analysis, let us consider data gathered in a $n \times p$ matrix $\mathbf{X}$. The rows of $\mathbf{X}$ are the transposes x_i' of the column vectors x_i ($i = 1, \ldots, n$) associated with n units or individuals. Each x_i corresponds to p measurements (variables). Principal component analysis is the more usual method to analyze (and visualize) this dataset (see Chapter 1). PCA can be replaced by variants such as correspondence analysis when the data are categorical (see Chapter 2) or more sophisticated methods in other special cases (see, for example, Chapter 5 for the case where the p measurements arise from the discretization of a function).

All of these methods depend upon singular value decomposition which leads to comprehensive simultaneous visualization of rows and columns of $\mathbf{X}$ (biplots according to Gabriel's terminology; see [GAB 71]). However, this kind of analysis uses only second order moments (inertia or variance), which is a serious limitation for full efficiency (e.g. [SIB 84]). Important aspects of the data structure are not likely to appear in any of the principal subspaces, as seen in Figure 3.1. When looking at this figure, imagine that the whole dimension is larger than 2 and the straight lines are parallel to subspaces of dimension 2. The first principal plane of PCA – roughly the horizontal axis in this example – is clearly unable to reveal the two clusters. These clusters could only appear by chance on further projections and the outlier cannot appear on any principal 2D subspace but only by projection on the oblique line.

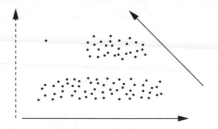

Figure 3.1. *What is an interesting projection?*

Actually, maximizing the variance of the display is likely to provide useful projections of the data, but obtaining such good projections is random. In the 1970s, with the increasing computational facilities, the idea emerged that good projections could be searched for in a more direct way.

The term projection pursuit has been coined by [FRI 74] to refer to a sequence of projections looking for hidden structures within a dataset; for that purpose, they optimize an index of heterogenity (see below). In parallel to the search for fixed interesting projections, the possibility of a dynamic exploration of the data has also been investigated [ASI 85]. The present chapter is mainly devoted to the search for good fixed views, but both approaches are complementary (e.g. [COO 95] and the example below in section 3.5).

When we look for an interesting display of the dataset, the first question is, of course: what is an interesting display? In Figure 3.1, for example, two directions of projection seem interesting: the projection on the vertical axis separates two groups and the projection on the oblique axis shows the presence of an isolated unit. The second question is: how can an interesting projection be characterized and searched for? A third question finally arises: can we rely on the projection which is thus found? Actually, if it is interesting to reveal a 'true' hidden structure, it is also important to be wary of an illusory one. Before discussing these points below, let us complete this section with two important references concerning the basic aspects of projection pursuit, both including an extensive discussion: [HUB 85, JON 87].

3.2.2. What is an interesting projection?

It is perhaps easier to answer the question: what is an uninteresting projection? An uninteresting projection is obviously a homogenous one. However, homogenity is not to be confused with no dispersion or even with small dispersion. On the one hand, the magnitude of dispersion depends on the measurement units; on the other hand, a larger dispersion can make some structural features clearer (this is the paradigm of PCA) but is not necessarily related to such features. A distribution is said to be homogenous if it is analogous to the distribution of an 'error' (a 'noise'). Various arguments lead us to consider that the 'most homogenous distribution' is the normal one (whatever its variance) from the historical arguments by Gauss to the argument of entropy: if the mean and the variance are given, the distribution with maximum entropy is the normal. In the same vein, Diaconis and Freedman [DIA 84] show that most projections of many high-dimensional clouds of points are nearly normal, which emphasizes that normality is not interesting since it is a banal feature.

It is worth noting that other distributions close to the normal can also be considered as 'standards of uninterestingness'. For example, Nason [NAS 01] advocates the Student distribution; Jones and Sibson [JON 87] note that the approach in [FRI 74] is implicitly related to a parabolic rather than a normal distribution. However, in this chapter, the normal distribution will be considered as the main standard of homogeneity. Interesting projections are therefore as far as possible from the normal. The problem is now to find a suitable measure of non-normality. This is introduced below and developed in section 3.3.

3.2.3. *Looking for an interesting projection*

Since the normal distribution is invariant under any affine transformation, a measure of non-normality should have the same property (method of type III) according to [HUB 85]. The simplest way to fulfill this requirement is to render the data spherical by means of the transformation $x_i \mapsto \mathbf{S}_n^{-1/2}(x_i - \overline{x}_n)\,(i = 1, \ldots, n)$ where $\overline{x}_n$ is the empirical mean and $\mathbf{S}_n$ is the empirical covariance matrix of the x_is. This transformation is carried out in the following without changing the notation for the units (unless otherwise stated).

The distance from normality can be measured by means of several indexes according to the kind of non-normality that is looked for. Most indexes proposed in the literature are described in section 3.3. Due to our visual perception, projections on a plane are the most useful [COO 93]. For this reason we focus on indexes for bi-dimensional distributions; most of them are obvious extensions of 1D indexes. They can also be generalized to higher order dimensions. Let $\Omega_p = \{(\mathbf{a}, \mathbf{b}) \in \mathbb{R}^p \times \mathbb{R}^p, \mathbf{a}'\mathbf{a} = 1, \mathbf{b}'\mathbf{b} = 1, \mathbf{a}'\mathbf{b} = 0\}$ (recall that a vector of $\mathbb{R}^p$ is assimilated to a column matrix and $\mathbf{a}'$ is the transpose of $\mathbf{a}$). A 2D index is a function $I : (\mathbf{a}, \mathbf{b}, \mathbf{X}) \mapsto \mathbb{R}$. When the index I has been defined, $I(\mathbf{a}, \mathbf{b}, \mathbf{X})$ is optimized with respect to $\mathbf{a}$ and $\mathbf{b}$ by means of classical algorithms. This provides a local optimum that depends on the initial $\mathbf{a}$ and $\mathbf{b}$.

We can look for several solutions associated with several choices of the initial values; the quality of the corresponding projections can be assessed both by the value of the index and by the apparent usefulness of the resulting displays. Authors of software propose various ways to deal with the whole process.

As well as the optimization of an index, another way to obtain interesting projections is to use a suitable generalization of PCA. Here again the aim is to find projections which are far from the normal. These methods of generalized PCA (GPCA) are presented in section 3.4. They are very simple to implement and provide a global optimum for any dimension of the projection subspace. Moreover, they are invariant under affine transformation and do not need previous spherization of the data.

3.2.4. *Inference*

An empirical index of non-normality is a statistic submitted to sampling variations; for fixed $\mathbf{a}$ and $\mathbf{b}$ the random variable $I(\mathbf{a}, \mathbf{b}, \mathbf{X})$ can take large values even if there is no structure in the data. The problem is even more serious with the maximum value of the index; the displays obtained by optimizing the index can be misleading by suggesting a structure where there is none. It is therefore important to test whether or not the graphical display is significant. For that, it is necessary to study the probability distribution of the maximum of I on Ω_p. This can be done in two ways:

– Generate a large number of n-samples from a spherical p-normal distribution to estimate critical values for the maximum of I. Most authors propose this simulation process, but the computation is extremely expansive for large n.

– Aim to derive the asymptotic distribution of the optimal value of the index. This is a difficult problem concerning the maximum of a stochastic process on Ω_p and the mathematical details are outwith the scope of this book. We merely refer to [LI 93, OHR 90, SUN 91] for the first solutions concerning 1D indexes, and we shall mention some other more general asymptotic testing procedures below.

3.2.5. *Outliers*

Most often, projection pursuit techniques deal with the search for a 'global' structure, in particular the search for clusters; this is the main purpose of the indexes described in section 3.3. However, the user is first interested in obtaining projections that display the outlying units if any are present in the dataset. On the one hand, detecting and positioning outliers is useful *per se*; on the other hand, outliers can seriously disturb the techniques based on the indexes of section 3.3 (the question of robustness will be evoked in this section). Pan *et al.* [PAN 00] examine the outlying character of the units one by one. A more global method is discussed in section 3.4. Concerning the detection of outliers, it is worth noting that inferential aspects are prominent in order to decide whether or not a unit should be considered as discordant. Pan *et al.* [PAN 00] stress that point; it will be also examined for the method described in section 3.4.

3.3. Some indexes of interest: presentation and use

Several families of projection indexes can be distinguished depending on whether the authors consider an entropy measure [COO 93, FRI 87, HAL 89], use a L^2 based distance [FRI 74, HUB 85, JON 87] or a Pearson chi-squared discrepancy measure [POS 90]. As mentioned above, all these indexes have to be calculated on sphered data.

3.3.1. *Projection indexes based on entropy measures*

3.3.1.1. *The Friedman–Tukey index*

The Friedman and Tukey original idea [FRI 74] consists of finding projection directions on which the observations are 'locally dense' and, eventually, form groups. To that aim, the index they propose is a measure of local density defined in the 2D case by:

$$I_{\mathrm{FT}}(\mathbf{a}, \mathbf{b}, \mathbf{X}) = \frac{1}{n^2 h^2} \sum_{i=1}^{n} \sum_{j=1}^{n} K\left(\frac{\mathbf{a}'(\boldsymbol{x}_i - \boldsymbol{x}_j)}{h}, \frac{\mathbf{b}'(\boldsymbol{x}_i - \boldsymbol{x}_j)}{h}\right)$$

where K is a bivariate kernel (in general equal to the product of univariate kernels) and h is a positive real number defining the window width. Denote the density of the projection of an observation x_i on the plane spanned by $\mathbf{a}$ and $\mathbf{b}$ by $f_{\mathbf{a},\mathbf{b}}$. Huber [HUB 85] and Jones and Sibson [JON 87] proved that the Friedman–Tukey index estimates, up to a proportionality factor:

$$\int_{\mathbb{R}^2} [f_{\mathbf{a},\mathbf{b}}(x,y)]^2 \, dx \, dy,$$

which is a monotonic function of the information or entropy of order 2 (according to the Rényi definition; see [RÉN 70]). Note that this information is not minimal for the Gaussian distribution but for a parabolic density.

3.3.1.2. *The Huber index*

This entropy index was proposed by Huber [HUB 85] and Jones and Sibson [JON 87]. It is defined in the 2D case by:

$$I_{\mathrm{ENT}}(\mathbf{a}, \mathbf{b}, \mathbf{X}) = \frac{1}{n} \sum_{i=1}^{n} \log \left(\frac{1}{nh^2} \sum_{j=1}^{n} K \left(\frac{\mathbf{a}'(x_i - x_j)}{h}, \frac{\mathbf{b}'(x_i - x_j)}{h} \right) \right)$$

and estimates:

$$\int_{\mathbb{R}^2} f_{a,b}(x,y) \log(f_{a,b}(x,y)) \, dx \, dy$$

which is the usual negentropy measure (of order 1 according to [RÉN 70]). Among the spherical absolutely continuous distributions on $\mathbb{R}^2$, the negentropy reaches its minimum value for a Gaussian distribution.

Both of the previous indexes require a kernel function K and an associated window width h for the bivariate density estimate of the projected data to be chosen. This question is addressed in [SIL 86, section 6.5.3] and [SCO 92, section 7.3.2], and considered in more details in [KLI 97, POL 95] who propose using a triweight kernel because the derived indexes are twice differentiable and easy to compute. Furthermore, according to [JON 87], these indexes are not too sensitive to the window width choice. However, a difficulty may occur if the window width is too small: numerical problems arise with the existence of local 'pseudo' maxima for the associated indexes.

Following a suggestion by Huber [HUB 85], Jones and Sibson [JON 87] propose an univariate index and its bivariate generalization based on the 3rd and 4th order moments of the distribution of the projected data. These indexes may be considered as 'approximations' of the entropy index. Nason [NAS 95] generalized this idea to 3D. The advantage of these indexes lies on their easy computation which does not heavily depend on the number of observations. However, their main drawback is their extreme sensitivity to a deviation from normality in the tails of the distribution. These indexes are therefore attracted by potential outliers or even by 'pseudo-outliers' [FRI 87].

3.3.1.3. *Computing procedures*

For all of the above-mentioned indexes, computing procedures have usually been developed by the authors themselves. Jones proposed Fortran procedures to compute the univariate and bivariate Friedman–Tukey indexes, as well as the Jones and Sibson indexes based on entropy or moments. Nason wrote a Splus function to calculate his 3D index based on moments. All these procedures are available on the web, notably on Nason's website [NAS 09]; the software XGobi and Xplore propose interactive vizualization of multivariate data in 1D, 2D or 3D. They incorporate the above-mentioned bivariate indexes to assist the exploration of the data. XGobi is free and available on the web [XGO 09] as well as the R-package xgobi which interfaces with the XGobi programs.

3.3.2. *Projection indexes based on L^2 distances*

Let us now consider some measures of distance in L^2-norm between the density of the projected data and the Gaussian density. Developing the density of the projected observations on an orthogonal polynomial basis, these indexes can be expressed as a sum of squared coefficients of the development.

3.3.2.1. *The Friedman index or the index based on Legendre polynomials*

In the 2D case, the index is calculated in the following way. Let Φ (respectively ϕ) be the cumulative distribution function (respectively the density) of the standardized 1D Gaussian distribution and $T : \mathbb{R} \rightarrow \mathbb{R}$ the transformation defined by $T(x) = 2\Phi(x) - 1$. Let us consider the transformation of the projected data:

$$(T(\mathbf{a}'\boldsymbol{x}_i), T(\mathbf{b}'\boldsymbol{x}_i)) = (2\Phi(\mathbf{a}'\boldsymbol{x}_i) - 1, \ 2\Phi(\mathbf{b}'\boldsymbol{x}_i) - 1)$$

where $g_{\mathbf{a},\mathbf{b}}$ denotes the density associated with the transformed projected data distribution and $f_{\mathbf{a},\mathbf{b}}$ is the density associated with the untransformed projected data. Let us consider the Legendre polynomials $P_k, k \in \mathbb{N}$, defined by:

$$\begin{cases} P_0(x) & = & 1 \\ P_1(x) & = & x \\ k\,P_k(x) & = & (2k-1)xP_{k-1}(x) - (k-1)P_{k-2}(x) \quad \text{for } k \geq 2 \end{cases}$$

and such that:

$$\int_{-1}^{1} P_k(x)P_l(x)dx = \begin{cases} 0 & \text{if } k \neq l \\ 2/(2k+1) & \text{if } k = l \end{cases} \quad \text{with } (k,l) \in \mathbb{N}.$$

The Friedman bivariate index [FRI 87] is defined by:

$$I_{\mathrm{FRI}}(\mathbf{a}, \mathbf{b}, \mathbf{X}) = \sum_{k=0}^{J}\sum_{l=0}^{J-k} \frac{(2k+1)(2l+1)}{4}\left[\frac{1}{n}\sum_{i=1}^{n} P_k(T(\mathbf{a}'\mathbf{x}_i))P_l(T(\mathbf{b}'\boldsymbol{x}_i))\right]^2 - \frac{1}{4}$$

where J is the development order of the density $g_{\mathbf{a},\mathbf{b}}$ in the Legendre polynomials basis.

Cook *et al.* [COO 93] proved that this index estimates:

$$\int_{\mathbb{R}^2} [f_{\mathbf{a},\mathbf{b}}(x,y) - \phi(x)\phi(y)]^2 \, \frac{1}{4\phi(x)\phi(y)} \, dx \, dy = \int_{-1}^{+1} \int_{-1}^{+1} g_{\mathbf{a},\mathbf{b}}^2(x,y) \, dx \, dy - \frac{1}{4}$$

for large J; more precisely, the index estimates the part of the right-hand side integral corresponding to the J first terms of the development of $g_{\mathbf{a},\mathbf{b}}$ in the Legendre polynomials basis. Because of the factor $w(x,y) = 1/(4\phi(x)\phi(y))$ in the left-hand side integral, the Friedman index is sensitive to deviations from normality in the tails of the projected data distribution. Hall [HAL 89] and Cook *et al.* [COO 93] proposed similar indexes with different factors w in order to reduce sensitivity to these deviations.

3.3.2.2. *The Hall index and the Cook et al. index or indexes based on Hermite polynomials*

Following the same principle as Friedman [FRI 87], Hall [HAL 89] and Cook *et al.* [COO 93] proposed indexes based on developments of the projected data density $f_{\mathbf{a},\mathbf{b}}$ in the Hermite polynomials basis.

The Hall index is an estimator $I_{\text{HAL}}(\mathbf{a}, \mathbf{b}, \mathbf{X})$ of:

$$\int_{\mathbb{R}^2} [f_{\mathbf{a},\mathbf{b}}(x,y) - \phi(x)\phi(y)]^2 \, dx \, dy,$$

or rather an estimator of the part corresponding to the first J terms of the development of $f_{\mathbf{a},\mathbf{b}}$ in the Hermite polynomials basis.

In a similar way, the Cook *et al.* index is an estimator $I_{\text{COO}}(\mathbf{a}, \mathbf{b}, \mathbf{X})$ of:

$$\int_{\mathbb{R}^2} [f_{\mathbf{a},\mathbf{b}}(x,y) - \phi(x)\phi(y)]^2 \, \phi(x)\phi(y) \, dx \, dy.$$

The factor $w(x,y) = \phi(x)\phi(y)$ gives more weight to deviations from normality at the center of the distribution.

For the indexes $I_{\text{FRI}}, I_{\text{HAL}}, I_{\text{COO}}$, we have to choose the number of terms J in the orthogonal polynomials basis . This choice has been studied by Cook *et al.* [COO 93], Sun [SUN 93] and Polzehl and Klinke [POL 95], but their conclusions differ substantially implying that the question is far from being resolved.

3.3.2.3. *Computing procedures*

The library *Interactive Projection Pursuit* (IPP) by Sun, for Splus projection pursuit, implements the univariate and bivariate Friedman indexes together with significance tests. It is free and available on Sun's website [SUN 09]. The software XGobi and Xplore incorporate the three indexes based on orthogonal polynomials. Note that in XGobi the Hall index is referred to as the Hermite index while the Cook *et al.* index is referred to as the 'natural' Hermite index.

3.3.3. *Chi-squared type indexes*

Posse [POS 90, POS 95a, POS 95b] proposed and studied a chi-squared type bivariate index defined as follows. The projection plane is divided into 48 boxes which are pieces of rings. On the one hand, the boxes are limited by 4 straight lines crossing at the origin at angles of 45 degrees; on the other hand, they are limited by 5 circles centered on 0 with radius $ir/5$, $i = 1, \ldots, 5$. Other choices for the number of classes are possible but this particular one is advocated by the author; moreover he recommends $r^2 = 2 \ln 6$ in order to have approximately balanced classes. If $\mathbf{a}$ and $\mathbf{b}$ are two orthogonal vectors of the projection space corresponding to two of the previously mentioned straight lines, a first index $I(\mathbf{a}, \mathbf{b}, \mathbf{X})$ [POS 90] is obtained from the usual Pearson chi-squared by summing the 48 terms of the form (observed frequency – theoretical frequency)^2/(theoretical frequency). The observed frequency of a class is the number of observations projected in this class and the theoretical frequency is the probability of this class for the spherical Gaussian distribution (recall that the data have been sphered).

For a given projection plane, the index $I(\mathbf{a}, \mathbf{b}, \mathbf{X})$ depends on the choice of $\mathbf{a}$ and $\mathbf{b}$, i.e. it depends on a rotation in this plane. In order to obtain an index that depends only on this plane, Posse [POS 95a] proposes replacing the use of one $I(\mathbf{a}, \mathbf{b}, \mathbf{X})$ by an average of indexes obtained through a series of rotations. In practice, from an arbitrary starting position, it is enough to calculate the mean of 9 indexes, associated with rotation angles $j\pi/36$ ($j = 0, \ldots, 8$), respectively. This new index, denoted below by I_{POS}, is efficient in many examples. Among its manifest properties, it is not sensitive to outliers, a property which is not shared by many indexes. The problem of significance levels is addressed in [POS 95a].

The MatLab library Computational Statistics [MAR 01] contains a function called 'csppeda' for computational statistics projection pursuit exploratory analysis which computes the index I_{POS}. This library is free and available on the website StatLib.

3.3.4. *Indexes based on the cumulative empirical function*

All previous indexes measure a kind of non-normality of the projected data through an estimate of their density function. All these estimates involve a tuning parameter: bandwidth for the kernel (section 3.3.1), order of truncation for polynomial expansions (section 3.3.2) and number of classes for the histogram (section 3.3.3). The value of this tuning parameter can significantly affect the behavior of the index. To eliminate the tuning problem, Perisic and Posse [PER 05] propose indexes based on the empirical cumulative distribution function (ECDF) of the projected data instead of the density.

Let us focus on 2D projections. A problem arises: while an index should be affine invariant, the ECDF depends on the quadrants defined by the axes, i.e. depends on the choice of a given rotation in the projection plane. As in the previous section, affine invariance is obtained by averaging over all possible rotations in this plane. Fortunately, it is enough in practice to consider only a few rotations (say 8 or 16). Let R denote the number of rotations, $(\mathbf{a}, \mathbf{b})$ an orthonormal basis of the plane of projection and $(\mathbf{a}_r, \mathbf{b}_r)$ the basis obtained by rotation of angle $2\pi r/R$ in this plane ($r = 0, \ldots, R-1$). Let F_n denote the ECDF of the (projected) data and Φ the standard normal distribution function. Then I_{KS} is defined by:

$$I_{\mathrm{KS}}(\mathbf{a}, \mathbf{b}, \mathbf{X}) = \frac{1}{R} \sum_{r=0}^{R-1} \max_i |F_n(\mathbf{a}_r' \boldsymbol{x}_i, \mathbf{b}_r' \boldsymbol{x}_i) - \Phi(\mathbf{a}_r' \boldsymbol{x}_i) \Phi(\mathbf{b}_r' \boldsymbol{x}_i)| .$$

Let Φ denote the standard normal distribution function and $G_n(s, t)$ the ECDF of the projected data coordinate transformed by Φ. I_{CvM} is defined by:

$$I_{\mathrm{CvM}} = \frac{1}{R} \sum_{r=0}^{R-1} [G_n(\Phi(\mathbf{a}_r' \boldsymbol{x}_i), \Phi(\mathbf{b}_r' \boldsymbol{x}_i)) - \Phi(\mathbf{a}_r' \boldsymbol{x}_i) \Phi(\mathbf{b}_r' \boldsymbol{x}_i)]^2$$

where KS (for Kolmogorov–Smirnov) and CvM (for Cramer–von Mises) are obviously related to the distance involved between the normal and the empirical distribution of the data. Perisic and Posse [PER 05] show that these indexes perform well compared to various others, with a slight advantage for I_{CvM}. They also propose two other indexes which look more specifically for asymmetry in the data distribution. The authors do not mention a ready-to-use computational procedure, but provide an algorithm for fast evaluation of the bivariate ECDF.

3.4. Generalized principal component analysis

3.4.1. *Theoretical background*

Generalized principal component analysis (GPCA) provides an alternative way to draw out a structure from a noise. Suppose that the mean of $\boldsymbol{x}_i$ is zero and express the theoretical covariance matrix of $\boldsymbol{x}_i$ as:

$$\Sigma = \Sigma_B + \Sigma_W \tag{3.1}$$

where Σ_B is the variance of the 'structural' part while Σ_W is the variance of the noise. Σ_W will be assumed non-singular. Considering first a 1D projection, it is natural to look for the linear combination, say $\mathbf{a}'\boldsymbol{x}_i$, which realizes the maximum of the structural variance versus the variance of the noise, i.e the maximum of the ratio $\mathbf{a}'\Sigma_B\mathbf{a}/\mathbf{a}'\Sigma_W\mathbf{a}$.

According to a well-known property of the eigenvectors and eigenvalues, this maximum is reached when $\mathbf{a}$ is an eigenvector of $\Sigma_W^{-1}\Sigma_B$ (or, equivalently, $\Sigma_W^{-1}\Sigma$) associated with the largest eigenvalue. If $\mathbf{u}$ is the Σ_W^{-1}-normed eigenvector of $\Sigma_B\Sigma_W^{-1}$ (or $\Sigma\Sigma_W^{-1}$) associated with the largest eigenvalue, $(\mathbf{a}'\boldsymbol{x}_i)\mathbf{u} = (\mathbf{u}'\Sigma_W^{-1}\boldsymbol{x}_i)\mathbf{u}$ is the Σ_W^{-1}-orthogonal projection of $\boldsymbol{x}_i$ on the subspace spanned by $\mathbf{u}$. (Recall that, for a positive definite matrix $\mathbf{M}$, the $\mathbf{M}$ norm of a vector $\boldsymbol{x}$ is $\boldsymbol{x}'\mathbf{M}\boldsymbol{x}$ and the vectors $\boldsymbol{x}$ and $\boldsymbol{y}$ are $\mathbf{M}$-orthogonal if and only if $\boldsymbol{x}'\mathbf{M}\boldsymbol{y} = 0$.) This gives the best 1D projection and the process can be continued to obtain projections on k-dimensional subspaces: the best k-dimensional projection of the units $\boldsymbol{x}_i$ is the Σ_W^{-1}-orthogonal projection on the subspace spanned by the k eigenvectors associated with the k largest eigenvalues of $\Sigma_B\Sigma_W^{-1}$ (or $\Sigma\Sigma_W^{-1}$).

This is a generalized PCA with metric Σ_W^{-1}. At this point, the dimension k is arbitrary. However, it is worth noting that, if $\mathrm{rank}(B) = q$, i.e. if all the interesting structural features lie in a q-dimensional subspace (the image of Σ_B), this subspace is spanned by the eigenvectors associated with the q strictly positive eigenvalues of $\Sigma_B\Sigma_W^{-1}$ (or the q eigenvalues of $\Sigma\Sigma_W^{-1}$ strictly larger than 1).

In practice, neither Σ_B, Σ_W or even the reality of model (3.1) are known. If the units were divided into known groups with different means and the same variance, the problem above would be the problem of discriminant factor analysis: the matrix Σ_W could be estimated as the within-group covariance matrix and Σ_B as the between-group variance. However, the problem we have in mind is very different: it is not known that the units are divided into groups and, even if this is the case, the units are not assigned to one of them (moreover the number of groups is unknown). If the 'true' structure is not known in practice, the user knows what kind of structure they are looking for. This is the key to the method that will allow the necessary estimation to be calculated. In section 3.4.2, some heuristic solutions proposed in the literature are reviewed. Developments of two of these solutions are given in section 3.4.3, including inferential aspects.

To deal with inference, it is necessary to base the model upon a probabilistic model. We provide the model employed here since it can help to understand the underlying principles of the method. The $\boldsymbol{x}_i$ are assumed to be random vectors, independently and identically distributed according to the mixture distribution:

$$\int \mathcal{N}_p(\boldsymbol{x}, \Sigma_W) \, dP(\boldsymbol{x}). \tag{3.2}$$

This means that each x_i is the sum of a random vector x whose probability distribution is P and a random normal $N_p(0, \Sigma_W)$ (the noise). Within this framework, P is the structural part of the distribution. Acording to previous assumptions, P is centered on zero; if the variance of P is Σ_B, the variance of mixture equation (3.2) is given by equation (3.1); if the whole distribution P lies in a q-dimensional subspace, say F, the rank of Σ_B is q. The aim of the analysis consists of finding a projection on F – or at least a good approximation of F – since this subspace contains all the information on the structural features of the dataset. In the special case where P is discrete with k possible values, equation (3.2) is a model for the repartition into k groups with the same variance Σ_W; the k means belong to a q-dimensional subspace with $q \leq k - 1$. (For the use of finite mixture models in supervised classifications, see Chapter 8.)

3.4.2. *Practice*

3.4.2.1. *Displaying outliers*

Outliers are (a small number of) units that are discared from the main more or less homogenous set of data. In the presence of outliers, the 'natural variability' – that of noise – is the variability of the data bulk. Within the framework of model (3.2), a robust estimate of the covariance matrix (say $\mathbf{V}_n$) is an estimate of Σ_W, while the ordinary empirical variance $\mathbf{S}_n$ is an estimate of Σ. The principle described above can then be implemented with $\mathbf{S}_n$ and $\mathbf{V}_n$ in place of Σ and Σ_W. The idea is founded upon work by Yenyukov [YEN 88]; Caussinus and Ruiz-Gazen [CAU 90, CAU 93, CAU 95] provided several developments (section 3.4.3).

3.4.2.2. *Groups and complex structures*

It is tempting to look for an estimate of Σ_W in the form of a local variance:

$$\mathbf{S}_W = \frac{1}{n(n-1)} \sum_{i=1}^{n-1} \sum_{j=i+1}^{n} \delta(\mathbf{X}, i, j) \, (x_i - x_j)(x_i - x_j)'$$

where δ can depend on the whole dataset $\mathbf{X}$ and the unit numbers (i, j). In practice, δ is chosen to cut off (or at least underweight) distant units in order to ensure the local character of $\mathbf{S}_W$; for example, if the units are divided into groups, two close units are expected to belong to the same group. (Note that $\mathbf{S}_W$ is the empirical variance $\mathbf{S}_n$ if $\delta = 1$.) Various choices of δ have been proposed. Art *et al.* [ART 82] propose a 'near neighbor' choice with δ equal to 1 or 0 depending on whether or not

$$\|x_i - x_j\|_{\mathbf{S}_W^{-1}} = (x_i - x_j)' \mathbf{S}_W^{-1} (x_i - x_j)$$

is one of the m smallest among these distances.

It is natural to measure the distance through $\mathbf{S}_W^{-1}$ as a Mahalanobis distance, but this necessitates an iterative computation. The optimal value for m is briefly discussed. This value m is further discussed by Burtschy and Lebart [BUR 91] who show that the method is actually a contiguity analysis where the graph of proximities is drawn from the data. These authors recommend choosing m close to $n/10$; they also mention the possibility of choosing δ as a smooth function (as below with a different choice). Lebart [LEB 01] returns to this question and suggests making m vary from 0 to n. Caussinus and Ruiz [CAU 90] suggest a smooth non-iterative version $\delta = K(\|x_i - x_j\|^2_{\mathbf{S}_n^{-1}})$ where K is a kernel function.

For this choice of δ, they give practical and theoretical developments concerning, in particular, the cases where the method is able to meet its goal [CAU 93, CAU 95]; tests are provided in [CAU 03b]. The following section accounts for these developments. Further theoretical results can be found in [TYL 09].

3.4.3. *Some precisions*

Let us consider the matrices:

$$\mathbf{S}_n = \frac{1}{n} \sum_{i=1}^{n} (x_i - \overline{x}_n)(x_i - \overline{x}_n)'$$

$$\mathbf{V}_n(\beta) = \frac{\sum_{i=1}^{n} \exp(-\frac{\beta}{2} \|x_i - \overline{x}_n\|^2_{\mathbf{S}_n^{-1}})(x_i - \overline{x}_n)(x_i - \overline{x}_n)'}{\sum_{i=1}^{n} \exp(-\frac{\beta}{2} \|x_i - \overline{x}_n\|^2_{\mathbf{S}_n^{-1}})}$$

$$\mathbf{T}_n(\beta) = \frac{\sum_{i=1}^{n-1} \sum_{j=i+1}^{n} \exp(-\frac{\beta}{2} \|x_i - x_j\|^2_{\mathbf{S}_n^{-1}})(x_i - x_j)(x_i - x_j)'}{\sum_{i=1}^{n-1} \sum_{j=i+1}^{n} \exp(-\frac{\beta}{2} \|x_i - x_j\|^2_{\mathbf{S}_n^{-1}})}$$

$$\mathbf{U}_n(\beta) = \left(\mathbf{V}_n(\beta)^{-1} - \beta \mathbf{S}_n^{-1}\right)^{-1}$$

where $\overline{x}_n = 1/n \sum_{i=1}^{n} x_i$, $\|x\|^2_{\mathbf{M}} = x'\mathbf{M}x$ and β is a real parameter. In $\mathbf{V}_n(\beta)$ it can be advisable to replace $\overline{x}_n$ by a robust estimate of the mean. In the following, β will be omitted when not necessary and $\mathbf{V}_n(\beta)$, $\mathbf{T}_n(\beta)$ and $\mathbf{U}_n(\beta)$ are written $\mathbf{V}_n$, $\mathbf{T}_n$ and $\mathbf{U}_n$, respectively. It must be stressed however, that (1) all these matrices depend on such a tuning parameter, (2) a suitable choice of β is important but guidelines can be provided and (3) this choice can be very different according to the matrix involved. We consider various GPCA formulated as in section 3.4.1 with $\mathbf{S}_n$ or $\mathbf{U}_n$ instead of $\mathbf{S}$ and $\mathbf{V}_n$, or $\mathbf{T}_n$ instead of $\mathbf{S}_W$.

The theoretical results below are derived within the framework of model (3.2). It is said that a method 'works' if, when n tends to infinity, the principal q-dimensional subspaces of the GPCA tends to the subspace containing the probability P, i.e. the method works if the projections are asymptotically on the 'good' subspace. A common property of all these GPCA is invariance under any affine transformation. This means that the same graphical display is obtained for the raw units and for any affine transformation of them, including non-orthogonal transformations. This emphasizes that the display focuses on the structural aspects of the data. In practice, the invariance ensures that the question of centering, standardizing or sphering the data is irrelevant.

3.4.3.1. *Displaying outliers*

A GPCA with metric $\mathbf{V}_n^{-1}$ can be used to display possible outliers; the centered x_i are $\mathbf{V}_n^{-1}$-orthogonally projected onto the principal subspaces of $\mathbf{S}_n\mathbf{V}_n^{-1}$. It can be proven that the method works under mild assumptions for a small value of β. All concrete examples that have been processed confirm this theoretical result; moreover, they have shown that β equal to 0.05 is generally a good choice for this parameter, with great stability of displays around this value.

Caussinus *et al.* [CAU 02, CAU 03a] give a testing procedure for the dimension of the projection, with tables of critical values. Let H_0 be the null hypothesis of normality and H_q the hypothesis that there exist outliers whose means belong to a q-dimensional subspace of $\mathbb{R}^n$. If H_0 is true, all the eigenvalues of $\mathbf{S}_n\mathbf{V}_n^{-1}$ are close to $\beta+1$. Under H_q, the q largest eigenvalues tend to be larger than $\beta+1$ while the others are still close to $\beta+1$. The qth eigenvalue is then used to test H_0 against H_q, the critical values being derived from the asymptotic distribution of the matrix $\mathbf{S}_n\mathbf{V}_n^{-1}$ under H_0.

3.4.3.2. *Displaying groups and complex structures*

For displaying groups or some other interesting structures, a similar method can be used with $\mathbf{T}_n^{-1}$ instead of $\mathbf{V}_n^{-1}$. The centered x_i are $\mathbf{T}_n^{-1}$-orthogonally projected onto the principal subspaces of $\mathbf{S}_n\mathbf{T}_n^{-1}$. It can be proven that the method works under fairly mild assumptions concerning the probability distribution P of model (3.2), with a value of β which is not too small. From the different examples that have been processed, β should be close to 2. In practice, it is advised to perform several displays with β varying from 1.5 to 2.5. (In most examples which have been considered, p is less than 8; it may be useful to widen the range for β if p is larger.) Caussinus *et al.* [CAU 03b] discuss a testing procedure for the dimension of the projection; the tests rely on the eigenvalues of $\mathbf{S}_n\mathbf{T}_n^{-1}$ which are now to be compared to $\beta+1/2$. The derivation of the asymptotic distribution of $\mathbf{S}_n\mathbf{T}_n^{-1}$ allows them to compute tables of critical values for large n.

The previous technique is very sensitive to the presence of outliers, which is not surprising since an outlier is a small group. However, it is not useful to exhibit outliers at this step since they are detected by the simplest method of section 3.4.3.1. This is even unfortunate since the projection subspace is therefore attracted by the outliers which can mask (and, in practice, does mask) other interesting features. It can be focused on these other structural features by means of a robust GPCA which rests on a robust estimate of the variance instead of $\mathbf{S}_n$.

In the present context, the most natural choice of such an estimate is the matrix $\mathbf{U}_n$ whose properties are investigated in [RUI 96]. The centered x_i are then $\mathbf{T}_n(\beta_2)^{-1}$

orthogonally projected onto the principal subspaces of $\mathbf{U}_n(\beta_1)\mathbf{T}_n^{-1}(\beta_2)$. Here, β_2 is to be chosen according to the previous suggestion (close to 2). β_1 should be larger than in section 3.4.3.1, say between 0.1 and 0.5 according to the expected number of outliers, β_1 increasing with this number (see [RUI 96]). Note that the number of expected outliers can be suggested by the technique of section 3.4.3.1.

3.4.3.3. *Biplots*

Projection pursuit aims to discover a structure within the set of units, that is the rows of the data matrix $\mathbf{X}$. A plot of the projected observations is therefore the first step of the analysis. However, it is also interesting to know which (combination of) measures are responsible for the visualized data structure. To this purpose, biplots [GAB 71] or some variants have been extensively used with PCA and correspondence analysis. They provide a simultaneous representation of rows and columns of $\mathbf{X}$, useful for understanding the relationship between the measurements and the projection of the units. Biplots can be used in a similar way with GPCA (see [GAB 02] for complete discussion and several examples).

Let $\mathbf{M}$ denote the metric (in practice $\mathbf{V}_n^{-1}$ or $\mathbf{T}_n^{-1}$) and $\mathbf{U}$ the matrix whose columns are the $\mathbf{M}$-normed eigenvectors of $\mathbf{S}_n\mathbf{M}$. The jth variable is then represented on the same graphical display as the units by vectors whose coordinates are the elements of the jth row of $\mathbf{U}$. In the whole p-dimensional space, the (ordinary) scalar product of the vectors representing the ith unit and the jth variable is the entry (i, j) of $\mathbf{X}$, i.e. the jth measurement on the ith unit. In the projection space, this scalar product is an approximation of the corresponding entry. Thus, a variable points to the direction of the units for which it takes large values.

It is important to note that the variable representation is not affine invariant. In practice, it is advisable to represent standardized variables. The length of a representing vector is then a good indicator of the overall contribution of the corresponding variable to the projection subspace, i.e. an indicator of its importance with respect to the graphical display of the units. The length of a vector associated with a standardized variable can vary from zero (no contribution, the variable is 'orthogonal' to the projection subspace) to one (maximum contribution, the variable is in the projection subspace).

3.5. Example

We consider the Lubischew [LUB 62] dataset which consists of $n = 74$ insects and $p = 6$ morphological measures. These observations are divided into three groups: group 1 (respectively, 2 and 3) is made of observations 1–21 (respectively, 22–43 and 44–74). This dataset has been studied several times within the projection pursuit context. In the present section, it illustrates the comparison of four 2D projection pursuit techniques. Three are based upon indexes implemented in different software, while the fourth is based upon GPCA implemented in R. The indexes we consider are listed as follows:

1) The natural Hermite index I_{COO} of order 4 is implemented using the software XGobi. Figure 3.2 illustrates the research of an optimum; it displays the temporal evolution of the index together with the 2D projection associated with the optimum found.

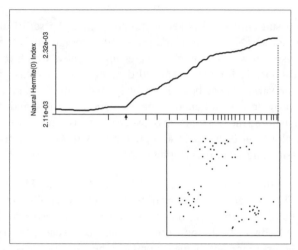

Figure 3.2. *Cook* et al. *index and associated projection for the Lubischew dataset (software XGobi)*

2) The Friedman index I_{FRI} is implemented in the Splus library Interactive Projection Pursuit (see section 3.3.2.3). Figure 3.3 gives nine projections associated with optima of I_{FRI} of order 4. The first projection at the top-left corner corresponds to an optimum computed from an initial random plane, while the other planes are orthogonal to the previous planes in order to eliminate the structures which have already been discovered. For details of the methodology, see [FRI 87].

3) The chi-squared index I_{POS} is implemented in the Matlab library Computational Statistics. Figure 3.4 illustrates the procedure by displaying a projection associated with an optimum of the Posse index.

4) The final procedure is the robust GPCA proposed by [CAU 90] and implemented in R. Figure 3.5 is a biplot which gives the projection of the labeled units obtained with $U_n(0.1)$ and $T_n(2)$, together with the representation of the six variables (arrows).

For each procedure, projections which discover the group structure of the dataset are yielded. These graphics are therefore very similar but the way they are obtained depends on the procedures. Procedure (4) provides one display corresponding to a global optimum once the parameters β_1 and β_2 have been fixed. On the contrary, the

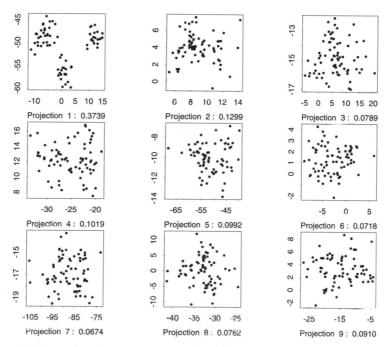

Figure 3.3. *Projection obtained by optimization of the Friedman index for the Lubischew dataset (IPP library in Splus)*

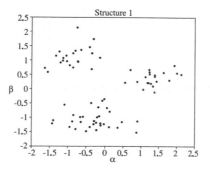

Figure 3.4. *Projection obtained by optimization of the Posse index for the Lubischew dataset (Computational Statistics library in Matlab)*

first three procedures depend on the choice of the initial projection. This choice may be manual in (1) or automatic in (2) and (3). The user of (1) is invited to look at the points cloud rotating to choose the initial projection, while procedures (2) and (3) randomly initiate the first projection plane.

The advantage of a dynamic method such as (1) is that it leads to projecting data on many directions with a lot of interactivity. The coupling 'Grand-tour/Projection

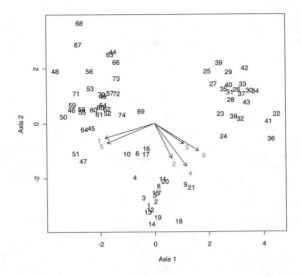

Figure 3.5. *Principal plane (1, 2) of robust GPCA with $U_n(0.1)$ and the metric $T_n^{-1}(2)$ for the Lubischew dataset (software R); individuals are represented by their labels, variables by arrows*

Pursuit' proposed by XGobi leads to dynamic visualization of the points cloud with the possibility of changing the index and its tuning parameters in real time.

On the curve displaying the temporal evolution of the index (top of Figure 3.2), the arrow indicates the instant when the user decided to optimize the index. When the local optimum is attained, the search stops and the obtained scatterplot is given at the bottom of Figure 3.2. The manual search for an initial interesting projection plane may however be tedious, especially when the number of dimensions is high. Methods such as (2) and (3) which are automatic do not suffer from this drawback and the user obtains an immediate result. Nevertheless, the result may depend heavily upon the initial choice and it is advisable to run the procedure several times to obtain different projections. Note that for the present example, however, procedure (2) turns out to be very stable.

The importance of controlling the significance of a projection has already been stressed in section 3.2.4. Procedure (2) is associated with a critical value computed by simulation, beyond which the projection is significant (at level 5%). In the example, this value for I_{FRI} is 0.1574 and only the first projection, which displays the structure in three groups, is significant. Tests are also possible for the index I_{POS} but they are not available in the implemented procedure. Concerning GPCA, a testing procedure based on the eigenvalues is given in [CAU 03b] with tables of asymptotic critical values for the non-robust GPCA. For Lubischew data, only the first two eigenvalues are significant; the conclusion is therefore similar to that obtained for (2).

As previously mentioned, many authors (e.g. [FRI 87, NAS 01]) raised the problem of sensitivity of the indexes to outliers. When outliers are present, they attract the projections associated with optimal indexes in such a way that other interesting structures can be masked. In order to study their sensitivity to spurious data, procedures (1)–(4) have been performed on a dataset which corresponds to the Lubischew data, except that one of the variables for the first individual has been replaced by a discordant variable. For this example, whose ambition is necessarily limited, finding the structure of the three groups using procedures (2) or (3) is not perturbed by the presence of the outlying observation. The group structure is however more difficult and takes more time to be recovered using procedure (1). Concerning procedure (4), this example confirms that the use of a robust covariance estimator leads to a method insensitive to outliers.

In general, it seems preferable to first detect the possible outlying observations, by using for example the method proposed in section 3.4.3.1 and to remove them, or rather, to make the projection pursuit procedure robust in a proper way. It is noticeable that procedure (2) displays the outlier on a second projection which becomes significant (Figure 3.6).

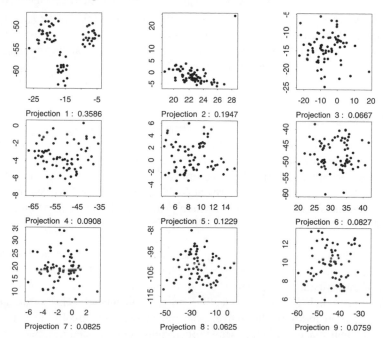

Figure 3.6. *Projections obtained by optimization of the Friedman index for the modified Lubischew data (library IPP of Splus)*

Finally, let us briefly comment upon the biplot in Figure 3.5, where the length of the arrows representing the variables has been multiplied by 2 to improve readability. Beyond the three groups, a glance at the display teaches us that:

– variables 2 and 3 are less responsible for the unit display than the others;

– group 1 is characterized by positive values of all the centered variables;

– individuals in group 2 have large negative values of the centered variables number 1 and 5; and

– individuals in group 3 are characterized by negative values of the centered variables 2, 3, 4 and 6.

3.6. Further topics

3.6.1. *Other indexes, other structures*

Not all indexes introduced in the literature are presented here. Some of them are still aimed at the search for groups (e.g. Eslava and Marriot [ESL 94]), emphasized in this chapter together with the search for outliers. However, the methods we have presented can allow the user to find other structures of interest, for example points concentrated around plane curves. Examples can be found in [CAU 95, CAU 03b, POS 95a]. A fairly similar example is 'the needle in a haystack' [FRI 87], where the problem is to detect a small number of points concentrated along a subspace. Some authors (e.g. [COO 93]) have proposed indexes which are aimed at discovering special kinds of structures such as holes, masses, etc.

3.6.2. *Unsupervised classification*

Projection pursuit aims to visualize a possible structure of a dataset, in particular the presence of groups. The purpose of classification is to split the dataset into groups and allocate the units to them. The two approaches are obviously complementary: visualizing helps to assess the effectiveness of the clusters and to determine their number. The linear combinations $a'x_i$ which provide a good visualization of groups are suitable candidates for clustering in a lower dimensional subspace.

Some authors have proposed using the first principal components to reduce the dimensionality of the clustering space. However, PCA is not designed for the purpose of classification. Bock [BOC 87] draws attention to this point and proposes an alternative which he calls 'projection pursuit clustering'. This is an iterative scheme where an index is alternatively optimized over all possible projections on low-dimensional spaces and all possible clusterings (with a fixed number of groups) within the projected subspace. As usual in clustering, the index is a within- versus between-variance index. Bolton and Krzanowski [BOL 03] give an algorithm of the same kind with an improved index and show examples where the technique works quite well. Peña and Prieto [PEÑ 01] look for 1D projections which minimize or maximize a kurtosis index (maximizing aims to detect groups while minimizing aims to find outliers). The presence of significant groups is controlled through the spacing of the projected points. The method gives good results for the examples which are presented, for a modest computational effort.

Clustering is the main motivation of the paper by Art *et al.* [ART 82] who propose data-based metrics for the input stage of a cluster analysis. In fact, data-based metrics are the key of GPCA. Caussinus and Ruiz-Gazen [CAU 07] discuss the complementarity of GPCA and clustering techniques. They revisit an example previously considered by Glover and Hopke [GLO 92] and Cook *et al.* [COO 04], and conclude:

– visual inspection of the data by means of GPCA is useful to get an idea of the number of possible homogenous classes and their major characteristics;

– clustering from the first principal components of GPCA rather than from the raw data improves efficiency in many circumstances (the sequence of eigenvalues provides a good guideline for choosing the number of components); and

– robustification of the GPCA eliminates discordant observations which are likely to spoil the display as well as the clustering.

3.6.3. *Discrete data*

Projection pursuit has so far been developed for numerical data. However, some recent papers attempt to provide parallel developments for discrete data.

Diaconis and Salzman [DIA 08] develop a notion of projection which is based upon the Radon transform. Let X be the set of all categories and Y a convenient class of subsets ('projection base'). If $f(x)$ is the frequency of category x belonging to X, the Radon transform at y in Y is the sum of $f(x)$ over X (as a simple example, among many other possibilities, the marginal frequencies of order k in a 2^q-contingency table are such projections). Within this framework, the most interesting projections are the less uniform. In fact, on the one hand, the uniform distribution on a finite set is that of maximum entropy. On the other hand, Diaconis and Salzman [DIA 08] establish that, for most datasets, most projections are nearly uniform. Their paper provides an application to illustrate the efficiency of the method. See also [AHN 03] for other examples and for visualizing and clustering aspects.

Caussinus and Ruiz-Gazen [CAU 06] propose a similar approach for numerical data by means of GPCA. Let Q discrete variables (each with any number of categories; the total number of categories will be denoted by J) be registered for each of n units. Caussinus and Ruiz-Gazen consider that the marginal distribution of these variables is not related to the structural part of the joint distribution; homogenity can therefore be characterized by the independence of the Q variables, the distribution with maximum entropy for fixed margins. Model (3.2) becomes

$$\int I(\boldsymbol{\pi})\, dP(\boldsymbol{\pi}), \tag{3.3}$$

where P is the structural part of the distribution and the noise I is the probability distribution, assuming the independence of the variables with marginal probabilities π. A typical example is the case where P is discrete and concentrated on $(m+1)$ points: model (3.3) turns out to be the latent class model. Within this framework, the distribution of interest is P; if it is concentrated on (or near) a subspace of $\mathbb{R}^J$, GPCA with a convenient metric $\mathbf{W}^{-1}$ will provide a good scatter plot for P. An estimate of $\mathbf{W}$ can be obtained from a latent class model. On the other hand, an interesting feature is related to Multiple Correspondence Analysis (MCA). Let us consider the indicator matrix of the dataset, whose elements are 0 apart from 1s in the positions to indicate the categories of response of each unit. MCA can be viewed as a GPCA of this dummy variables matrix, with a metric which is not an estimate of $\mathbf{W}^{-1}$, but is fairly close to such an estimate. This suggests that MCA could be fairly efficient in displaying the interesting latent structure.

3.6.4. *Related topics*

As indicated in its title, this chapter deals only with exploratory projection pursuit, i.e. methods to explore a dataset in order to find a hidden structure and visualize it on a low-dimensional graphical display. However, finding 'good' low-dimensional projections is useful with several other problems to struggle against the so-called 'curse of dimensionality'. Much attention has been paid to two questions of this kind: the estimation of a multidimensional density and the non-linear regression with many explanatory variables. They are beyond the scope of this text, but their importance should be stressed.

Since exploratory projection pursuit techniques are somewhat complementary to PCA, we have to mention some other techniques which are also complementary to PCA. A robust PCA is to some extent parallel but opposite to the GPCA of section 4.3.1 aimed at discovering outliers. Finally, there is a strong similarity between the GPCA solution of the projection problems addressed in this chapter and the independent components analysis (ICA) advocated for blind signal separation (e.g. [CAR 98, PHA 96]). The reader is referred to [TYL 09] for a recent survey of GPCA (referred to as 'Invariant Coordinate Selection') and ICA, including a broad generalization of previous theoretical results.

3.6.5. *Computation*

Concerning the available computing procedures, it is notable that they have generally been developed by the authors themselves and are often adapted to a particular operating system. The free software XGobi is adapted to the Unix system but is not compatible with the Microsoft Windows system. This shortcoming is corrected in the new software GGobi, but GGobi does not offer all the possibilities of XGobi in terms of available projection pursuit indexes. The procedures by C. Posse and J. Sun are free libraries developed for well-known commercial software (Matlab for C. Posse and Splus for J. Sun). The library by J. Sun is written in Fortran 77 and is only portable by default on some Unix systems.

Interest in exploratory methods is currently progressing, especially within the field of data mining and, thanks to computer science improvements, exploratory projection pursuit is possible even for very large datasets. It is therefore legitimate to expect that the exploratory procedures described in the present paper will soon be integrated as standards in all the well-known statistical software.

3.7. Bibliography

[AHN 03] AHN J., HOFMANN H., COOK D., "A projection pursuit method on the multidimensional squared contingency table", *Computational Statistics*, vol. 18, p. 605–626, 2003.

[ART 82] ART D., GNANADESIKAN R., KETTENRING J. R., "Data-based metrics for cluster analysis", *Utilitas Mathematica*, vol. 21A, p. 75–99, 1982.

[ASI 85] ASIMOV D., "The grand tour: a tool for viewing multidimensional data", *SIAM Journal on Scientific and Statistical Computing*, vol. 6, num. 1, p. 128–143, 1985.

[BOC 87] BOCK H., "On the interface between cluster analysis, principal component analysis, and multidimensional scaling", H. BOZDOGAN H., GUPTA A., Eds., *Multivariate Statistical Modeling and Data Analysis*, p. 17–34, D. Reidel Publishing Company, 1987.

[BOL 03] BOLTON R., KRZANOWSKI W., "Projection Pursuit clustering for exploratory data analysis", *Journal of Computational and Graphical Statistics*, vol. 12, num. 1, p. 121–142, 2003.

[BUR 91] BURTSCHY B., LEBART L., "Contiguity analysis and projection pursuit", GUTTIERREZ R., VALDERRAMA M., Eds., *Applied Stochastics Models and Data Analysis*, p. 117–128, World Scientific, Singapour, 1991.

[CAR 98] CARDOSO J. F., "Blind signal separation: statistical principles", LIU R.-W., TONG L., Eds., *Proceedings of the IEEE, Special Issue on Blind Identification and Estimation*, vol. 90, p. 2009–2026, 1998.

[CAU 90] CAUSSINUS H., RUIZ A., "Interesting projections of multidimensional data by means of generalized principal component analyses", *COMPSTAT 90*, p. 121–126, Physica-Verlag, Heidelberg, 1990.

[CAU 93] CAUSSINUS H., RUIZ-GAZEN A., "Projection pursuit and generalized principal component analyses", S. MORGENTHALER ET AL., Ed., *New Directions in Statistical Data Analysis and Robustness*, p. 35–46, Birkhäuser Verlag, Basel Boston Berlin, 1993.

[CAU 95] CAUSSINUS H., RUIZ-GAZEN A., "Metrics for finding typical structures by means of principal component analysis", Y. ESCOUFIER ET AL., Ed., *Data Science and its Applications*, p. 177–192, Academic Press, 1995.

[CAU 02] CAUSSINUS H., HAKAM S., RUIZ-GAZEN A., "Projections révélatrices contrôlées – recherche d'individus atypiques", *Revue de Statistique Appliquée*, vol. L, num. 4, p. 81–94, 2002.

[CAU 03a] CAUSSINUS H., FEKRI M., HAKAM S., RUIZ-GAZEN A., "A monitoring display of multivariate outliers", *Computational Statistics and Data Analysis*, vol. 44, num. 1–2, p. 237–252, 2003.

[CAU 03b] CAUSSINUS H., HAKAM S., RUIZ-GAZEN A., "Projections révélatrices contrôlées – groupements et structures diverses", *Revue de Statistique Appliquée*, vol. LI, num. 1, p. 37–58, 2003.

[CAU 06] CAUSSINUS H., RUIZ-GAZEN A., "Projection Pursuit approach for categorical data", GREENACRE M., J. BLASIUS J., Eds., *Multiple Correspondence Analysis and Related Methods*, p. 405–418, 2006.

[CAU 07] CAUSSINUS H., RUIZ-GAZEN A., "Classification and generalized principal component analysis", BRITO P., BERTRAND P., CUCUMEL G., DE CARVALHO F., Eds., *Selected Contributions in Data Analysis and Classification*, p. 539–548, Springer, 2007.

[COO 93] COOK D., BUJA A., CABRERA J., "Projection pursuit indexes based on orthonormal function expansions", *Journal of Computational and Graphical Statistics*, vol. 2, num. 3, p. 225–250, 1993.

[COO 95] COOK D., BUJA A., CABRERA J., HURLEY C., "Grand tour and projection pursuit", *Journal of Computational and Graphical Statistics*, vol. 4, num. 3, p. 155–172, 1995.

[COO 04] COOK D., CARAGEA D., HONAVAR H., "Visualization in classification problems", *Proceedings in Computational Statistics (COMPSTAT 2004)*, Berlin, Springer, p. 799–806, 2004.

[DIA 84] DIACONIS P., FREEDMAN D., "Asymptotics of graphical projection pursuit", *Annals of Statistics*, vol. 12, num. 3, p. 793–815, 1984.

[DIA 08] DIACONIS P., SALZMAN J., "Projection pursuit for discrete data", *IMS COLLECTIONS, Probability and Statistics: Essays in Honor of David A. Freedman*, vol. 2, p. 265–288, 2008.

[ESL 94] ESLAVA G., MARRIOTT F., "Some criteria for projection pursuit", *Statistics and Computing*, vol. 4, p. 13–20, 1994.

[FRI 74] FRIEDMAN J., TUKEY J., "A projection pursuit algorithm for exploratory data analysis", *IEEE Transactions of Computing*, vol. C-23, p. 881–889, 1974.

[FRI 87] FRIEDMAN J., "Exploratory projection pursuit", *Journal of American Statistical Association*, vol. 82, p. 249–266, 1987.

[GAB 71] GABRIEL K. R., "The biplot graphic display of matrices with application to principal component analysis", *Biometrika*, vol. 58, num. 3, p. 453–467, 1971.

[GAB 02] GABRIEL K., "Le biplot - outil d'exploration de données multidimensionnelles", *Journal de la Société Française de Statistique*, vol. 143, num. 3–4, p. 5–55, 2002.

[GLO 92] GLOVER D., HOPKE P., "Exploration of multivariate chemical data by projection pursuit", *Chemometrics and Intelligent Laboratory Systems*, vol. 16, p. 45–59, 1992.

[HAL 89] HALL P., "On polynomial-based projection indexes for exploratory Projection Pursuit", *Annals of Statistics*, vol. 17, num. 2, p. 589–605, 1989.

[HUB 85] HUBER P. J., "Projection Pursuit (with discussion)", *Annals of Statistics*, vol. 13, p. 435–525, 1985.

[JON 87] JONES M., SIBSON R., "What is projection pursuit? (with discussion)", *Journal of Royal Statistical Society A*, vol. 150, p. 1–38, 1987.

[KLI 97] KLINKE S., *Data Structures in Computational Statistics*, Physica-Verlag, 1997.

[LEB 01] LEBART L., "Classification et analyse de contiguïté", *La Revue de Modulad*, vol. 27, p. 1–22, 2001.

[LI 93] LI G.-Y., CHENG P., "Some recent developments in Projection Pursuit", *Statistica Sinica*, vol. 3, p. 35–51, 1993.

[LUB 62] LUBISCHEW A., "On the use of discriminant functions in taxonomy", *Biometrics*, vol. 18, p. 455–477, 1962.

[MAR 01] MARTINEZ W., MARTINEZ A., *Computational Statistics Handbook with MATLAB*, CRC Press, London, 2001.

[NAS 95] NASON G., "Three-dimensional projection pursuit", *Journal of Royal Statistical Society C*, vol. 44, p. 411–430, 1995.

[NAS 01] NASON G., "Robust projection indices", *Journal of Royal Statistical Society B*, vol. 63, p. 551–567, 2001.

[NAS 09] NASON G., 2009, http://www.stats.bris.ac.uk/~guy/Research/PP/PP.html.

[OHR 90] OHRVIK J., "Structure or noise", *Computational Statistics Quarterly*, vol. 6, p. 1–20, 1990.

[PAN 00] PAN J., FUNG W., FANG K., "Multiple outlier detection in multivariate data using projection pursuit techniques", *Journal of Statistical Planning and Inference*, vol. 83, p. 153–167, 2000.

[PEÑ 01] PEÑA D., PRIETO F., "Cluster identification using projections", *Journal of the American Statistical Association*, vol. 96, num. 456, p. 1433–1445, 2001.

[PER 05] PERISIC I., POSSE C., "Projection Pursuit indices based on the empirical distribution function", *Journal of Computational and Graphical Statistics*, vol. 14, num. 3, p. 700–715, 2005.

[PHA 96] PHAM D. T., "Blind separation of instantaneous mixture of sources via an independent component analysis", *IEEE Transactions of Signal Processing*, vol. 44, num. 11, p. 2768–2779, 1996.

[POL 95] POLZEHL J., KLINKE S., Experiences with bivariate projection pursuit indices, Report num. 373, Sonderforschungsbereich, University of Berlin, 1995.

[POS 90] POSSE C., "An effective two-dimensional projection pursuit algorithm", *Communication in Statistics: Simulation and Computation*, vol. 19, num. 4, p. 1143–1164, 1990.

[POS 95a] POSSE C., "Projection pursuit exploratory data analysis", *Computational Statistics and Data Analysis*, vol. 20, p. 669–687, 1995.

[POS 95b] POSSE C., "Tools for two-dimensional exploratory projection pursuit", *Journal of Computational and Graphical Statistics*, vol. 4, num. 2, p. 83–100, 1995.

[RÉN 70] RÉNYI A., *Probability Theory*, American Elsevier Publishing Company, New York, 1970.

[RUI 96] RUIZ-GAZEN A., "A very simple robust estimator for a dispersion matrix", *Computational Statistics and Data Analysis*, vol. 21, p. 149–162, 1996.

[SCO 92] SCOTT D., *Multivariate Density Estimation: Theory, Practice and Visualization*, Wiley, New York, 1992.

[SIB 84] SIBSON R., "Present position and potential developments: some personal views - multivariate analysis", *Journal of Royal Statistics Society A*, p. 198–207, 1984.

[SIL 86] SILVERMAN B., *Density Estimation for Statistics and Data Analysis*, vol. 26, Chapman and Hall, London, 1986.

[SUN 91] SUN J., "Significance levels in exploratory Projection Pursuit", *Biometrika*, vol. 78, num. 4, p. 759–769, 1991.

[SUN 93] SUN J., "Some practical aspects of exploratory Projection Pursuit", *SIAM Journal on Scientific Computing*, vol. 14, num. 1, p. 68–80, 1993.

[SUN 09] SUN J., 2009, http://sun.cwru.edu/~jiayang/nsf/ipp.html.

[TYL 09] TYLER D., CRITCHLEY F., DÜMBGEN L., OJA H., "Invariant coordinate selection (with discussion)", *Journal of the Royal Statistical Society*, 2009, to appear.

[XGO 09] XGOBI, 2009, http://lib.stat.cmu.edu/general/XGobi/.

[YEN 88] YENYUKOV I., "Detecting structures by means of Projection Pursuit", *COMPSTAT 88*, p. 47–58, Physica-Verlag, Heidelberg, 1988.

Chapter 4

The Analysis of Proximity Data

4.1. Introduction

Proximity measurements form the core information treated by many methods relevant either to statistical data reduction, exploratory data analysis or machine learning. Proximity data consist of measures of *similarity* or *observed dissimilarity* between elements of a set of n objects of interest (individuals, stimuli, etc.). We obtain a measurement $o_{j,k}$ of either the similarity or the observed dissimilarity between each pair of objects (j, k).

Proximity data may arise from computations carried out on a usual $n \times p$ multivariate data matrix, recording the values of p variables on n individuals. However, this dyadic information may also consist of measures of similarity or dissimilarity collected directly as the output of a comparative judgement study or some pair comparison experiment. In the former case, recourse to proximity data induces some loss of information since we ignore one member (most often variables or features) of the duality relationship coupling variables (or features) and individuals (or examples). The present chapter will favor the latter setting: featureless proximity measures.

Proximity data analysis produces graphical representations aimed at e.g. modeling subjective judgements of similarity to uncover eventual latent patterns structuring the mental representation of subjects. The production and the interpretation of such displays constitute a major challenge in contemporaneous cognitive sciences. As

---

Chapter written by Gerard D'AUBIGNY.

usual in pattern recognition, three main families of representation models may be distinguished:

– *Spatial* or *continuous* models display objects (stimuli) on a continuum, and identify them with points in a metric space. These methods interpret observed similarities as monotonic decreasing functions of the corresponding interpoint distances read on the solution metric space.

– *Combinatorial* or *discrete* models identify objects with elements of some particular subsets, hopefully related to natural classes determined by some observable features of the objects. Similarity is then interpreted as a weighted sum, describing shared or distinctive subsets. This class of models is discussed in Chapters 8–10.

– *Hybrid* models try to elaborate some useful combination of the two above-mentioned approaches. This class of models will not be discussed in this book; see [DEG 70, SAR 90] for a presentation and references.

Multidimensional scaling (MDS) methods originate from the USA in the 1950s with the breakthrough by Torgerson [TOR 52, TOR 58] and Shepard [SHE 62a, SHE 62b]. Both tried to supply psychologists with synthetic continuous descriptions of homogenous perceptive spaces (such as a 3D space for perception of colors). Such spatial (geometric) descriptions were expected to reveal the most salient dimensions used by subjects to discriminate stimuli. In psychometry, these methods are considered from experience suitable for analyzing perception and opinion data. The term *ordination methods* is prefered in biometry, and recent developments in machine learning and pattern recognition ascribes to MDS the status of a cornerstone for many proposals in *manifold mearning* (ML) or *non-linear dimensionality reduction* (NLDR). These developments are outwith the scope of the present introduction. They would justify in themself the addition of a new chapter to future editions of the book; see [LEE 07] for a comprehensive presentation and references.

Combinatorial methods – such as *hierarchical* (agglomerative) clustering (HAC) or *additive trees* (AT) – are of wider applicability than MDS. In theory, they are able to treat discrete structures, featuring many combinatorial aspects of more complex cognitive domains, such as hierarchical structures (e.g. families of phonemes) or phylogenetic structures, as well as the multidimensional features of stimuli (i.e. numerical characteristics). In psychometry, these methods are considered suitable for analyzing selection and cognition data.

As a visualization and synthesis tool, MDS is a data reduction method which makes it possible to summarize a dataset in a space of small dimensionality. Thus, the acronym MDS is often used in a generic way, to designate a class of data analysis methods aimed at preserving a notion of *geometric structure* of data when represented in a metric space of reduced dimensionality. In this situation, geometric structures are interpreted as (pre-)metric relations describable in terms of a set of observed dissimilarity data. It means that two points judged as near (respectively, far apart) in the observation space would appear as nearby (respectively, distant) locations in the representation space.

During a long period of time, the applications of the MDS methodology in psychology, in sensory analysis and in marketing confined this exploratory tool to small datasets, for which algorithmic questions remained subsidiary. As a consequence, and because of the complexity of the numerical problems involved, many algorithmic proposals suffered slow (sub-linear) convergence properties, converging towards only local optima and not global optima.

These realities drastically changed with the opening of new fields of application, e.g. in molecular biology [CRI 88, GLU 93, HAV 91, MOR 96], in geostatistics [SAM 92, SAM 94] and, more recently, in a great variety of computer engineering sub-domains, from artificial intelligence and pattern recognition to machine learning and data mining. Thus, feasibility questions evolved because the size of file to be treated in these fields were much larger and, at the same time, a much higher standard of numerical accuracy was required. From this point of view, we are still making progress. Our presentation will therefore neglect these very important questions, and we have tried to limit our discussion of numerical questions as much as possible. Our text is therefore more oriented towards the needs of social science users rather than those of computer scientists.

The contribution of computer scientists to the field is not limited to algorithmic problems. While often ignored by statisticians and psychologists, some work very similar to the metric MDS approach was completed during the same period of time by specialists of pattern recognition, under the heading of *vector quantization* (VQ). This research began with a paper by Sammon [SAM 69], who referred to the method as *non-linear mapping* (NLM).

This approach will be discussed, but a noticeable distinction between MDS and NLM should be noted, which concerns the observed data. Sammon does not tackle the problem of Euclidean embedding of one set of observed featureless dissimilarity data. He is interested in questions of data reduction while suffering minimal loss of information in the case where a great number of features have been observed, from which a Euclidean distance matrix has been calculated. This data generation process implies more structure in the observations than in the MDS setting.

Both approaches share a badness of fit function measuring the (weighted) discrepancy between observed and solution distances, and this common feature distinguishes them from Hilbertian methods used in the classical methods of multivariate statistical analysis. The former attempt to respect affine and metric (and generally, but not always, Euclidean) structures while the latter try to respect vector space and angular structures. From that point of view, MDS and VQ are two close examples of a *multidimensional quantification* (MQ) approach.

A more discrete analysis of MQ methods is needed to understand the sources of diversity of recent methodological proposals in this field. Preservation of a distance structure may be understood in at least two ways. On the one hand, we may impose that each observed dissimilarity judgement would be given an equal importance. Then, any MQ method should elaborate a *global isometry*, mapping the space spanned by observed data into the representation space. Such an approach characterizes so-called *isometric scaling* methods. On the other hand, we may prescribe only the respect of dissimilarities among *neighbor* points. In this case, the MQ method should take into account only *local isometry* relations, and secure an optimal preservation of a metric structure among *neighbors*. Two points, which are close in terms of their observed dissimilarity, remain close together in the representation space.

This point leads us to distinguish two groups of methods:

– *Global embedding* methods in which the mapping defining the embedding or the representation must be an *isometry* and so takes into account the (approximate) reconstitution of any observed dissimilarity, either small of large. Linear methods, such as PCoA (elaborated in Torgerson's proposal), and non-linear methods, such as MDS (elaborated in Shepard's proposal) or NLM (elaborated on Sammon's proposal), do belong to this class, in which we shall distinguish parametric methods from the semi-parametric (or ordinal) methods. The original (psychometric) approach of MDS clearly belongs to this first class i.e. the global isometry approach.

– *Local embedding* or topological methods in which the mapping defining the embedding or the representation may be a *local isometry* which takes into account only the (approximate) reconstitution of the observed dissimilarity among neighbor points.

This second class of methods, based on local isometry, has developed more recently in the computer science community under the headings of *topological mapping* and *manifold learning*. It aims to preserve the local topological or metric properties of eventually curvilinear structures, smooth enough to be embeddable on manifolds.

Depending on the collected information and on the goals assigned to the embedding, two sub-classes of methods need be contrasted. In the first class, the method focuses on the respect of the *typological* properties induced by such *neighborhood* relationships. Kohonen's *self-organizing maps* (SOM) [KOH 82] is the best well-known example of such an approach, but other methods of *topographic mapping* have also been designed, generally in the form of a specific architecture of a neural network. More statistically principled formulations of the problem have motivated the appearance of *generative topographic mapping* (GTM) [BIS 96, BIS 98, TIP 96], and more geometrically principled formulations of the problem have motivated the appearance of *curvilinear component analysis* [DEM 94, DEM 97], which can be seen as a variant of MDS. See the literature for a discussion of these approaches, e.g. [LEE 07].

This chapter is restricted to the core MDS (in a narrow sense) methodological aspects. In this narrow sense, MDS is dedicated to the analysis of one two-way table of observed dissimilarity data. Wider definitions of MDS exist, and some authors (e.g. [GOW 96]) would consider Chapters 1–7 to be discussing some MDS methods. A more limited focus seems inconsequential because many more comprehensive presentations of MDS (in a wider sense) now exist, which try to list a great number of existing methods (e.g. [BOR 81, COX 94, DAV 83, EVE 97, KRU 78, SCH 81, SHE 72a]). Carroll and Arabie [CAR 80] give a useful typology of problems encountered in (wide sense) MDS, and a panorama of MDS methods for constructing Euclidean configurations may be found in [LEE 82]. This work is extended to non-Euclidean configurations in [ARA 91]. These papers are usefully complemented by recent books e.g. [LEE 07, PEK 05].

This exposition of MDS methods draws inspiration (and some hypothetical examples) from [AUB 93, BOR 97, CAR 80, COX 94, EVE 97]. Four introductory examples are first presented. Section 4.2 then presents the MDS problem and introduces some notation. The following section focuses on the resolution of the *metric embedding problem* in the classical case. Section 4.4 presents first the MDS model in a parametric setting, and then two extensions of this method to a semi-parametric setting. The first generalization concerns a typical situation met by applications to sensory analysis: the measurement scale of observed dissimilarities is ordinal. The second generalization tries to fulfill the needs expressed in molecular biology, when theory imposes *a priori* known boundary constraints to the domain of variation of interpoint distances in the solution display. The final section is devoted to an example and presents some assistance tools, usable in the interpretative stage in any application of the MDS methodology.

4.2. Representation of proximity data in a metric space

4.2.1. *Four illustrative examples*

The varied applications of MDS entail distinct algorithms, as illustrated by the following examples. The first two compare a small number of objects, a characteristic common to a great majority of applications of MDS in psychology and more broadly in social sciences. This is to be contrasted e.g. to the usual set-up of spectroscopic analysis illustrated by example 4.3, since protein molecules typically contain a few hundred atoms, and their visualization requires more effective optimization methods.

EXAMPLE 4.1.– One of the most famous applications of MDS is due to Shepard [SHE 62a, SHE 62b] which builds a representation of data collected by Ekman [EKM 54] in a 2D Euclidean space. During a psychophysics experiment on color perception, Ekman submitted a series of pairs of stimuli differing in hue to a sample of 31 subjects. Each subject was asked to grade every pair from zero (no similarity) to four (identity). The average score attributed to each pair (j, k) and the associated observed dissimilarity were computed. It is striking that the Shepard algorithm recovered a 2D display which positioned colors approximately along the famous color circle. Previous analysis, applying classical factor analysis to the matrix S of 'similarity perceived between colors' required seven common factors.

EXAMPLE 4.2.– Romesburg [ROM 84] looked into votes of 15 members of the congress in the state of New Jersey on approval of 19 texts devoted to environmental matters. The observed dissimilarity o_{jk} counts the number of time the two representatives j and k disagreed, listed in Table 4.1. The MDS algorithm provides an illuminating result, distinguishing the votes of republican representatives from those of democrats. Only one exception is noticeable, for republican number 12. One more interesting feature is the location of representatives 2 and 4 who abstained from voting more often than others (respectively, 9 and 6 times).

	c1	c2	c3	c4	c5	c6	c7	c8	c9	c10	c11	c12	c13	c14
c2	8	.	.	.	.	.	.	.	.	.	.	.	.	.
c3	15	17	.	.	.	.	.	.	.	.	.	.	.	.
c4	15	1	9	.	.	.	.	.	.	.	.	.	.	.
c5	10	13	16	14	.	.	.	.	.	.	.	.	.	.
c6	9	13	12	12	8	.	.	.	.	.	.	.	.	.
c7	7	12	15	13	9	7	.	.	.	.	.	.	.	.
c8	15	16	5	10	13	12	17	.	.	.	.	.	.	.
c9	16	17	5	8	14	11	16	4	.	.	.	.	.	.
c10	14	15	6	8	12	10	15	5	3	.	.	.	.	.
c11	15	16	5	8	12	9	14	5	2	1	.	.	.	.
c12	16	17	4	6	12	10	15	3	1	2	1	.	.	.
c13	7	13	11	15	10	6	10	12	13	11	12	12	.	.
c14	11	12	10	10	11	6	11	7	7	4	5	6	9	.
c15	13	16	7	7	11	10	13	6	5	6	5	4	13	9

Table 4.1. *Vote disagreements among members of the New Jersey congress over 19 questions concerning environment matters*

Figure 4.1a shows the MDS representation of the Ekman data, adorned with interpretative aids not produced by the package. Segments linking points to their two nearest neighbors have been drawn (according to Ekman). This makes the latent continuous and non-linear pattern apparent, which is assumed to setup this display. Moreover, it was also attempted to localize parts of this arc associated with each color. Lastly, every point is labeled with an exogenous information: the *a priori* known wavelength of each color.

Figure 4.1b is no more naive since the MDS program output has been completed by point labels which also use an exogenous information, namely the political belonging of each representative (R for republican and D for democrat), to emphasize the partition of the representation space induced by partisanship, from a discriminant analysis point of view.

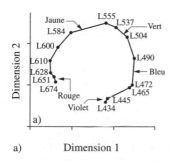
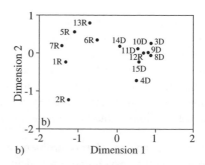

a)   b)
Dimension 1   Dimension 1

Figure 4.1. (a) Perceived similarity between colors which correspond to wavelengths varying in the interval from 434 nm to 674 nm [EKM 54]; (b) frequencies of vote disagreement among 15 members of the New Jersey congress over 19 questions concerning environmental matters [ROM 84]

EXAMPLE 4.3.– A classical problem of molecular biology (e.g. [CRI 88, GLU 93, HAV 91, MOR 96]) considers proteins with molecules formed by n atoms (see Figure 4.2a). Pairwise distances between atoms are measured by some nuclear magnetic resonance (NMR) system. Typically, the dissimilarities so obtained concern only a portion (generally about one-third) of $n(n-1)/2$ atomic distances which should be measured. Despite the existence of missing data, a representation of the complete molecule is produced into $\mathbb{R}^3$. The form estimated from the observed distance matrix is fundamental to the understanding of the molecule's working and the protein's properties. An intensive research field has emerged in biochemistry during the last 20 years, which focuses on the 2D and 3D representations of molecules, under the heading of *conformation problem*.

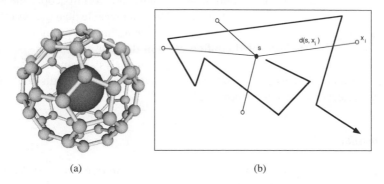

(a)   (b)

Figure 4.2. (a) Representation of an endohedral molecule reconstructed from measurements of inter-atomic distances by NMR (only one-third of the measurements have been made) and (b) drawing of a fictitious path and adjustment of its location related to four landmarks

EXAMPLE 4.4.– Another classical problem, now relevant to engineering, concerns the navigational and surveying facility known as *global positioning system* (GPS). Figure 4.2b illustrates this geographical positioning method, which assumes that two sets of objects are available. The first set refers to landmarks, reference marks or other observed marks on the shore, denoted $j \in \mathcal{J}$. The second set comprises the measurement spots, $i \in \mathcal{I}$. The vector $\mathbf{o}_i \triangleq \{o_{ij}; j \in \mathcal{J}\}$ collects the gap separating its position from the location of landmarks, measured in terms of (pre-)distances (or dissimilarities) and makes it possible to deduce its location. This approach assumes that when measurement errors are neglected, $o_{ij} \approx d_{ij} \triangleq \|\mathbf{x}_i - \mathbf{s}_j\|$ where $\mathbf{x}_i$ and $\mathbf{s}_j$ denote the locations of the object i and the landmark j, respectively.

An ordinal analog of this problem, which motivated intensive and continued research activity in psychometrics and in econometrics, is preference data analysis. The observed dissimilarity o_{ij} measures the (relative) intensity of the preference expressed by the subject i about the object j. The more object j draws closer to the *ideal point* which represents a judge (subject) i, the stronger is the preference of i for this object j. The resulting affine (often Euclidean) description methods have developed in social sciences under the heading of *multidimensional unfolding distance model*, also known among econometricians as *(geometric) individual choice modeling*.

REMARK 4.1.– Example 4.1 illustrates a situation where the analyzed dissimilarities were calculated as averages of a sample of 31 raw individual dissimilarities, demonstrating many sources of variation and error classically assignable to human judgement. The resulting inaccuracy was aggravated here by the use of a (rough) five levels *Likert scale* for expressing judgement. This feature is shared by a great majority of applications of the MDS approach to sensor data, and this must be contrasted with the much higher standard of measurement accuracy carried out in the NMR spectroscopy setting. Moreover, behavioral scientists require less accurate results than biochemists, referred to in example 4.3. Briefly, the spectroscopic measure of observed dissimilarities is less prone to bias and less changeable than the observations produced by human beings. As a consequence, we are inclined to promote parametric methods to generate the estimated configuration in the case of example 4.3, and to adopt *non-parametric* or *semi-parametric* methods of analysis in applications such as examples 4.1 and 4.2. This allows data transformation of observed dissimilarity and reduces the effect of potential distortions.

4.2.2. Definitions

DEFINITION 4.1.– *A real matrix* $\mathbf{O} = (o_{jk})$, *which is square and of order n, is an observed dissimilarity (or predistance) matrix if it satisfies the following properties 1 and 2. It is said to be semi-proper if its diagonal terms vanish, that is if it also verifies property 3. It is called proper when its diagonal terms are the only null terms. Moreover, this is a pseudo-distance matrix if its terms also verify property 4. A pseudo-distance matrix which is semi-proper is a semi-distance matrix. It is a distance matrix if it is moreover proper:*

1) it is symmetric, i.e. if $o_{jk} = o_{kj}$, $\forall j, k \in \{1, \cdots, n\}$;

2) its terms are positive, i.e. if $o_{jk} \geq 0$, $\forall j, k \in \{1, \cdots, n\}$;

3) $o_{jj} = 0$, $\forall j \in \{1, \cdots, n\}$;

4) triangle inequality: $o_{ik} \leq o_{ij} + o_{jk}$, $\forall i, j, k \in \{1, \cdots, n\}$.

MDS models seldom assume that the observed dissimilarity is a distance function. As a consequence, they allow an admissible class of transformations, such as the affine functions (when interval scale measurements are assumed) or the monotonic functions (when ordinal scale measurements are assumed). We refer to any dissimilarity matrix derived from the observed matrix by an admissible transformation as a *disparity matrix*. Later, any disparity matrix which results from an admissible recoding of the observed dissimilarity matrix $\mathbf{O} = (o_{jk})$, is denoted by $\mathbf{T} = (t_{jk})$. Fixing the class of admissible transformations retained for an investigation is the researcher's responsibility.

4.2.2.1. *Euclidean embedding of a proximity network*

The above-mentioned applications in molecular biology, in chemistry and in physics highlighted a specific field of research, tightly related to multidimensional quantization. The problem then consists of representing *with high accuracy* some molecular structure in 2D or 3D based on the observed inter-atomic distance measures and on complementary conformational constraints, such as the chirality constraints expressed in terms of distances [CRI 88, GLU 93, HAV 91]. Most often, the number of observed inter-atomic distances is limited, and the dissimilarity matrix $\mathbf{O}$ is not fully observed. This feature is specific to research focusing on this application domain, referred to as the *Euclidean distance matrix completion* (EDMC) problem: for a given *sparse* symmetric matrix $\mathbf{O}$, find an imputation of the missing elements in a way to secure the Euclideanicity of the resulting matrix [ALF 97, ALF 98, TRO 97].

Let us define a network as a pair $\mathcal{N} = (G, o)$, where $G = (\mathcal{I}, \mathcal{E})$ is a graph with a finite set of nodes, $card(\mathcal{I}) = n$ and any edge $e \in \mathcal{E} \subseteq \mathcal{I} \times \mathcal{I}$ is equipped with a non-negative weight o_e.

DEFINITION 4.2.– $\mathcal{N}$ *is q-embeddable if there exists a mapping* $\Upsilon : \mathcal{I} \mapsto E \backsim (\mathbb{R}^q, \| - \|)$ *such that:*

$$d_{jk}(\Upsilon) \triangleq \|\Upsilon(s_j) - \Upsilon(s_k)\| = o_{jk} \quad \forall e = (j, k) \in \mathcal{E}.$$

$\mathcal{N}$ *admits a Euclidean embedding if there exists some integer* $q > 0$ *for which* $\mathcal{N}$ *is q-embeddable.*

Whatever the integer $q > 0$ is, Saxe [SAX 79] showed that the q-embedding problem is NP-difficult if the observed dissimilarity o is integer valued. If the real matrix $\mathbf{O}$ is symmetric, with non-negative terms and a zeroed diagonal, the smallest integer q for which there exist points $\mathbf{x}_j \in E \backsim (\mathbb{R}^q, \| - \|)$ such that

$\|\mathbf{x}_j - \mathbf{x}_k\| = o_{jk}$, is called the *embedding dimensionality* of $\mathbf{O}$. Barvinok [BAR 95] shows that if a proximity network is embeddable, then it is q^*-embeddable for $q^* = (\sqrt{8\,card(\mathcal{E}) + 1}\,)/2$.

4.2.2.2. *A hypothetical counter-example*

Recall that a square distance matrix $\mathbf{D} = (d_{ik})$ of order n is Euclidean if there exists a configuration of points $\mathbf{X} = \{\boldsymbol{x}_j \in \mathbb{R}^p, j \in \{1, \cdots, n\}\}$ such that d_{jk} is equal to the Euclidean distance separating the points $\boldsymbol{x}_j$ and $\mathbf{x}_k$. We regroup the $\boldsymbol{x}_j$ to form the lines of a matrix $\mathbf{X}$ of order $n \times p$, called the *configuration matrix*, from which we derive the Euclidean distance matrix $\mathbf{D}(\mathbf{X}) = (d_{jk}(\mathbf{X}))$. Let us denote the set of such matrices by $\mathcal{D}_n(p)$.

The following matrix (Table 4.2) violates the positivity postulate of a distance, in such a way that no similarity mapping can transform it in a distance matrix. It does not allow us to embed the five objects in a Euclidean space.

o_{ik}	1	2	3	4		t_{ik}	1	2	3	4
2	0.2					2	5.0			
3	1.2	0.2				3	6.0	5.0		
4	0.2	3.2	0.2			4	5.0	8.0	5.0	
5	−1.8	−0.8	−1.8	−0.8		5	3.0	4.0	3.0	4.0

Table 4.2. *The raw dissimilarities o_{ik} and the disparities t_{ik} obtained by adding the additive constant a= 4, 8*

When an observed dissimilarity matrix $\mathbf{T}$ and a reference dimensionality p are given, the embedding problem to be solved in classical metric geometry consists of establishing whether $\mathbf{T}$ belongs to $\mathcal{D}_n(p)$ or not. In the affirmative, we need to generate a configuration matrix $\mathbf{X}$, of order $n \times p$, such that $\mathbf{T} = \mathbf{D}(\mathbf{X})$. However, no embedding problem has to be solved in the above toy example, since there does not exist a real constant α for which the matrix $\mathbf{T} = \alpha\mathbf{O}$ is a distance matrix.

We denote by δ_{ik} the Dirac symbol which equals 1 iff (if and only if) $i = k$ and 0 otherwise. Then, we may render all the dissimilarities positive by adding a large enough constant (here $a > 1.8$) to every off-diagonal term in $\mathbf{O}$. Yet, the matrix $\mathbf{T}$ with generic term $t_{ik} = o_{ik} + a(1 - \delta_{ik})$ is not necessarily a distance matrix, as we may check for $a = 2.8$. When looking for the triplet for which the triangle inequality is the more strongly violated, we notice that $o_{45} + o_{52} = -1.6 \leq 3.2 = o_{42}$ and so, at worst, the disparities t_{ik} realize a triangle equality for $a = 4.8$, since $t_{45} + t_{52} = -1.6 + 2a \geq 3.2 + a = t_{42}$. The transformed $\mathbf{T}$ generated in this way is a distance matrix.

4.3. Isometric embedding and projection

The metric embedding problem stated here is: when is $\mathbf{T}$ a Euclidean distance matrix? This question has been answered independently by Schoenberg [SCH 35] and Young and Householder [YOU 38]. Its use for applications follows the equivalent proposals of Torgerson [TOR 52, TOR 58], which differ from Schoenberg and Young and Householder by his choice of an origin in the representation space. Torgerson chooses the barycenter, and not one of the observed objects. A complementary question stated by this approach consists of finding the value of a which minimizes the algebraic dimension of the representation space of $\mathbf{T}$.

This approach is known by a great number of names: *classical scaling* [MAR 78] or *classical multidimensional scaling* (CMDS) in the statistical literature; and *principal coordinate analysis* (PCoA) after the comprehensive work of Gower [GOW 66], demonstrating its tight links with principal component analysis. *Triplet's analysis* is the French contribution to the field (e.g. [BEN 73, CAI 76]), which emphasizes that a (little more general) full presentation of Torgerson's approach involves the knowledge of a triplet $(\mathcal{I}, W_{\mathcal{I}}, \mathbf{O}_{\mathcal{I} \times \mathcal{I}})$ where $\mathcal{I}$ is a finite set of objects weighted by a non-negative measure $W_{\mathcal{I}}$ and $\mathbf{O}_{\mathcal{I} \times \mathcal{I}}$ is a matrix of pairwise dissimilarities.

PCoA generates a representation of the objects $i \in \mathcal{I}$ in the form of a *configuration matrix* $\mathbf{X}$, of order $n \times q$, the ith row of which gives the vector of coordinates $\boldsymbol{x}_i$ of the point of $\mathbb{R}^q$, associated with the object i in the solution representation of $\mathcal{I}$. $\mathbf{D}(\mathbf{X})$ yields a set of pairwise Euclidean distances which reproduces the observations $\mathbf{O}$ when this matrix is Euclidean and of rank q [MAR 78], and estimates them otherwise.

In the Euclidean case, the columns of $\mathbf{X}$ are interpretable as *variables* (features) – the principal components of $\mathbf{X}$ – and each row of $\mathbf{X}$ regroups the values of these variables – the *principal coordinates* – corresponding to an element of $\mathcal{I}$. This principal coordinates representation matrix may be made conspicuous in the set of possible Euclidean representations of $\mathcal{I}$ in a space of dimension q, because of its duality relationship with PCA. We have the following proposition.

PROPOSITION 4.1.– *The application of PCA (based on covariances) to the columns of any matrix $\mathbf{Z}$ of order $n \times m$, which describes the configuration of points induced by $\mathbf{X}$, admits the principal coordinates axes as their principal axes.*

The well-known relationship between inner product and distances specific to Euclidean geometry constitutes the fundamental principle of PCoA (see Figure 4.3):

$$b_{jk} = h_j^2 + h_k^2 - d_{jk}^2. \tag{4.1}$$

Consider a set $\{w_j \in]0, 1], j \in \{1, \cdots, n\}\}$ of weights defined on $\{1, \cdots, n\}$ and such that $\sum_{j=1}^n w_j = 1$. Then when $\mathbf{D}$ is an Euclidean distance matrix, the convention

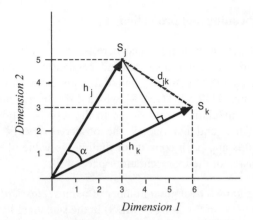

Figure 4.3. *Illustration of the relationship between scalar product and Euclidean distance in a 2D space*

adopted by Torgerson [TOR 52] consists of choosing the barycenter $\overline{x} = \sum_j w_j x_j$ as the origin of the data cloud of weighted points $\{(x_j, w_j) \in \mathbb{R}^q, j \in \{1, \cdots, n\}\}$, when Schoenberg prefers to choose one of these n points. Torgerson's proposal allows the following classical computations:

$$b_{jk} \triangleq < x_j - \overline{x}, x_k - \overline{x} >= h_j h_k \cos(\alpha) \tag{4.2}$$

where $h_j^2 = b_{jj} = ||x_j - \overline{x}||^2 = d_{j.}^2 - \frac{1}{2}d_{..}^2, d_{j.}^2 = \sum_k w_k d_{jk}^2$ and $d_{..}^2 = \sum_j w_j d_j^2$.

Let us denote the line of constant functions on $\mathcal{I}$ by Δ_{1_n} and the projector into the subspace D_p-orthogonal to Δ_{1_n} by Q_1. Then equation (4.2) can be written in matrix form:

$$b_{jk} = h_j^2 + h_k^2 - d_{jk}^2 \Leftrightarrow B = -\frac{1}{2}Q_1 H Q_1 \triangleq \tau(H) \tag{4.3}$$

where H is the Hadamard square of the distance matrix:

$$h_{jk} = d_{jk}^2 \Leftrightarrow H \triangleq D \odot D \tag{4.4}$$

and $Q_1 = I_n - 1_n 1_n' D_p$ is a centering operator (regarding the barycenter) which gives its name to the *double centering* operator τ. The notation τ is traditional and honors the pioneering contribution of Torgerson [TOR 52, TOR 58] aimed at the exploitation and diffusion of this result, classic in applied multivariate statistics. Moreover, note that $h_{jj} = 0 \neq h_j^2$.

Let $\mathcal{S}_n^q$ be the set of square positive semi-definite matrices of order n and rank q. We then have the following theorem [SCH 35, TOR 52, YOU 38].

THEOREM 4.1.– $O \in \mathcal{D}_n(q)$ *iff* $B_O \triangleq \tau(O \odot O)$ *is an element of* $\mathcal{S}_n^q$. *In this case, every factorization* $B_O = XX'$ *provides one configuration matrix* X, *of order* $n \times q$, *such that* $D(X) = O$.

It is therefore sufficient to compute the spectral decomposition $\mathbf{B_O} = \mathbf{UD}_\lambda\mathbf{U'}$ and to generate $\mathbf{X} \triangleq \mathbf{UD}_\lambda^{1/2}$. This elegant and computationally inexpensive solution has a weakness: it is *algebraic*, since based on the computation of inner products from distances, and this feature makes it a highly non-robust approach. Its use should be reserved for cases where the observed distance matrix is Euclidean and free of large measurement errors. Actual practice (especially in biometry) suffers numerous rash applications of PCoA, where the consequences of the non-Euclideanicity of the observed dissimilarity matrix are underestimated.

4.3.1. *An example of computations*

Let us consider the following table, which gives the observed dissimilarities between four objects:

$$\mathbf{O} = \begin{pmatrix} 0 & 4.05 & 8.25 & 5.57 \\ 4.05 & 0 & 2.54 & 2.69 \\ 8.25 & 2.54 & 0 & 2.11 \\ 5.57 & 2.69 & 2.11 & 0 \end{pmatrix} \Rightarrow \mathbf{H} = \mathbf{O} \odot \mathbf{O} = \begin{pmatrix} 0 & 16.40 & 68.06 & 31.02 \\ 16.40 & 0 & 6.45 & 7.24 \\ 68.06 & 6.45 & 0 & 4.45 \\ 31.02 & 7.24 & 4.45 & 0 \end{pmatrix}.$$

The matrix obtained by the Torgerson's double centering operation uses:

$$\mathbf{Q}_1 = \begin{pmatrix} +3/4 & -1/4 & -1/4 & -1/4 \\ -1/4 & +3/4 & -1/4 & -1/4 \\ -1/4 & -1/4 & +3/4 & -1/4 \\ -1/4 & -1/4 & -1/4 & +3/4 \end{pmatrix}$$

to obtain the pseudo-inner-products matrix $\mathbf{B} = -\frac{1}{2}\mathbf{Q}_1\mathbf{H}\mathbf{Q}_1$:

$$\mathbf{B} = - \begin{pmatrix} 2.52 & 1.64 & -18.08 & -4.09 \\ 1.64 & -0.83 & 2.05 & -2.87 \\ -18.08 & 2.05 & 11.39 & 4.63 \\ -4.09 & -2.87 & 4.63 & 2.33 \end{pmatrix}$$

and the spectral decomposition $\mathbf{B} = \mathbf{U'D}_\lambda\mathbf{U}$ leads to:

$$\mathbf{U} = \begin{pmatrix} 0.77 & 0.04 & 0.50 & -0.39 \\ 0.01 & -0.61 & 0.50 & 0.61 \\ -0.61 & -0.19 & 0.50 & -0.59 \\ -0.18 & 0.76 & 0.50 & 0.37 \end{pmatrix} \quad \text{and} \quad \mathbf{D}_\lambda = \begin{pmatrix} 35.71 & 0.00 & 0.00 & 0.00 \\ 0.00 & 3.27 & 0.00 & 0.00 \\ 0.00 & 0.00 & 0.00 & 0.00 \\ 0.00 & 0.00 & 0.00 & -5.57 \end{pmatrix}.$$

The content of $\mathbf{D}_\lambda$ shows that the observed dissimilarity was neither Euclidean nor of full rank. If we retain a representation of the objects in the real vector space spanned by the two first principal coordinates, we generate the configuration matrix $\mathbf{X}_{[2]} = \mathbf{U}_{[2]} \times \mathbf{D}_{\sqrt{\lambda}}$ with components:

$$\mathbf{X}_{[2]} = \begin{pmatrix} 0.77 & 0.04 \\ 0.01 & -0.61 \\ -0.61 & -0.19 \\ -0.18 & 0.76 \end{pmatrix} \times \begin{pmatrix} 5.98 & 0.00 \\ 0.00 & 1.81 \end{pmatrix} = \begin{pmatrix} 4.62 & 0.07 \\ 0.09 & -1.11 \\ -3.63 & -0.34 \\ -1.08 & 1.38 \end{pmatrix}.$$

4.3.2. *The additive constant problem*

Adding a constant to the off-diagonal terms in the observed dissimilarity matrix yields a makeshift method of rendering the Euclidean embedding of $\mathcal{I}$ feasible. This method requires some caution, and Messick and Abelson [MES 56] demonstrate it on a toy example. They show that the choice of this constant induces deformations of the estimated configurations of points and may change the order of the eigenvalues held back, if non-necessarily positive constants are allowed.

Moreover, the analysis of the triplet $(\mathcal{I}, W_{\mathcal{I},\mathbf{T}})$ depends on the selected additive constant a, since $t_{jk} = o_{jk} + (1 - \delta_j^k)a$ with $\delta_j^k = 1$ when $j = k$ and 0 otherwise. As a consequence, double-centering gives:

$$\mathbf{B_T} = \tau(\mathbf{T} \odot \mathbf{T}) = \mathbf{B_O} + 2a\mathbf{B}_{\sqrt{\mathbf{O}}} + \frac{a^2}{2}\mathbf{Q}_1$$

where $\mathbf{B}_{\sqrt{\mathbf{O}}} \triangleq \tau(\mathbf{O})$.

Cailliez [CAI 83] proves that the smallest additive constant which guarantees the Euclideanicity of $\mathbf{T}$ is the solution of an algebraic problem: the computation of the greatest eigenvalue of

$$\begin{pmatrix} 0 & 2\mathbf{B_O} \\ -1 & 4\mathbf{B}_{\sqrt{\mathbf{O}}} \end{pmatrix}.$$

For example, let us consider four points located on a circle. The observed dissimilarities are expressed in radians on the circle:

$$\mathbf{O} = \begin{pmatrix} 0 & \pi & \pi/4 & \pi/2 \\ \pi & 0 & 3\pi/4 & \pi/2 \\ \pi/4 & 3\pi/4 & 0 & 3\pi/4 \\ \pi/2 & \pi/2 & 3\pi/4 & 0 \end{pmatrix} \approx \begin{pmatrix} 0.0000 & 3.1416 & 0.7854 & 1.5708 \\ 3.1416 & 0.0000 & 2.3562 & 1.5708 \\ 0.7854 & 2.3562 & 0.0000 & 2.3562 \\ 1.5708 & 1.5708 & 2.3562 & 0.0000 \end{pmatrix}.$$

There exists no Euclidean embedding of these objects on the circle. Clearly, layman readers read 'bird's-eye' distances in the plane and (wrongly) infer that there exists Euclidean distances on the circle. These plane distances are not equal to the observed distances recorded in $\mathbf{O}$. In applying Cailliez's result, we compute the eigenvalue:

$$\lambda_{max} \begin{pmatrix} 0.00 & 0.00 & 0.00 & 0.00 & 3.16 & -5.48 & 2.24 & 0.08 \\ 0.00 & 0.00 & 0.00 & 0.00 & -5.48 & 5.63 & -1.47 & 1.31 \\ 0.00 & 0.00 & 0.00 & 0.00 & 2.24 & -1.47 & 2.55 & -3.32 \\ 0.00 & 0.00 & 0.00 & 0.00 & 0.08 & 1.31 & -3.32 & 1.93 \\ -1 & 0.00 & 0.00 & 0.00 & -2.55 & 2.95 & -0.98 & 0.59 \\ 0.00 & -1 & 0.00 & 0.00 & 2.95 & -4.12 & 1.37 & -0.20 \\ 0.00 & 0.00 & -1 & 0.00 & -0.98 & 1.37 & -2.55 & 2.16 \\ 0.00 & 0.00 & 0.00 & -1 & 0.59 & -0.20 & 2.16 & -2.55 \end{pmatrix} \approx 1.29.$$

We may then check that the matrix $\mathbf{B_{T_{1.29}}} = \tau(\mathbf{T}_{1.29} \odot \mathbf{T}_{1.29})$ derived from the affine transform:

$$\mathbf{T}_{1.29} = \begin{pmatrix} 0 & \pi & \frac{\pi}{4} & \frac{\pi}{2} \\ \pi & 0 & \frac{3\pi}{4} & \frac{\pi}{2} \\ \frac{\pi}{4} & \frac{3\pi}{4} & 0 & \frac{3\pi}{4} \\ \frac{\pi}{2} & \frac{\pi}{2} & \frac{3\pi}{4} & 0 \end{pmatrix} + 1.29 \begin{pmatrix} 0 & 1 & 1 & 1 \\ 1 & 0 & 1 & 1 \\ 1 & 1 & 0 & 1 \\ 1 & 1 & 1 & 0 \end{pmatrix}$$

is positive semi-definite, with rank $(n - 2) = 2$. We therefore generate a Euclidean representation of the four points in $\mathbb{R}^2$, located on a circle as expected.

This elegant method may be criticized on two respects. First, it is an algebraic approach, so is therefore not robust to measurement errors. Second, no control exists on dimensionality q, since the existence of a Euclidean embedding is proved only in a space of dimension at most $(n-2)$. The solution may therefore reveal itself useless in practice. Two alternative methods may be substituted, involving the use of an additive constant to apply PCoA to a transformed semi-definite positive matrix derived from **T**:

1) The first is an elementary method, and uses any *matrix exponential transformation* $\mathbf{A} \triangleq exp(\alpha \, \mathbf{O})$ of a symmetric matrix $\mathbf{O}$, with α real and strictly positive. Such a transformation is known to yield a symmetric positive semi-definite matrix, the eigenvalues of which are derived from those of $\alpha \, \mathbf{O}$ by exponentiation. The associated eigenvectors remain unchanged in this transformation of $\mathbf{O}$.

2) The second method is preferred because it is founded on the interpretation of this problem in terms of Euclidean embedding of a network. [AUB 89] proves that the adjoint of the operator τ defines $\mathbf{O}$ as the image of the positive semi-definite matrix $\mathbf{L}$, referred to as the *combinatorial Laplacian*, which is known to describe the network complexity:

$$\mathbf{L} = \mathbf{D}_{h_+} - \mathbf{H}, \quad \mathbf{D}_{h_+} \triangleq Diag(h_{j+}), \quad h_{j+} = \sum_{k \neq j} h_{jk}.$$

REMARK 4.2.– When interpreted in terms of matrix regularization, the additive constant plays the same part as the global shrinkage in *ridge* regression, while the Laplacian matrix plays the same part as the perturbation of **B** in an Empirical Bayes regression method. The compound solution $A \triangleq \exp(\alpha L)$ is interpreted in physics as the resolution of an heat diffusion equation, and this Markovian modeling strategy has recently been proposed in pattern recognition as a method of manifold learning [LAF 06]. In a recent past, the combinatorial (or Graph) Laplacian has also been used in machine learning for spectral-clustering or dimension reduction purposes (see [LEE 07] for a comprehensive discussion).

4.3.3. *The case of observed dissimilarity measures blurred by noise*

In the case of noisy dissimilarity data, Torgerson [TOR 58] and Gower [GOW 66] show that an approximative solution to the metric embedding problem may be designed as the following constrained optimization problem, formulated with the help of the Frobenius norm of a matrix:

$$(\mathcal{P}_1) \triangleq Min\{STRAIN \triangleq \|\mathbf{B} - \tau(\mathbf{T} \odot \mathbf{T})\|_F^2; \mathbf{B} \in \mathcal{S}_n^q\}. \qquad (4.5)$$

Once again, the solution is generated by algebraic means. Let us write the spectral decomposition of $\mathbf{B_O} = \tau(\mathbf{O} \odot \mathbf{O})$ as $\mathbf{B_O} = \mathbf{U}\mathbf{D}_\lambda^t\mathbf{U}$, in which the diagonal matrix $\mathbf{D}_\lambda$ is coded in order to satisfy $\lambda_1 \geq \lambda_2 \geq \cdots \geq \lambda_n$. We may then prove the following result, in which $\mathbf{D}_{\overline{\lambda}} = diag(\overline{\lambda}_j; j = 1, \cdots, q)$ and:

$$\overline{\lambda}_s = \begin{cases} \max(\lambda_s, 0) & \text{if } s = 1, \cdots, q \\ 0 & \text{if } s = (q+1), \cdots, n. \end{cases}$$

THEOREM 4.2.– *The matrix* $\overline{\mathbf{B}}_\mathbf{O} = \mathbf{U}\mathbf{D}_{\overline{\lambda}}^t\mathbf{U}$ *is a solution of problem (4.5), the global minimum of which is:*

$$m_q(\tau(\mathbf{O} \odot \mathbf{O})) = \sum_{j=1}^{q}(\lambda_j - \overline{\lambda}_j)^2 + \sum_{k=p+1}^{n}(\lambda_k)^2.$$

As a consequence, when modifying the loss function (4.5) as $\|\mathbf{B} - \tau(\mathbf{T}_a \odot \mathbf{T}_a)\|_F^2$ to insert an additive constant, we obtain a method to estimate the value a, adapted to the case where the observed dissimilarity data may be blurred by errors. This method, due to de Leeuw and Heiser [LEE 82], seems more satisfactory than its competitors proposed by Cooper [COO 72] and Saito [SAI 78], which determinate the additive constant in optimizing a criterion distinct from that optimized by the triplet analysis. So far, however, its implementation is numerically tricky. The contribution of Trosset [TRO 97] seems to furnish more satisfactory numerical results.

4.4. Multidimensional scaling and approximation

Nowadays, when interpreted as a statistical *data visualization method* suitable for dissimilarity measures, the name *multidimensional scaling* has become ambiguous because its use has a double meaning:

– In a *strict* sense, it refers to the family of techniques devoted to the representation of objects in a space of small dimensionality, constrained to respect as much as possible the information contained in an observed set of pairwise dissimilarity data. This definition covers the work of Shepard [SHE 62a, SHE 62b], de Leeuw *et al.* [LEE 82] and Cox *et al.* [COX 94]

– In a wider sense, the expression refers to any technique which generates some geometric representation of a set of objects, based on multivariate data. This definition encompasses certain clustering methods as well as most multivariate exploratory data analysis techniques, dotted with data visualization tools. PCA and CA are two examples of this class of statistical methods. Recent books by Ripley [RIP 96] or Gower [GOW 96] adopt such a meaning.

Another distinction in MDS is between *weighted* MDS, *replicated* MDS and MDS *on a single matrix* [YOU 94]. In replicated MDS, several matrices of proximity data can be analyzed simultaneously. The matrices are provided by different subjects or by a single subject observed at multiple times. A single scaling solution captures the proximity data of all matrices through separate metric or non-metric relationships for each matrix. This approach can take individual differences in response idiosyncrasy into account. In weighted replicated MDS e.g. in the *Individual Scaling* (INDSCAL) model [CAR 70], the dimensions in the scaling solution can be weighted differently for each subject or subject replication to model differences regarding the different dimensions.

Finally, there is the distinction between *deterministic* and *probabilistic* MDS. In deterministic MDS, each object is represented as a single point in multidimensional space (e.g. [BOR 97]), whereas in probabilistic MDS [MAC 89, RAM 77, TAK 78] each object is represented as a probability distribution in multidimensional space. In understanding the mental representation of objects, this latter approach is useful when representation of objects is assumed to be noisy (i.e. the presentation of the same object on every trial gives rise to different internal representations).

Our text assumes the strict meaning of MDS and focuses on the developments born in psychometry (and its competitors proposed with a different vocabulary in machine learning), and is often referred to as *vector quantization*. Moreover, related distance methods in *preference analysis* (e.g. [AUB 79, HEI 81]) and in *individual differences analysis* (a more debatable restriction) have been omitted. We refer the reader to [BOR 97, COX 94, DAV 83, EVE 97, TAK 77b, YOU 87] for a presentation of these closely related methods, and to [LEE 07] for a comparative study of recent developments in *non-linear data reduction* methodology.

4.4.1. *The parametric MDS model*

One way to analyze proximity data blurred by errors consists of looking for a configuration matrix $\mathbf{X}$ as the solution of a badness of fit problem which operates directly on the observed dissimilarity coefficients.

This approach departs from PCoA in a way to obtain direct control on the dissimilarity measurement, without the algebraic intermediation induced by the double-centering operator to obtain pseudo-inner products. The constrained optimization problem to be solved may be written:

$$(\mathcal{P}_2) \triangleq Min\{\text{STRESS} \triangleq \mathcal{E}(\mathbf{O}, \mathbf{D}(\mathbf{Z})); \mathbf{D}(\mathbf{Z}) \in \mathcal{D}_n(q)\} \tag{4.6}$$

where $\mathcal{E}(\mathbf{O}, \mathbf{T})$ is a regular discrepancy function. The two most often used criteria have been established by analogy to the OLS regression problem:

– Kruskal [KRU 64a, KRU 64b] formalized the heuristics originally proposed by Shepard [SHE 62a, SHE 62b]. He minimizes the mean square error function *raw* STRESS (STandardized REsidual Sum of Squares), interpretable as the (Frobenius) norm: $\mathcal{E}(\mathbf{O}, \mathbf{T}) = ||\mathbf{O} - \mathbf{T}||_F^2$.

– For numerical reasons, Takane *et al.* [TAK 77b] prefer to minimize the criterion function STRESS, which quantifies the mean squares discrepancy between squares of observed and estimated dissimilarities: $\mathcal{E}(\mathbf{O}, \mathbf{T}) = ||\mathbf{O} \odot \mathbf{O} - \mathbf{T} \odot \mathbf{T}||_F^2$.

In fact, both approaches are examples of a (weighted) mean squares criterion, corresponding to $\nu = \frac{1}{2}$ and $\nu = 1$, respectively:

$$\mathcal{E}(\mathbf{O}, \mathbf{T}) = \sum_{j<k} w_{jk}\vartheta_{jk} \quad \text{with} \quad \vartheta_{jk} = \left((o_{jk})^\nu - (t_{jk})^\nu\right)^2. \tag{4.7}$$

The introduction of weights w_{jk} acting on pairs of objects, and not on objects themseves, departs from the PCoA set-up. The weights w_{jk} are assumed positive, not uniformly null, and they allow missing data (in this case, $w_{jk} = 0$). They are also useful in situations where existing context effects lead to attribute differential weights to the pairs (j, k) of objects. For example, by taking $\nu = \frac{1}{2}$ and the weights $w_{jk} = (o_{jk})^{-1} \times \sum_{j<k}(o_{jk})^{-1}$, one may rediscover the *non-linear mapping* (NLM) method due to Sammon [SAM 69] (often cited and used in the pattern recognition community). One other particular case is the *stochastic* MDS problem due to Ramsay [RAM 77, RAM 78a, RAM 80, RAM 82]. Ramsay assumes that the observed dissimilarities are log-normal distributed and since, by continuity, $\ln(u) \sim (u)^\nu$ when ν tends to 0, Ramsay's solution appears as a limiting case of optimization problem (4.7).

From a numerical analysis point of view, the MDS problem is non-linear in the parameters and a difficult non-convex global optimization problem. A great number of numerical strategies have therefore been experimented with, but no general algorithm exists that can outclass competitors in any situation. Each subproblem has its own idiosyncrasies, on which specific effective numerical methods have developed (e.g. Guttman [GUT 68], de Leeuw [LEE 77], Glunt *et al.* [GLU 93], Groenen [GRO 93, GRO 95] and Pham and Le Thi [PHA 97, LET 98]). The difficulty of this question is attested by the results of a series of comparative studies (e.g. Machmouchi [MAC 92], Kearsley *et al.* [KEA 94], Li *et al.* [LI 95] and Basalaj [BAS 01]). In any case, convergence to a local optimum *only* can be proved, and there is a consensus to say that convergence of MDS algorithms is generally slow. Moreover, it is usually difficult to know which numerical approach has been implemented in a commercial package. It is therefore good practice to experiment with various packages and hence numerical methods.

The following presents only a limited number of numerical methods, selected for their historical value or their pedagogical interest. Presentation is moreover limited to their methodological features, without discussing the implementation aspects.

4.4.2. *The Shepard founding heuristics*

This section presents one iteration of the heuristics proposed by Shepard in his pioneering articles [SHE 62a, SHE 62b], to illustrate a toy dissimilarity demonstrated by Rabinowitz [RAB 75]. The objects are candidates in the US presidential election. The ranks produced by a subject, when asked to rank pairs of politicians with respect to their global similarity, are listed in the upper triangular part of Table 4.3. The lower triangular part yields hypothetical inter-point distances.

Code	Name	H	M	P	W	N
H	Humphrey	—	1	5	7	6
M	McGovern	8.06	—	2	10	8
P	Percy	11.18	6.32	—	9	4
W	Wallace	7.28	9.49	7.62	—	3
N	Nixon	9.06	7.81	4.12	3.61	—

Table 4.3. *US presidential elections hypothetical data [RAB 75]: the upper triangular part reports the ranks produced by a subject and the lower triangular part yields one set of hypothetical inter-point distances*

Let us assume that a bidimensional representation is required. The first stage of the iteration entails the choice of an initial configuration. This choice is not devoid of consequences, since it influences the convergence speed of the iteration. Figure 4.4a is an example of initialization, which locates the points at the vertices of a simplex.

The first iteration may be summarized by its input $\mathbf{X}_{old}$ and its output $\mathbf{X}_{new}$:

$$\mathbf{X}_{old} = \begin{pmatrix} -6 & -3 \\ -2 & 4 \\ 4 & 2 \\ 1 & -5 \\ 3 & -2 \end{pmatrix} \quad \Rightarrow \quad \mathbf{X}_{new} = \begin{pmatrix} -4.82 & -1.84 \\ -2.29 & 3.60 \\ 3.25 & 2.78 \\ 0.81 & -6.16 \\ 3.06 & -2.38 \end{pmatrix}.$$

Shepard heuristics may be presented in the language of (elastic) networks: each object $i \in \mathcal{I}$ is associated with a node of the graph $(\mathcal{I}, \mathcal{E})$ and each step of the iteration modifies the weight of edges. The weight of edge e computed at iteration ν is a function of the discrepancy between the observed dissimilarity o_e and the estimated distance $d_e(\mathbf{X}^{(\nu)})$. Shepard interprets this quantity as a force, either *attractive* or *repulsive* depending on the value of the computed discrepancy. Shepard heuristics

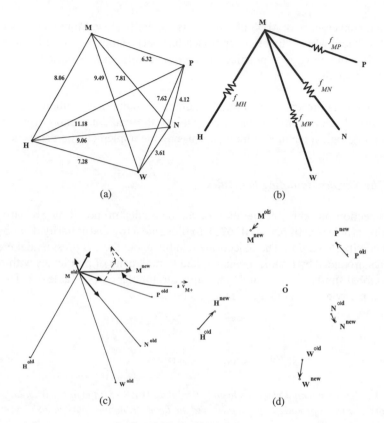

Figure 4.4. *US presidential elections data [RAB 75]. Four stages of one iteration of the Shepard heuristics: (a) initial configuration; (b) balance of forces to which the candidate McGovern is submitted; (c) resultant of the forces to which the candidate McGovern is submitted; (d) balance of the translations to which the configuration is submitted during one iteration*

therefore anticipate the ideas founding the *neural nets* methodology in looking for equilibrium states of a set of nodes binded by elastics, which submit their neighbors to forces proportional to these discrepancies.

Figure 4.4b illustrates the situation in the case of candidate McGovern. Its location is changed during each iteration, in function of the balance of forces applied to it by the other candidates. $\mathbf{X}^{(\nu)}$ is estimated with help of disparities $t_{jk}^{(\nu)} = m^{(\nu)}(o_{jk})$, in computing the quantity $e_{jk}^{(\nu)} = t_{jk}^{(\nu)} - d_{jk}(\mathbf{X}^{(\nu)})$.

More precisely, given a configuration $\mathbf{X}$, each object j is represented by the point with coordinates x_j and we reduce the discrepancy e_{jk} in moving the point j in the direction of point k. e_{jk} determines both the orientation and the intensity of the shifting vector from the current locations of j and k. Shepard uses the correction factor:

$$f_{jk} = \frac{e_{jk}}{d_{jk}(\mathbf{X})} = 1 - \frac{t_{jk}}{d_{jk}(\mathbf{X})}$$

and the contribution of point k to the location shift of point j cumulates motions v^s along each axis s. For each pair (j, k),

$$v_{jk}^s = f_{jk}(x_j^s - x_k^s) \quad s = 1, 2, \cdots, n.$$

The resultant shift of location associated with each point j is obtained by averaging:

$$v_{j.}^s = \frac{1}{n} \sum_{k=1}^{n} v_{jk}^s = \frac{1}{n} \sum_{k=1}^{n} (1 - \frac{t_{jk}}{d_{jk}(\mathbf{X})})(x_j^s - x_k^s)$$

with the convention $\frac{t_{jj}}{d_{jj}(\mathbf{X})} = 0$ for all j. The new coordinate of the representative point of object j on the axe s at the end of iteration $(\nu + 1)$ is:

$$x_j^{s(\nu+1)} = x_j^{s(\nu)} + v_j^{s(\nu)}.$$

Globally, each iteration establishes a balance of forces exerted on every node by the set of other nodes. The resultant of these forces at each point (Figure 4.4c) is calculated, and its resulting shift of position is deduced. Figure 4.4d shows the global transformation of candidate locations undergone by the configuration $\mathbf{X}^\nu$ during iteration $\nu : \mathbf{X}^\nu \mapsto \mathbf{X}^{\nu+1}$. [AUB 98] proposes a justification of Shepard heuristics, based on an analogy with the electric network theory. This may be complemented from an numerical analysis point of view by the following observation.

LEMMA.– *The gradient of the energy function* $\mathcal{E}(\mathbf{X}) = \sum_j \sum_k (t_{jk} - d_{jk}(\mathbf{X}))^2$ *is a n-dimensional real vector with sth coordinate*

$$\frac{\partial}{\partial x_j^s} \mathcal{E}(\mathbf{X}) = 2n v_j^s.$$

This means that Shepard heuristics may be interpreted as a gradient method with constant step size:

$$x^{s(\nu+1)} = x^{s(\nu)} - \frac{1}{2n} \left[\frac{\partial}{\partial x_j^s} \mathcal{E}(\mathbf{X}) \right]^{(\nu)}$$

and ignores the complexities due to the non-differentiability of $\mathcal{E}(\mathbf{X})$ in points $\mathbf{X}^{(\nu)}$ for which $d_{jk}(\mathbf{X}^{(\nu)}) = 0$ and $j \neq k$. From this numerical point of view, Kruskal's improvements [KRU 64a, KRU 64b] based on an effective management of the step size $\alpha(\nu)$ give a clear rationality to his choice of the raw STRESS, as an energy function to minimize. More recent contributions to this algorithmic matter, in particular de Leeuw and Heiser [LEE 80, LEE 82] in continuation of Guttman [GUT 68], may be interpreted as elaborations and refinements of this initial gradient iterative method. According to Kruskal:

$$x^{s(\nu+1)} = x^{s(\nu)} - \alpha(\nu) \left[\frac{\partial}{\partial x_j^s} \mathcal{E}(\mathbf{X}) \right]^{(\nu)},$$

which generates a configuration $\mathbf{X}$ solution of the problem:

$$Min\{\mathcal{E}(\mathbf{X}) , \ \mathbf{X} \in \mathcal{M}_{p \times q}(\mathbb{R}) , \ \mathbf{X} \text{ centered}\}.$$

The generalization for $\mathcal{E}(\mathbf{X})$ to the case where the observations are weighted is immediate: let w define a weight function on $\mathcal{I} \times \mathcal{I}$, such that $w_{jk} = w_{kj}$ and $w_{jj} = 0$ for all j and k, $j \neq k$.

DEFINITION 4.3.– *If $(\mathcal{I}, O_{\mathcal{I} \times \mathcal{I}}, W_{\mathcal{I} \times \mathcal{I}})$ is the triplet of interest, the energy function:*

$$\mathbf{X} \rightarrow \mathcal{E}(\mathbf{X}) = \sum_j \sum_k w_{jk}(t_{jk} - d_{jk}(\mathbf{X}))^2$$

is named the raw STRESS and the disparity $t_{jk} = m(o_{jk})$ is a transformation of the observed dissimilarity o_{jk} by a monotone increasing function m.

4.4.3. *The majorization approach*

Although the analytical definition of the raw STRESS function can appear to be rather simple, its minimization is a difficult numerical problem of global optimization [GRO 93]. Since the adaptation of Shepard's heuristics to a gradient method by Kruskal (1964), a huge number of numerical methods have been experimented upon to find more effective algorithms for the minimization of the STRESS function, from quasi-Newton iterations to D-C (difference of convex) programming [LET 98], simulated annealing [MAC 92, MAC 94], neural networks [NAU 01, WEZ 01, WEZ 04] or genetic algorithms [N'G 95, EVE 01, VAR 06].

Some of them have been submitted to a critical comparison by Kearsley *et al.* [KEA 94]. They are plagued by the same weaknesses: in most cases, the STRESS function has many local minima, and its minimization problem is high dimensional. Moreover, this criterion is not a convex function of X when the dimensionality of the solution space is fixed, and its non-differentiability cannot be ignored. However, [GRO 95] proves that if $d_{jk}(\mathbf{X})$ is a Minkowski ℓ_p-metric such that $p > 1$, gradient optimization methods can be applied in vicinities of local minima where the raw STRESS is differentiable. A Minkowski ℓ_p-metric in $\mathbb{R}^q$ is written:

$$d_{jk}(\mathbf{X}) = \left[\sum_{s=1}^{q} |x_j^s - x_k^s|^p\right]^{1/p}.$$

The best compromise for dissimilarity matrices of order $n \leq 100$, easily available in commercial packages (e.g. SPSS) and free software (e.g. $\mathcal{R}$), is considered to be the majorization method [LEE 82, LEE 88]. This was developed as an extension of the work of [GUT 68] to the case where the weights w_{jk} are not uniform. This choice is based on a comparative experiment designed by [BAS 01], which yields majorization and *simulated annealing* algorithms as the two winners of the contest. Note that the treatment of a massive data file also necessitates the use of incremental methods (e.g. [ELH 98, TIP 96]).

Returning to the Majorization algorithm, [LEE 82, LEE 88] show that the raw STRESS can be written:

$$\mathcal{E}(\mathbf{X}) = \eta_O^2 + \eta^2(\mathbf{X}) - 2\rho(\mathbf{X}) = \eta_O^2 + 2\mathcal{DC}(\mathbf{X}) \quad \eta_O^2 \triangleq \sum_{j=1}^{n-1}\sum_{k=j+1}^{n} w_{jk}o_{jk}^2.$$

The η_O^2 part is fixed by data, and can be assumed equal to 1 without loss of generality. The $\mathcal{DC}(\mathbf{X}) = \frac{1}{2}\eta^2(\mathbf{X}) - \rho(\mathbf{X})$ is the active part, and

$$\eta^2(\mathbf{X}) \triangleq \sum_{j=1}^{n-1}\sum_{k=j+1}^{n} w_{jk}d_{jk}^2(\mathbf{X}) = \text{tr}(\mathbf{B}_X\mathbf{G})$$

where $\mathbf{B}_X \triangleq \mathbf{X}\mathbf{X}'$ and

$$\rho(\mathbf{X}) \triangleq \sum_{j=1}^{n-1}\sum_{k=j+1}^{n} w_{jk}t_{jk}d_{jk}(\mathbf{X}) = \sum_{j=1}^{n-1}\sum_{k=j+1}^{n} \frac{w_{jk}t_{jk}}{d_{jk}(\mathbf{X})} \times d_{jk}^2(\mathbf{X}) = tr(\mathbf{B}_X\mathbf{L}(\mathbf{X})).$$

$\eta(\mathbf{X})$ and $\rho(\mathbf{X})$ are two seminorms in $\mathcal{M}_{n,r}(\mathbb{R})$. We can show [LEE 77] that the metric MDS problem is equivalent to the convex maximization problem:

$$(P_{\max}) \quad \max\left\{\frac{\rho(\mathbf{X})}{\eta(\mathbf{X})} : \eta(\mathbf{X}) \neq 0\right\} \Leftrightarrow \max\{\rho(\mathbf{X}) : \eta(\mathbf{X}) \leq 1\}. \tag{4.8}$$

Any configuration matrix $\mathbf{X}$, minimizer of the STRESS function, is therefore a solution of the stationary equation:

$$\mathbf{X} = \mathbf{G}^+\mathbf{L}(\mathbf{X})\mathbf{X}. \tag{4.9}$$

First, an initialization step generates a starting configuration matrix $\mathbf{X}^{(0)}$, which is then updated at each iteration ν in the following way:

$$\mathbf{X}^{(\nu+1)} = \mathbf{G}^+\mathbf{L}(\mathbf{X}^{(\nu)})\mathbf{X}^{(\nu)} \tag{4.10}$$

where $\mathbf{G}$ and $\mathbf{L}(\mathbf{X})$ are two symmetric, positive semi-definite matrices, with generic terms:

$$g_{jk} = \begin{cases} -2\,w_{jk} & \text{if } j \neq k \\ \sum_{k,k\neq j} w_{jk} & \text{if } j = k \end{cases}$$

and

$$l_{jk}(\mathbf{X}) = \begin{cases} -2\,w_{jk}\,\frac{t_{jk}}{d_{jk}(\mathbf{X})} & \text{if } j \neq k \\ \sum_{k,k\neq j} w_{jk}\,\frac{t_{jk}}{d_{jk}(\mathbf{X})} & \text{if } j = k, \end{cases}$$

respectively. Both $\mathbf{G}$ and $\mathbf{L}(\mathbf{X})$ are a graph Laplacian, and $\mathbf{G}^+$ is a generalized inverse of $\mathbf{G}$.

The majorization algorithm generates a sequence $\{\mathbf{X}^{(\nu)}, \nu \in \mathbb{N}\}$, associated with a non-increasing sequence of values $\{\mathcal{E}(\mathbf{X}^{(\nu)}), \nu \in \mathbb{N}\}$ of the STRESS function, and converges (to a local minimum). A more detailed analysis of the convergence of this algorithm [LEE 77, LEE 80, LEE 82, LEE 88] shows that the Guttman iteration (4.10) converges to a connected set of local stationary points if any pair of distinct objects $(i, k), i \neq k$ fulfills the condition $w_{jk}t_{jk} > 0$. It has been shown [LET 98, PHA 97] that the majorization algorithm solves the DC (also known as difference of convex functions) problem:

$$(P_{\min}) \quad \min\{\mathcal{DC}(\mathbf{X}) : \mathbf{X} \in \mathcal{M}_{n,r}(\mathbb{R})\} \tag{4.11}$$

since this active part of the STRESS function may be written as the difference of two convex functions. Many good properties of this algorithm are therefore inherited from those of the wider class of DC programming methods.

The interpretation of the matrices $\mathbf{G}$ and $\mathbf{L}(\mathbf{X})$ is a classic in multivariate data analysis, since they define a Euclidean geometry on two affine spaces in duality. The kernel $\mathbf{G}$ induces a global geometry which depends on weights w_{jk} only, while the kernel $\mathbf{L}(\mathbf{X})$ also depends on the current configuration $\mathbf{X}$ due to the non-linearity of the problem. Note that $\mathbf{L}(\mathbf{X})$ accounts for the discrepancy which exists between the disparity t_{jk} and the distance $d_{jk}(\mathbf{X})$ in the current configuration $\mathbf{X}$, in the form of a weighted ratio. As a consequence, we observe in practice that an affine transformation of the data may produce a great perturbation of the solution, and may take a long time to be corrected by the algorithm.

4.4.3.1. *The influence of the choice of a discrepancy function*

Raw STRESS is the most elementary MDS badness of fit function, and it is easily extendable to the case where pairs (j, k) are differentially weighted. This makes the integration of missing pair comparisons possible. Moreover, since distances verify $d_{jk}(\alpha\mathbf{X}) = \alpha d_{jk}(\mathbf{X})$, $\mathcal{E}(\mathbf{X})$ can be made arbitrarily small by alternating an update of the configuration $\mathbf{X}$ and an update of disparities $t_{jk} = \alpha o_{jk}$ with α and $\mathbf{X}$ shrinking gradually. This deficiency can be overcome by transforming $\mathcal{E}(\mathbf{X})$. A first method consists of normalizing $\mathcal{E}(\mathbf{X})$ by the mean squares semi-norm of disparities $\eta_{\mathbf{T}}^2$. The result is known as the *normalized STRESS*:

$$\mathcal{E}_{norm}(\mathbf{X}) = \frac{\mathcal{E}(\mathbf{X})}{\eta_{\mathbf{T}}^2} \quad \text{with} \quad \eta_{\mathbf{T}}^2 \triangleq \sum_{j=1}^{n-1} \sum_{k=j+1}^{n} w_{jk} t_{jk}^2 \qquad (4.12)$$

and was proposed [KRU 64a, KRU 64b] to normalize by the semi-norm $\eta^2(\mathbf{X})$. This normalization factor defines what this author calls the *Stress formula 1*, or STRESS1 loss function, which is written STRESS1 $= \sqrt{\mathcal{E}_{Krus1}(\mathbf{X})}$, with:

$$\mathcal{E}_{Krus1}(\mathbf{X}) = \frac{\mathcal{E}(\mathbf{X})}{\eta^2(\mathbf{X})} \quad \text{and} \quad \eta^2(\mathbf{X}) = \sum_{j=1}^{n-1} \sum_{k=j+1}^{n} w_{jk} d_{jk}^2(\mathbf{X}). \qquad (4.13)$$

Basalaj [BAS 01] proves that for a given configuration $\mathbf{X}$, $\mathcal{E}_{norm}(\mathbf{X})$ and $\mathcal{E}_{Krus1}(\mathbf{X})$ are equivalent if we allow for an optimal scaling $t_{jk} = \alpha\, o_{jk}$ of dissimilarities in each case and, as a consequence, at the respective optima α_{norm}^* and α_{Krus1}^*:

$$\mathcal{E}_{norm}(\alpha_{norm}^*\mathbf{X}) = \mathcal{E}_{Krus1}(\alpha_{Krus1}^*\mathbf{X}) \triangleq \mathcal{E}^*(\mathbf{X}) = 1 - \left(\frac{\rho(\mathbf{X})}{\eta(\mathbf{X}).\eta_{\mathbf{O}}}\right)^2. \qquad (4.14)$$

This common index is scale free and varies in $[0, 1]$ in such a way that lower values of $\mathcal{E}^*(\mathbf{X})$ correspond to better output configurations $\mathbf{X}$.

The original definition of the raw STRESS function suffers another disadvantage: errors associated with large and small dissimilarities are penalized equally, making it difficult to preserve local properties of data. Some contributions focus on the weight function w as the means to counteract this weakness. For example, take $w_{jk} = \alpha g(t_{jk})$ where g is monotone decreasing and α is a normalizing constant. The simplest choice is $g(t) = t^{-1}$, since weights do not depend on $\mathbf{X}$. A more demanding choice consists of constraining the weight function to depend on the current configuration matrix $\mathbf{X}$ since, in this case, specific MDS algorithms have to be designed. Two successful approaches of this type are Sammon *non-linear mapping* [SAM 69] and Demartines *curvilinear component analysis* [DEM 94]. The former sets $w_{jk} = (\alpha d_{jk}(\mathbf{X}))^{-1}$ normalized by $\alpha = \sum_{j<k} d_{jk}(\mathbf{X})$, while the latter uses $w_{jk} = g(d_{jk}(\mathbf{X}))$ and g is a fixed monotone non-increasing function. A specific algorithm had to be developed in each case since the weight function depends on $\mathbf{X}$.

A second modification of the MDS problem is the result of preprocessing the observed dissimilarities and/or the output distances to obtain better convexity properties of the induced numerical problem. The loss function is then written:

$$\mathcal{E}_{\phi,\psi}(\mathbf{X}) = \sum_j \sum_k w_{jk} (\phi(t_{jk}) - \psi(d_{jk}(\mathbf{X})))^2. \qquad (4.15)$$

The most significant example of such a modification is due to Takane *et al.* [TAK 77b], and results from the choice $\phi = \psi = 2$. This leads to the evaluation of discrepancies between *squared* disparities and *squared* Euclidean distances, and consequently to the definition of a new badness of fit index called the SSTRESS (squared STRESS): SSTRESS $= \mathcal{E}_{2,2}(\mathbf{X})$. In this case, the optimization problem to be solved simplifies because of the good convexity properties of squared Euclidean distances.

Optimizing SSTRESS necessitates a specific algorithm. The initial algorithm [TAK 77b] has now been superseded by the Hendrickson [HEN 95] algorithm which is more efficient with large datasets. We are now able to display, in a reasonable computing time, proteins containing up to 124 amino acids i.e. 777 atoms (objects). A second efficient optimizer of SSTRESS has been proposed by Moré and Wu [MOR 96]. This algorithm is innovative for two reasons: first the DC problem is solved by a *trust regions* algorithm; second, the iteration is protected against risk of capture by a local minimum by smoothing the loss function with a Gaussian convolution Kernel. Although time consuming, this method has been successfully applied to the 216 atoms constituting molecules.

Fitting squared distances preferably to distances gives an augmented weight to large dissimilarities, with consequences discussed above. Some authors tried to reduce this risk by normalizing the energy function. For example, Niemann [NIE 80] proposed improving the Sammon non-linear mapping algorithm by optimizing the energy function:

$$\mathcal{E}_{Niem}(\mathbf{X}) = \frac{\mathcal{E}_{2,2}(\mathbf{X})}{\sum_{j<k} w_{jk} d_{jk}^4(\mathbf{X})} \quad \text{and} \quad w_{jk} = d_{jk}^{-2}(\mathbf{X}).$$

4.4.3.2. *Initialization*

It has become common practice to use the PCoA output configuration of the desired dimensionality as an initialization of the MDS algorithm. This practice seems debatable, since it increases the number of cases where the iteration becomes trapped by a local optimum. Naud [NAU 01] shows that the energy function is more prone to violate the expected decreasing monotonicity feature in the neighborhood of local optima. One of the main conclusions of the numerical experiment conducted by Basalaj [BAS 01] is that PCoA and MDS global optima often yield output configurations showing statistically significant differences.

Many users start with a PCoA solution and then omit it to check if the obtained optimum is global or local. Acting so, they quite often retain MDS solutions relatively similar to the PCA, because they become trapped by a local optimum. This mistaken practice reinforced the misleading statement that both methods often give similar solutions. Diffusing such a misconception would lead to underestimating the non-negligible improvements brought by the adoption of MDS algorithms in preference to PCoA when the observed dissimilarities are not Euclidean.

When using a MDS algorithm, it is reasonable to multiply initializations and choose them to be as diverse as possible. This can be done manually (e.g. [AUB 89]), but recourse to stochastic strategies enables the job to be completed automatically. Simulated annealing strategies (e.g. [AUB 98, BAS 01, MAC 92, MAC 94]) and genetic algorithms (e.g. [LER 95, N'G 95, ZIL 07]) offer two promising and complementary ways to generate random initial configurations. They are not a panacea since their use increases computing time, especially for genetic algorithms. When making comparisons, it should always be remembered that multiplying initial configurations will increase the necessary computing time.

4.4.4. *Extending MDS to a semi-parametric setting*

We now consider the often-met situation (e.g. in sensor analysis) where the information collected is subjective and assumed invariant by isotonic (monotone and bijective) transformation. A related and more general situation occurs in molecular spectrography, where theory dictates variation boundaries to all or part of the inter-atomic distances measured by nuclear magnetic resonance (NMR). For computing purposes, we are given a class of admissible transformations of $\mathbf{O}$ in which the MDS algorithm looks for an optimal representative in a least mean squares sense. The MDS algorithm generates an output configuration matrix $\mathbf{X}$, and estimates a disparity matrix $\mathbf{T}$ in the class of admissible functions of $\mathbf{O}$.

4.4.4.1. *The ordinal multidimensional scaling method*

The ordinal multidimensional scaling method or *non-metric multidimensional scaling* (NMDS) method assumes that isotonic transformations of $\mathbf{O}$ are admissible, and associate with every pair $e = (j, k)$ the disparity $t_e = m(o_e)$, where m is an admissible real function. In practice, *ex æquo* pairs – $e \neq e'$ such as $o_e = o_{e'}$ – necessitate specific attention. See [AUB 75] or [GUT 68] for a discussion of this technical point.

Each iteration of an NMDS algorithm adds to the corresponding iteration in metric MDS, a step devoted to the resolution of an isotonic regression problem. This is inexpensive when solved in a least mean squares sense [BAR 72] which is expressed (for iteration τ):

$$\mathbf{T}^{(\tau)} = \operatorname{argmin}\left\{||\mathbf{\Theta} - \mathbf{D}(\mathbf{X}^{(\tau)})||_F^2; \ \theta_e = m(o_e) \ \forall e \in \mathcal{I} \times \mathcal{I}\right\}. \qquad (4.16)$$

The NMDS solution so obtained is often barely discernible from the MDS solution when the number of objects is large. The observed differences are then mainly due to existing outliers. As a matter of fact, the analysis of small samples (typical of sensor analysis and behavioral sciences) has showed that NMDS algorithms are much more robust than their parametric analog.

4.4.4.2. *An example of isotonic regression*

This section presents the isotonic regression algorithm proposed by Barlow *et al.* [BAR 72], applied to the Rabinowitz data on the US presidential candidate election. The block algorithm described by [BAR 72] operates gradually, and its sequence of steps is reported in the columns of Table 4.4: each step compares the current values of distance d_e with those $d_{e'}$ associated with values $o_{e'}$ smaller than o_e. They remain unchanged as long as they respect the order of the o_es. In case of order violation, the responsible distances are averaged with preceding contiguous values until constitution of an amalgamation block associated with an average which respects (in a wide sense) the order induced by observations o_{ik}.

Pairs	o_{ik}	d_{ik}	1	2	3	4	5	6	7	8	t_{ik}
Humphrey–McGovern	1	7.8	7.8	5.5	3.93	3.38	3.38	3.38	3.38	3.38	3.38
McGovern–Percy	2	3.2	3.2	5.5	3.93	3.38	3.38	3.38	3.38	3.38	3.38
Nixon–Wallace	3	0.8	0.8	0.8	3.93	3.38	3.38	3.38	3.38	3.38	3.38
Nixon–Percy	4	-1.7	1.7	1.7	1.7	3.38	3.38	3.38	3.38	3.38	3.38
Humphrey–Percy	5	9.1	9.1	9.1	9.1	9.1	8.5	8.13	6.68	5.8	5.32
Humphrey–Nixon	6	7.9	7.9	7.9	7.9	7.9	8.5	8.13	6.68	5.8	5.32
Humphrey–Wallace	7	7.4	7.4	7.4	7.4	7.4	7.4	8.13	6.68	5.8	5.32
McGovern–Nixon	8	2.3	2.3	2.3	2.3	2.3	2.3	2.3	6.68	5.8	5.32
Percy–Wallace	9	2.3	2.3	2.3	2.3	2.3	2.3	2.3	2.3	5.8	5.32
McGovern–Wallace	10	2.9	2.9	2.9	2.9	2.9	2.9	2.9	2.9	2.9	5.32

Table 4.4. *US presidential elections data: disparities t_{ik} resulting from an isotonic regression of distances d_{ik} on dissimilarities o_{ik}*

For example, in step 5, amalgamating the pairs (Humphrey, Nixon) and (Humphrey, Wallace) would not be sufficient to break the violation of the order constraints induced by the corresponding observed dissimilarities 6 and 7, since step 4 attributes the value 8.5 to the pairs (Humphrey, Nixon) and (Humphrey, Percy). We need to amalgamate these three pairs to obtain an amalgamation block with average value $8.13 > 3.38$, respecting the order fixed by o. One output of the iteration is the disparity t, which constitutes a monotonic transformation of o realizing a least mean squares isotonic regression of d on o [BAR 72]. This output is visualized in Figure 4.5.

4.4.4.3. *Extending NMDS to more general boundary constraints*

In molecular biology, spectrographic measurements usually dictate structural stability constraints to molecules, which assign variation boundaries to admissible inter-atomic distances:

$$l_e \le d_e(\mathbf{X}) \le u_e, \ \forall e \in \mathcal{I} \times \mathcal{I}.$$

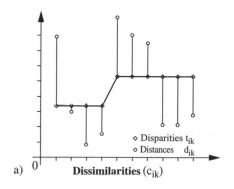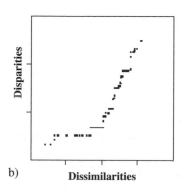

a)   **Dissimilarities (c_{ik})**   b)   **Dissimilarities**

Figure 4.5. *Two examples of isotonic regression applied to distance table: (a) applied to the US presidential elections data and (b) applied to the moral concepts data submitted to a multidimensional scaling in q = 2 dimensions*

These structural stability constraints include the ordinal MDS case, since the lower and upper boundaries may be fixed by exogenous information and not deduced from the order of observed dissimilarities. The resulting *block constraints* are conventionally defined as:

$$\mathbf{D}(\mathbf{X}) \in [\mathbf{L}, \mathbf{U}] \leftrightarrow l_e \leq d_e(\mathbf{X}) \leq u_e, \ \forall e \in \mathcal{I} \times \mathcal{I}.$$

When the badness of fit index $\mathcal{E}(\mathbf{O}, \mathbf{D}(\mathbf{Z}))$ retained is the raw STRESS function, the optimization problem under boundary constraints to be solved is written:

$$(\mathcal{P}_3) \triangleq \min\{\mathcal{E}(\mathbf{O}, \mathbf{D}(\mathbf{Z})); \mathbf{D}(\mathbf{Z}) \in [\mathbf{L}, \mathbf{U}] \cap \mathcal{D}_n(p)\}. \qquad (4.17)$$

Glunt *et al.* [GLU 93] produce a sub-gradient algorithm, relevant to the Guttman algorithm. More recently, several optimization methods based on criteria analogous to the SSTRESS function also appeared [LET 98, TRO 97], written:

$$(\mathcal{P}_3) \triangleq \min\{\mathcal{E}'(\mathbf{D}(\mathbf{Z})); \mathbf{D}(\mathbf{Z}) \in \mathcal{D}_n(p)\} \qquad (4.18)$$

where

$$\mathcal{E}'(\mathbf{D}(Z)) = \sum_{e=1}^{n(n-1)/2} q_e \vartheta_e(\mathbf{Z})$$

and where

$$\vartheta_e(\mathbf{Z}) = \min_e \left\{ \frac{d_e^2(\mathbf{Z}) - l_e^2}{l_e^2}, 0 \right\} + \max_e \left\{ \frac{d_e^2(\mathbf{Z}) - u_e^2}{u_e^2}, 0 \right\}.$$

The results of the first published numerical experiments, cross-validated using numerous molecular biochemistry datasets [MOR 96], seem promising.

4.5. A fielded application

This section is devoted to data published by Tournois and Dickes [TOU 93] and reported in the form of the summary Table 4.5. A random sample of 64 subjects produced their own dissimilarity judgements on 17 moral concepts (objects) compared on subjective grounds. Each subject was invited to subjectively sort these concepts into a number of classes fixed by themselves. The observed (compound) dissimilarity matrix was deduced from the 64 partitions so obtained.

	a	b	c	d	e	f	g	h	i	j	k	l	m	n	o	p
a	128															
b	66	125														
c	84	109	38													
d	132	132	136	149												
e	121	130	127	138	33											
f	123	110	115	124	73	73										
g	118	102	122	135	124	124	111									
h	119	103	126	140	118	112	110	65								
i	140	73	133	127	117	124	102	102	93							
j	144	124	132	140	125	129	106	104	104	113						
k	130	114	133	145	122	118	115	62	51	96	88					
l	89	115	99	115	122	117	102	89	91	109	110	94				
m	154	111	145	127	140	140	125	115	118	105	34	98	126			
n	131	89	137	128	129	129	125	82	63	84	116	50	103	96		
o	20	115	79	72	144	133	133	128	132	134	154	143	101	144	123	
p	28	133	69	85	128	118	116	118	124	137	145	134	86	155	137	39

Table 4.5. *Moral concepts data [TOU 93]: lower triangular part of the (symmetric) observed dissimilarity matrix*

4.5.1. *Principal coordinates analysis*

The moral concepts dissimilarity data (Table 4.5) is first submitted to a principal coordinates analysis (PCoA) of the triplet $(\mathcal{I} = \{a, b, \cdots, p\}, \frac{1}{16}\mathbf{I}_{16}, \mathbf{O}_{\mathcal{I} \times \mathcal{I}})$. Table 4.6 lists the histogram of eigenvalues and a $q = 2$ dimensional PCoA solution (on the first principal plane), displayed in Figure 4.6a.

Note that the spectrum of obtained eigenvalues contains some negative values. If not considered negligible, their existence indicates that the observed dissimilarity function was not Euclidean.

Adding an additive constant, as discussed in section 4.3.2, allows us to correct for these negative eigenvalues. Here, their absolute value may be considered small enough regarding the roughness of observed dissimilarity judgements. This interpretation can be checked by comparing PCoA results to those of a more robust method, namely non-metric multidimensional scaling, assuming an ordinal scale of measurement of dissimilarities.

Negative eigenvalues	Eigenvalues	% of eigenvalues	Positive eigenvalues
	39610.52	63.043	***
	21973.6	19.401	*********************
	14685.08	8.666	****************
	11765.55	5.562	************
	5869.46	1.384	******
	4664.44	0.874	*****
	3309.55	0.440	***
	2265.36	0.206	**
	1910.19	0.147	**
	1554.1	0.097	**
	637.74	0.016	*
	512.53	0.011	*
	42.09	7.118E–05	
	–17.08	1.172E–05	
*	–544.42	0.012	
*	–1024.53	0.042	
**	–1569.74	0.099	

Table 4.6. *Spectral decomposition of the table derived by double centering of the scrutinized observed dissimilarity matrix*

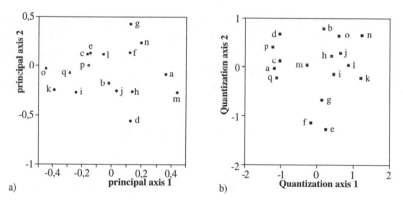

Figure 4.6. *Moral concepts data: comparison of solution displays obtained by PCA and MDS with dimension q = 2: (a) plane spanned by the first two principal coordinates and (b) plane produced by semi-parametric multidimensional scaling*

The displays were produced with the help of a commercial package. Many choices are available e.g. use of SAS (Statistical Analysis System), SPSS (Statistical Programs for Social Sciences) and SYSTAT (System for Statistics) in parallel, to vary the algorithm used. Moreover, several initial configurations were submitted to control the risks of sub-optimal solution corresponding to some local optimum. The configuration displayed in Figure 4.6b and those to be presented in the following were produced by SYSTAT.

4.5.2. *Dimensionality for the representation space*

Determining the correct dimensionality of the representation space is relatively standard in the PCoA case because, as in the PCA case, output sub-spaces of smaller dimension are *embedded* in higher dimensional sub-spaces and obtained by orthogonal projection.

Figure 4.6a illustrates a resulting noticeable feature of Euclidean distances evaluated on the PCoA q-dimensional output configurations $\mathbf{X}_q$; it satisfies the *inner* approximation property:

$$d_e(\mathbf{X}_q) \le d_e(\mathbf{X}_{(q+1)}) \le t_e, \quad \forall e \in \mathcal{I} \times \mathcal{I}, \quad \forall q \quad 1 \le q \le (n-1).$$

As a data reduction method, PCoA is tautological and outputs distances which are biased (shrunk) estimates of the observed dissimilarities, due to the contracting feature of projection operators:

$$d_e(\mathbf{X}_q) = o_e - \epsilon_e, \quad \epsilon_e \in \mathbb{R}^+ \quad \forall e \in \mathcal{I} \times \mathcal{I}.$$

In contrast, MDS yields *bilateral* approximations of observed disparities, produced by fitting a falsifiable model. As a consequence, Figures 4.7a and 4.7b look very different. In both cases, the output distances correspond to an optimum established in 2D, and noticeable differences are attributable to the inner approximation feature of PCA. This does not minimize the raw STRESS, but only the discrepancy between derived Torgerson transforms by:

$$\text{STRAIN} = \sum_{jk} \left(b_{\mathbf{O},jk} - b_{jk}(\mathbf{X}) \right).$$

Figure 4.7b illustrates the fact that MDS allows overestimations of disparities by a Euclidean distance as well as underestimations. Moreover, Figure 4.7b plots disparities, and not dissimilarities, on the X-axis as an ordinal version of MDS has been used. The disparity used in this plot is the output monotone increasing transformation of the observed dissimilarity.

Since the solution MDS algorithms cannot be obtained as orthogonal projections on embedded affine manifolds, the optimal configuration in q dimensions does not necessarily contain the optimal configuration in $(q-1)$ dimensions. This is illustrated here on the moral concepts dataset by submitting the outputs of MDS in 2D and 3D to a canonical analysis (CA). As expected, the results of CA substantiate the general coherence of the two solutions. At the same time, however, it appears that the vector space spanned by the columns of $\mathbf{X}_2$ is not strictly embedded in that spanned by the columns of $\mathbf{X}_3$. Table 4.7 and Figure 4.8 show that the multiple correlation of each column vector in $\mathbf{X}_2$ on $\mathbf{X}_3$ differs from 1.

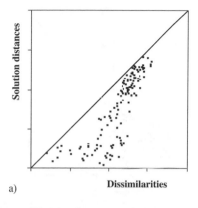
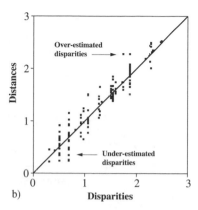

a) **Dissimilarities**   b) **Disparities**

Figure 4.7. *Moral concepts data: comparison of observed dissimilarity approximations by distances derived by PCoA and by NMDS with q = 2, respectively: (a) PCoA calculates a projection on an affine sub-space, therefore approximating from below the observed dissimilarities and (b) MDS yields a two-sided approximation, and does not define an orthogonal projector on an affine sub-space*

	Dim2(1)	Dim2(2)
Multiple correlation	0.998	0.989
Squared multiple correlations	0.996	0.978
Ajusted $R^2 = 1 - \frac{(n-1)(1-R^2)}{ddl}$ $(n = 17, ddl = 13)$	0.995	0.973

	Regression coefficient $B = (X'X)^{-1}X'Y$		Standardized regression coefficient	
	Dim2(1)	Dim2(2)	Dim2(1)	Dim2(2)
Dim(1)	1.085	−0.019	0.997	−0.025
Dim(2)	0.037	1.050	0.025	0.988
Dim(3)	−0.067	−0.043	-0.031	−0.028

Table 4.7. *Moral concepts data: results of a multivariate regression of the 2D MDS solution on the 3D solution*

4.5.3. The scree test

An important issue in both PCoA and MDS is choosing the number of dimensions for the scaling solution. A configuration with a high number of dimensions achieves low stress values, but cannot easily be comprehended by the human eye and is apt to be determined more by noise than by the essential structure in the data. On the other hand, a solution with too few dimensions might not reveal enough of the structure in the data.

A well-known method to select the dimensionality is *the scree test* (also known as the elbow test), the methodological status of which differs in PCoA and in MDS for part of the non-embeddability of solutions space in the latter case. In MDS, the raw STRESS (or any other badness of fit criterion) is plotted against the dimensionality. Ideally, this choice is visually obvious from the *elbow* in the scree plot indicating when, after a certain number of dimensions, the stress is no longer reduced

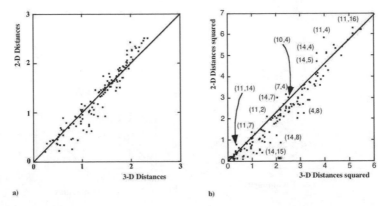

Figure 4.8. *Moral concepts data: comparison of distances, obtained by multidimensional scaling in p = 3 (in X-axis) and in p = 2 dimensions (in Y-axis): (a) comparison of distances and (b) comparison of squared distances (pairs which do not respect an embedding of 2D and 3D solutions are marked by their label)*

substantially. However, in many datasets, the energy function decreases smoothly with increasing dimensionality making the choice of appropriate dimensionality very difficult with this method.

A more salient indicator for the appropriate dimensionality can be obtained by *cross-validation*. The idea is to test how the configuration optimized to model the proximity data for one group of subjects can generalize to the proximity data of a different group of subjects. This method is only feasible if integrated to the design of a rich enough data collection, however. Lee [LEE 01] and Oh and Raftery [OH 01] have independently explored these issues, promising more principled techniques of dimensionality determination based on balancing the trade-off between model fit and model complexity.

The application of the scree test to the moral concepts data necessitates the construction of the optimal configuration in a sequence of choices of dimension q. The resulting diagram must be interpreted with caution. In first approximation, Table 4.8 seems to indicate that a 3D solution would be preferable to a 2D solution. Whatever the choice, the end of this chapter uses the output solution in 2D.

Dimensions	STRESS	Barchart diagram
1	0.30149	*********************************
2	0.11812	*************
3	0.07897	*********
4	0.03003	***
5	0.01788	**

Table 4.8. *Scree test applied to the moral concepts data; computations were made using the SYSTAT package*

4.5.4. *Recourse to simulations*

Whatever the method adopted to fix an optimal dimensionality able to secure a fair representation of the observed proximity data, two difficulties must be faced. First, stating that the best solution necessitates q dimensions is not sufficient to determine if it is good. Can a STRESS equal to 0.11812 be judged close enough to zero to justify 2D? Kruskal [KRU 64a, KRU 64b] produced a series of rules of thumb, applied by many users but hard to justify on theoretical grounds. Others prefer to base their choice on the argued interpretability of the solution.

One way to resolve this dilemma consists of making decision based on simulations: for each sample size, we submit a random display of a large number of uniformly distributed points in a space of fixed dimensionality to MDS. This makes it possible to establish abacuses of the resulting values of the energy function of interest. Several abacuses of this type have been published in the psychometric literature for the STRESS function [CLA 76, KLA 69, SHE 72b, SPE 72, SPE 74b, SPE 74a, SPE 78, STE 69]. Their practical use is based on the idea that informative dimensions would result in values of the STRESS which depart from the value generally obtained on randomly (structureless) generated data.

The simulation of non-uniform configurations may be carried out with the help of probabilistic dissimilarity models (developed by e.g. [MAC 81, RAM 78a, RAM 80, RAM 82, TAK 77a, TSU 96]). This does not seem to have been undertaken on a large scale, however.

4.5.5. *Validation of results*

4.5.5.1. *The Shepard diagram*

The relation existing between observed dissimilarities and output distances produced by the ordinal MDS program is plotted in Figure 4.9 for the moral concepts data. This display is known as the *Shepard diagram* in the psychometric literature, while it is of common use in simple curvilinear regression. The main information given by this compound display for the moral concepts data is that the MDS algorithm could approximate large dissimilarities easier than small dissimilarities (by Euclidean distances in a 2D solution space).

Consulting and validating this type of plot may be tricky because it combines two sources of approximation: the first arises from the estimation of an isotonic function and the second concentrates on a metric MDS problem. The first component is represented in Figure 4.5b for the moral concepts data. It focuses on a non-parametric estimation problem. The isotonic regression of a set of distances on the observed dissimilarities and the predicted distances are the disparities in NMDS. The second component is a metric MDS component which takes the disparities for fixed, and fits

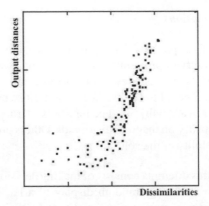

Figure 4.9. *Shepard diagram obtained for moral concepts data: output interpoint distances computed over the optimal plane and solution of the ordinal MDS problem*

a Euclidean distance to them. The smaller the badness of fit function, the more points concentrate along the first bisecting line. Results are presented in Figure 4.10b.

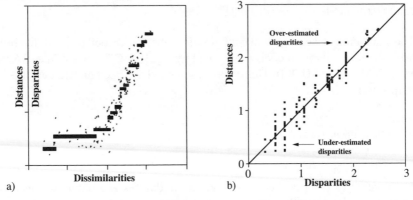

Figure 4.10. *Moral concepts data: decomposition of the Shepard diagram in semi-parametric MDS. (a) A monotonic transformation maps observed dissimilarities into disparities and (b) a transformation maps distances in the plane configuration into optimal disparities, derived by an ordinal MDS algorithm*

For the moral concepts data, Figure 4.10b shows that the isotonic regression step partially corrected the differential treatment of small and large dissimilarities by the metric MDS step. Fitting distances to the optimal disparities exhibits equally dispersed residuals for small and large disparities. Note that the gap between points and the bisecting line allow us to appreciate visually the size of the terms $\ell_{jk}(\mathbf{X}) = -2w_{jk}t_{jk}d_{jk}(\mathbf{X})^{-1}$ for which $t_{jk}d_{jk}(\mathbf{X})^{-1} \neq 1$, where $\ell_{jk}(\mathbf{X}) \neq g_{jk}$.

Any pair $e = (j, k)$ makes the function $\mathcal{E}(\mathbf{X})$ different from zero at the optimum if $t_e \neq d_e(\mathbf{X})$, while the solution $\mathbf{X}$ globally satisfies the fixed point equation $\mathbf{X} = \mathbf{G}^+\mathbf{L}(\mathbf{X})\mathbf{X}$.

On the other hand, Figure 4.10a describes the component of the solution which informs us about its agreement to the pre-order defined on the set of pairs $e = (j, k)$ by the observed dissimilarities. Each landing observable on the plot 4.10a corresponds to one inequality violated by the output distances. During the isotonic regression step, the algorithm generates a block of minimal width into which distances are averaged to obtain equal disparities out of respect for the corresponding monotonicity constraints.

The number of landings and their width are informative for two reasons. First, the less numerous and widespread the observed landings are, the more the isotonicity constraint is satisfied. Second, each landing corresponds to (approximately) equal distances in the output configuration, i.e. to artifactual regular geometric patterns such as isosceles or equilateral triangles, squares, rhombuses, etc. An interactive diagnostic tool would be useful to identify and characterize such patterns in the representation, as well as pairs creating the problem.

4.5.5.2. Tucker index of metric concordance

Features shown by a Shepard diagram are usually left unquantified. The Tucker index [TUC 51] may be used for this purpose; several interesting properties of this index are therefore discussed here as is its relation to the STRESS function.

Let E denote a representation space and $\mathcal{I}$ a set of n elements i represented as n points in E. The mapping $\Upsilon : \mathcal{I} \mapsto E$ associates with each $i \in \mathcal{I}$ its image $x_i \in E$. Neither E nor $\mathcal{I}$ is assumed *a priori* supplied with a geometric structure. We only assume that we are given the observed dissimilarity o on $\mathcal{I}$ and distance d on E.

DEFINITION 4.4.– *We refer to Tucker's Index of metric concordance as the cross-product coefficient:*

$$I_{CM}(\Upsilon) = \sum_{j=1}^{n} \sum_{k<j} o(j, k)d(\Upsilon(j), \Upsilon(k)). \qquad (4.19)$$

This is also definable as $I_{CM}(\Upsilon) = \sum_{j,k} o(\Upsilon^{-1}(j), \Upsilon^{-1}(k))d(j, k)$ since Υ is one-to-one. A reliable representation therefore corresponds to a large value of the index. The way in which we define o and d depends upon the problem dealt with. Moreover, if there exists a mapping Υ which preserves the dissimilarity order, then the maximization of I_{CM} enables us to estimate it. As a matter of fact, Goodhill *et al.* [GOO 95, GOO 96] prove the following result.

LEMMA.– *Let $\{x_i; i \in \mathbb{N}\}$ and $\{y_i; i \in \mathbb{N}\}$ be two real positive, monotone increasing sequences which admit a finite number of non-zero terms, and let π denote a permutation defined on $\mathbb{N}$. Then:*

1) $\pi = $ identity maximizes $I_{CM}(\pi) = \sum_{i=1}^{\infty} x_i y_{\pi(i)}$ w.r.t. π, and

2) the maximization of $I_{CM}(\pi)$ provides the permutation π^*, such as:

$$x_j < x_k \Rightarrow y_{\pi^*(j)} \le y_{\pi^*(k)} \quad and \quad y_{\pi^*(j)} < y_{\pi^*(k)} \Rightarrow x_j \le x_k.$$

A straightforward application of this lemma to multidimensional quantification problems is expressed as follows.

COROLLARY.– *Let $(\mathcal{I}, o)$ and (E, d) be two pre-metric spaces of equal cardinality. The bijection Υ^* which maximizes $I_{CM}(\Upsilon)$ belongs to the set $\mathcal{B}$ of one-to-one mappings $b : \mathcal{I} \mapsto E$ such that for all $s_1, s_2, s_3, s_4 \in \mathcal{I}$ and $x_1, x_2, x_3, x_4 \in E$:*

1) $o(s_1, s_2) < o(s_3, s_4) \Rightarrow d(b(s_1), b(s_2)) < d(b(s_3), b(s_4))$, and

2) $d(x_1, x_2) < d(x_3, x_4) \Rightarrow o(b^{-1}(x_1), b^{-1}(x_2)) < o(b^{-1}(x_3), b^{-1}(x_4))$.

In particular, in the case of metric MDS, $\mathcal{I}$ and F are identified, o is an observed dissimilarity function and d is a Euclidean distance function. In the case of the Sammon non-linear mapping method, both o and d are assumed to be Euclidean distance kernels. In both cases, these properties of the Tucker index I_{CM} yield some theoretical foundation to past practice which used the Mantel index – a rediscovery of I_{CM} – as a mean of quantification of the quality of representations produced by these methods. A series of concurrent index measuring quality of MDS solutions exist; see Goodhill *et al.* [GOO 95, GOO 96] or Bezdek *et al.* [BEZ 95] for discussions and references.

We limit ourselves to the following observation: equation (4.14) establishes a useful link between the Tucker index and the energy function minimized in MDS (either metric or non-metric) and in NLM. It shows that the normalized Tucker index

$$T_{CM} = \frac{I_{CM}(\mathbf{O}, \mathbf{D}(\mathbf{X}))}{\sqrt{I_{CM}(\mathbf{O}, \mathbf{O}) I_{CM}(\mathbf{D}(\mathbf{X}), \mathbf{D}(\mathbf{X}))}}$$

is well suited for descriptive purposes since it may be interpreted as a cosine and varies in $[-1, 1]$, with the null value as the non-informative case.

The relation with the normalized STRESS is complete when the Tucker index is extended to the case of non-trivial weights w_{jk}. Then, its normalized form for the pair $(\mathbf{t}, \mathbf{d}(\mathbf{X}))$ is written:

$$T_{CM}(\mathbf{t}, \mathbf{d}(\mathbf{X})) \triangleq \frac{< \mathbf{t}, \mathbf{d}(\mathbf{X}) >_{\mathbf{D}_W}}{||\mathbf{t}||_{\mathbf{D}_W} \times ||\mathbf{d}(\mathbf{X})||_{\mathbf{D}_W}}. \qquad (4.20)$$

Its evaluation at the optimum of a MDS algorithm yields descriptive information on the global reliability of the output configuration, in terms of metric recovery. Moreover, the denominator of $c(\mathbf{t}, \mathbf{d}(\mathbf{X}))$ is permutation invariant and so its numerator, i.e. the Tucker (or Mantel) index I_{CM}, may be used as a permutation test statistic useful for evaluating metric recovery.

4.5.5.3. *Looking for influential elements*

The Shepard diagram can help to spot pairs of objects which take a prominent part in the development of the iteration in a representation space of a given dimensionality. The importance of this spotting results from the fact that MDS algorithms not only output some optimal configuration $\mathbf{X}$ but conjointly output the associated optimal metric, induced by $\mathbf{L}(\mathbf{X})$. Specific statistics may be computed to quantify these notions, but their discussion is outwith the introductory scope of this text. Nevertheless, let us note that we work towards two complementary ends: the search of both influential pairs of points and – a more tricky problem – influential points.

4.5.6. *The use of exogenous information for interpreting the output configuration*

Another important issue is the interpretation of the output configuration resulting from MDS procedures. Once an optimal configuration has been generated, we need to interpret the components of the resulting plot(s). This work is made easy when some geometric structure organizes the plot, such as a clustering tendency or the belongingness to some (curvi) linear manifold. This is the case for the Ekman data, for which the output configuration in Figure 4.1a depicted a bidimensional structure distributing the colors along the arc of a circle. A great number of proposals have appeared recently, which try to capitalize on the existence of differentiable geometric structures to analyze the metric structure of their distribution support. See [LEE 07] for an overview of this exciting research area on *manifold learning*.

Finding such endogenous structures is mainly an informal data mining activity. Each knowledge discovery resulting from this exploratory process therefore entails some form of confirmation by way of heuristic or principled means. For example, an often-used method [KRU 78] consists of submitting the data $\mathbf{O}$ to some (hierarchical) clustering algorithm and then drawing the output (hierarchy of) partitions on the plot(s) obtained by an MDS algorithm. However, this *endogenous validation* method may encounter problems, because these two approaches are somehow antinomic. As a matter of fact, labeled hierarchies of partitions are isomorphic to ultrametric distances, and Holman [HOL 72] proved that the Euclidean embedding of n objects may necessitate $n-1$ dimensions if their interpoint distances are ultrametric. This has to be contrasted with the wish for few dimensions that justifies the use of MDS.

MDS and HCA have different privileged domains. The MDS assumes the existence of a continuum of values into which the objects can be embedded. HCA assumes a combinatorial structure, more able to describe aggregative features. In the first case, the form of the histogram of observed dissimilarities tends to be unimodal and skewed toward small values, while the ultrametric dissimilarities tend to produce repartitions skewed toward large values and sometimes multimodal.

In other situations, both approaches may reveal themselves to be complementary, and this kind of assistance to interpretation becomes effective. This is the case for the example of the votes of US representatives (Figure 4.1). Here, the representation of partitions produced independently by a clustering method on the MDS plot adds endogenous information and helps the interpretation.

Reality is often somewhere in between the two models, and we hope to obtain an approximate view of *homogenous groups* to appear in a low dimensional plot issued from an MDS method, interpreted using *typological* aims.

4.5.6.1. *Calibration of a configuration and observed illustrative variable*

The moral concepts database is special because the dissimilarity data was completed by a series of complementary variables measured on each concept (object), usable for interpretive purposes. Since they did not participate in the elaboration of MDS output, it is usual to call them *illustrative elements*. Here, two sources of *exogenous information* exist. The first is a numerical variable: a sample of 176 subjects scored the moral gravity of the 17 moral concepts; column 3 of Table 4.9 provides the average score attributed to each concept.

Name	Moral concept	Mean gravity	Object	Modality	Dim. X1	Dim. X2
a	Crime	6.525	Situation	Instrumental	−1.174	−0.032
b	To dishonour	5.881	Behavior	Cognitive	0.178	0.774
c	Hate	5.997	Situation	Affective	−1.020	0.118
d	To hate	5.857	Behavior	Affective	−1.031	0.663
e	Inequality	6.051	Situation	Cognitive	0.250	−1.290
f	Injustice	6.358	Situation	Cognitive	−0.146	−1.155
g	Intolerance	6.267	Situation	Cognitive	0.122	−0.685
h	Cowardice	5.881	Situation	Affective	0.404	0.217
i	Dishonesty	6.080	Situation	Instrumental	0.462	−0.148
j	To be disrespectful	5.756	Behavior	Cognitive	0.649	0.282
k	Distrust	4.540	Situation	Affective	1.206	−0.242
l	Lies	5.790	Situation	Instrumental	0.867	0.028
m	Nuisance	6.159	Situation	Instrumental	−0.259	0.035
n	To distrust	4.005	Behavior	Affective	1.221	0.648
o	To cheat	5.761	Behavior	Instrumental	0.602	0.622
p	To kill	6.540	Behavior	Instrumental	−1.216	0.394
q	Violence	6.403	Situation	Instrumental	−1.115	−0.229

Table 4.9. *Moral concepts data: available exogenous information [TOU 93]. The column 'Mean gravity' has been derived from empirical measurements and the columns 'Object' and 'Modality' result from subject matter arguments*

A linear regression of the exogenous 'gravity' variable on the two variables defining a basis of the representation space for the moral concepts data has been used. It yields a *calibration* method, often useful in assisting the interpretation of the MDS solution. Table 4.10 reports the results of the multiple linear regression of the gravity variable on the MDS output dimensions.

Variable	Coefficient	STD error	STD coeff	Tolerance	T	P(2 tail)
Constant	5.874	0.108	0.000	—	54.190	0.000
Dim(1)	−0.563	0.133	−0.713	1.000	−4.223	0.001
Dim(2)	−0.338	0.186	−0.306	1.000	−1.811	0.092

		Analysis of variance			
Source	Sum-of-squares	DF	Mean-square	F-ratio	P
Regression	4.217	2	2.108	10.557	0.002
Residual	2.796	14	0.200	—	—

Warning: Case 14 is an outlier (studentized residual = −3.268)

Table 4.10. *Moral concepts data [TOU 93]: multiple linear regression of the (available) exogenous variable 'Mean gravity' on the solution dimensions of the MDS problem*

For example, the squared multiple correlation coefficient equals 0.775 on these data, and it is known with a standard error 0.447. This highlights the interest, but a restricted one, in this indicator for calibration purposes. In general, the criticisms of the artificiality of representation axes may sometimes be removed when they reveal themselves highly (curvi) linearly correlated with some exogenous variable.

As discussed by [GOW 96] for the definition of biplots in MDS, the situation is more complex than linear regression can accommodate. Since MDS is *not* a projection method, linear regression only uses a part of the information contained in the variable gravity. This may be completed by using the values of the variable 'gravity' as labels of corresponding points. This yields Figure 4.11b where the association between the exogenous variable and vectors lying in the representation space are now free of *a priori* linearity constraints.

This labeling shows that the projected 'gravity' axis suffered violations of the order induced by this variable, and was not of much assistance in a reliable interpretation of the plot. A more advanced research was imperative. Note that the labeling of points did act as a third dimension variable added to the plot, making it reminiscent of the response surfaces methodology based on the set of measurement locations fixed by the MDS output configuration.

This approach showed us that quadratic regression models could help. The results are not shown here, however, because it is observed above that a better idea would be to retain three dimensions in applying MDS to the moral concepts data. Note also that drawing segments between points on the plot of Ekman data (Figure 4.1) was a way to approximate a non-linear manifold supporting an informative structure assigned to the data.

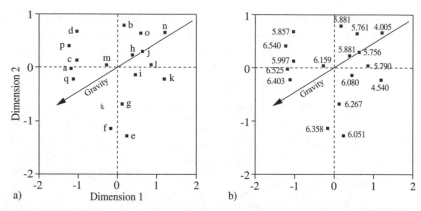

Figure 4.11. *Moral concepts data: introduction of an illustrative calibration based on exogenous information. (a) Use of linear regression of the 'mean gravity' variable and (b) use of non-linear calibration consisting of points labeling with help of the values of an exogenous variable 'gravity average'*

4.5.6.2. *Calibration of a configuration and derived illustrative variables*

A variant of the above method has been used on the plot of votes of New Jersey representatives (see Figure 4.1). It consists of defining regions of the MDS configuration, induced by some qualitative exogenous variable. An example is the partisan adherence of representatives. This *regional interpretation* is non-linear, meaning that two geographically nearby (respectively, distant) points can belong to distinctive (respectively, identical) regions. This approach differs from typological approaches, since it is based on the *facet theory* of Guttman [LEV 85].

This can also be brought into play on the moral concepts data, with the two facets 'Object' and 'Modality' reported in Table 4.9. Variants of the analysis of variance (ANOVA) may be used for scaling purposes, to quantify certain aspects of this qualitative method. Concretely, any facet defines a set of elements, which corresponds to some conceptual category. For example, the 17 moral concepts data can be characterized by two facets depending on (1) whether the concept refers either to situations (described by nouns) or to behaviors (described by verbs) and (2) whether the modality distinguishes cognitive concepts from affective and instrumental concepts.

Using facets to interpret a plot remains informal and assumes that we dispose of a theory which fixes the data generation process. In Figure 4.12a, two regions partition the space in function of the facet 'Object'. A horizontal line separates the two regions, and an ANOVA (not reported here) highlighted a statistically significant effect on dimension 2 and a non-significant effect on dimension 1. The facet 'Modality' defines regions too complex to be describable by a single ANOVA factor.

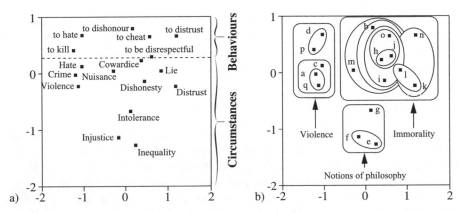

Figure 4.12. *Moral concepts data: introduction of a regional interpretation of the display. (a) Partition of space in two zones contrasted by dimension 2(Y-axis) and (b) drawing of embedded partitions produced by hierarchical clustering (Ward criterion)*

A more exploratory technique has been used in Figure 4.12b for interpreting the 2D MDS configuration for the moral concepts data. It shows the results of a hierarchical clustering algorithm which minimizes variance according to the Ward criterion [CHA 81] and applied to table **X**. The resulting dendrogram is given in Figure 4.13.

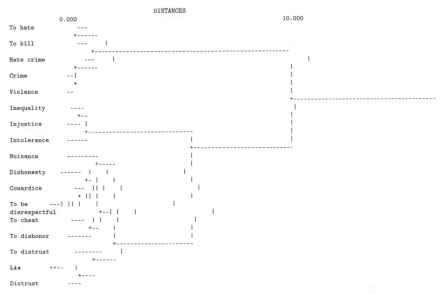

Figure 4.13. *Moral concepts data: hierarchical clustering based on the Ward criterion*

Both approaches agree on these data, so that hierarchical clustering produces groups with a simple spatial contiguity structure which define, as expected, embedded zones without overlapping. This yields a simple and useful method to assist in the interpretation of the MDS optimal configuration.

4.5.6.3. *The positioning of illustrative (new) points*

As illustrated in Figure 4.14, a dual source of interpretive aids in MDS is furnished by so-called *illustrative* (or *out-of-sample* or *supplementary*) objects. We test working hypotheses by adding some illustrative points to the output display. Illustrative points are defined as objects *not* having participated in the construction of the MDS output configuration. Illustrative is therefore opposed to *active*. Active (or *learning*) objects form the dissimilarity matrix $\mathbf{O}_A$ from which the MDS solution configuration $\mathbf{X}_A$ of rank q is derived. Positioning illustrative (out-of-sample) objects consists of deriving the matrix $\mathbf{X}_S$ of their coordinates in the affine space spanned by $\mathbf{X}_A$.

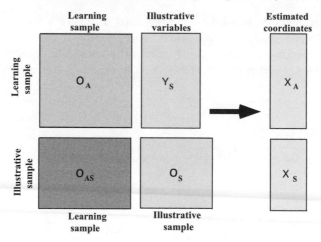

Figure 4.14. *A summary of observations available for a complete MDS study*

This is possible if either observed dissimilarities between the illustrative object(s) and the active objects are available, or substantive arguments permit the simulation of such cross-dissimilarities. In both cases, let us consider that a sample of $n = n_A + n_S$ objects has been partitioned in n_A active and n_S illustrative objects. Let us also consider that two dissimilarity matrices are given: the active matrix $\mathbf{O}_A$ and the illustrative *cross-dissimilarity* matrix $\mathbf{O}_{SA}$ of order $n_S \times n_A$, the general term $o_{j_s k_a}$ which measures the dissimilarity between an illustrative object j_s and an active object k_a. Positioning illustrative objects on the plot (e.g. [AUB 89, GOW 68, TRO 06]) differs from representing p illustrative variables previously measured on the n objects of interest, and forming some illustrative data table Y_S of order $n_A \times p$. In PCoA, they induce two dual problems.

An analytic solution to the 'adding a point' problem in PCoA is due to [GOW 68] and [WIL 70]. A matrix formulation has been given by [MAI 79] for one point and extended to n_S illustrative points in MDS by [AUB 84, AUB 88, AUB 89].

In PCoA, a weight w_i is usually imposed on each object. Let us impose a null weight to illustrative objects while respecting the normalization $\sum_i w_i = 1$, and submit the n objects (active and illustrative) to a PCoA. This bipartition induces the following decompositions in blocks:

$$\mathbf{D}_n = \begin{pmatrix} \mathbf{D}_{n_A} & \mathbf{0} \\ \mathbf{0} & \mathbf{0} \end{pmatrix}, \mathbf{H} = \begin{pmatrix} \mathbf{H}_{11} & \mathbf{H}_{12} \\ \mathbf{H}_{21} & \mathbf{H}_{22} \end{pmatrix}, \mathbf{B} = \begin{pmatrix} \mathbf{B}_{11} & \mathbf{B}_{12} \\ \mathbf{B}_{21} & \mathbf{B}_{22} \end{pmatrix}, \mathbf{Q} = \begin{pmatrix} \mathbf{Q}_1 & \mathbf{0} \\ -\mathbf{Q}_{21} & \mathbf{I}_{n_S} \end{pmatrix}.$$

The matrix $\mathbf{Q}_{21} = \mathbf{1}_{n_S} \mathbf{1}'_{n_A} \mathbf{D}_{n_A}$ is derived from available data, as well as the active part of $\mathbf{B} = \mathbf{Q}\mathbf{H}\mathbf{Q}'$. In fact, only the block $\mathbf{B}_{22}$ depends on unavailable data, since $\mathbf{H}_{22}$ is an unknown part of $\mathbf{H} = (-\frac{1}{2}o_{jk}^2)$. The important point to note is that it does not intervene in the necessary computations. As a matter of fact, we can prove the following result [AUB 84, AUB 88, AUB 89, MAI 79].

PROPOSITION 4.2.– *If* $\mathbf{X}_A$ *is the output configuration and solution of the PCoA problem and if* $\mathbf{D}_\lambda$ *is the associated diagonal matrix of eigenvalues, then the coordinates of illustrative objects form the columns of the matrix* ${}^t\mathbf{X}_S = \mathbf{D}_\lambda^{-1} {}^t\mathbf{X}_A \mathbf{D}_{n_A} \mathbf{B}_{12}$.

This solution is only valid for PCoA, but Pekalska *et al.* [PEK 99] show how to extend this method to the positioning of illustrative points on the solution space of a MDS problem or of a NLM problem. First, they notice that if n_A objects form the sample of active points and if $\mathbf{X}_A$ denotes the configuration matrix of order $n_A \times q$ derived from $\mathbf{O}_A$, then since $\mathbf{O}_A$ has full rank in NLM, there exists a linear mapping with matrix $\mathbf{M}_A$ of order $n_A \times q$ such that $\mathbf{X}_A = \mathbf{O}_A \times \mathbf{M}_A$. $\mathbf{M}_A$ must therefore be computed first, and the cross-dissimilarities between elements of $\mathbf{I}_S$ and $\mathbf{I}_A$ must be made available to form a matrix $\mathbf{O}_S$ of order $n_S \times n_A$. Then, adding the n_S illustrative objects only requires the computation of $\mathbf{X}_S = \mathbf{O}_S \times \mathbf{M}_A$. The case of MDS is more general, but if both matrices $\mathbf{O}_A$ and $\mathbf{O}_S$ are available, the above method applies under a rank condition: $\mathbf{X}_A$ should verify $rank(\mathbf{X}_A) \leq rank(\mathbf{O}_A)$.

4.5.7. *Introduction to stochastic modeling in MDS*

Our presentation of the MDS methodology basically considers it as a family of exploratory tools classifiable among *dimension reduction* methods. This is not totally accurate since there exist circumstances where it is reasonable to consider observed dissimilarities as fallible measurements of some latent distances. In such situations, recourse to stochastic modeling gives the opportunity to estimate the variability of the observed dissimilarities and of the resulting location of points in the MDS output configuration. Inferences of various kinds become available, e.g.

adding confidence regions to each point on the solution plot. This add-on extra seems especially promising for shape analysis and face recognition problems [BRO 06].

Prime work on this topic is due to MacKay and Zinnes [MAC 81]. These authors attach to each object a latent location vector $\xi \in \mathbb{R}^q$ assumed Gaussian, and estimate its mathematical expectation by a vector $x \in \mathbb{R}^q$. Various models of dependence are possible to constrain the coordinates of ξ, which conduct to model inner product matrices as Wishart distributed random matrices. The (image) distribution law of the corresponding Euclidean distance matrix $\Delta \triangleq (d_{jk}(\xi))$ follows. It is not standard, and leads to complex computations. This approach has been reworked and improved by Tsuchiya [TSU 96].

Ramsay [RAM 77, RAM 78b, RAM 78a, RAM 80, RAM 82] developed a second class of models, adopted also by [TAK 77a, TAK 78]. It consists of modeling the repartition law of latent distances Δ from scratch. Dissimilarities o are then interpreted as fallible observations blurred by random errors. The Ramsay model is of semi-parametric type, and considers disparities $t_{jk} = m[ln(o_{jk})]$ as realizations of a family of independent random variables, normally or (preferably) log-normally distributed:

$$T_{jk} \sim LG(ln[d_{jk}(\xi)], \sigma_{jk}^2). \tag{4.21}$$

As argued by Ramsay, one justification for this logarithmic transformation is based on the empirical establishment that the measurement accuracy of observed dissimilarities tends to decrease with their size. It seems that more technical reasons can also enter into play when we integrate the relative feature of dissimilarity judgments, and decide to treat them as a peculiar type of *compositional data*.

Regardless, the variance term of the model must be designed. Ramsay considers the case where a sample of N respondents replicated the experiment and produced their own dissimilarity data matrix. The retained variance model therefore assigns a specific parameter to each (group of) respondent(s) and is written:

$$\sigma_{jk}^2 = \sigma_i^2 \times \gamma_{jk}^2, \quad i \in \{1, \cdots, N\}. \tag{4.22}$$

This model allows individual (or group) variations in the level of measurement accuracy. Moreover, Ramsay proposes choosing between three different types of transformation of dissimilarities into disparities:

$$t_{jk}^i = m^i[\ln(o_{jk}^i)] = \ln(o_{jk}^i) + \nu_i, \tag{4.23}$$

$$t_{jk}^i = m^i[\ln(o_{jk}^i)] = \beta_i \ln(o_{jk}^i) + \nu_i, \tag{4.24}$$

$$t_{jk}^i = m^i[\ln(o_{jk}^i)] = S_i(\ln(o_{jk}^i)) + \nu_i. \tag{4.25}$$

Models (4.23) and (4.24) are parametric. Model (4.25) is non-parametric and integrates a monotonic spline function S_i to smooth the transformation obtained during the isotonic regression step of each iteration, rendering it less prone to be trapped by local optima. A detailed analysis of this non-parametric add-on in an exploratory MDS setup is available in [ELH 88].

Model (4.24), taken as an example, can be written:

$$\beta_i \ln[o_{jk}^i] + \nu_i = \ln[d_{jk}^i(\xi)] + \epsilon_{jk}^i. \qquad (4.26)$$

Ramsay and Takane developed maximum likelihood numerical methods to estimate the parameters of these models and, in particular, the estimates $d_{jk}^i(\mathbf{X})$ of distances $d_{jk}^i(\xi)$. The choice of the correct model in this context is not straightforward, and is based on arguments of asymptotic behavior. Oh and Raftery [OH 01] and Lee [LEE 01] contributed MDS models in a Bayesian context. This set-up yields more principled probabilistic tools aimed at model specification and choice of dimensionality. No asymptotics are used. However, the two underlying probabilistic models differ from McKay and Zinnes, from Ramsay and from one another. A careful comparison of these models remains to be made.

4.6. Bibliography

[ALF 97] ALFAKHI A. Y., KHANDANI A., WOLKOWICZ H., An interior point method for the Euclidean Distance Matrix Completion problem, Research Report num. CORR 97-09, Department of Combinatorics and Optimization, University of Waterloo, Canada, 1997.

[ALF 98] ALFAKHI A. Y., WOLKOWICZ H., On the embedability of weighted graphs in Euclidean spaces, Research Report num. CORR 98-12, Department of Combinatorics and Optimization, University of Waterloo, Canada, 1998.

[ARA 91] ARABIE P., "Was Euclid an unnecessarily sophisticated psychologist?", *Psychometrika*, vol. 56, p. 567–587, 1991.

[AUB 75] AUBIGNY (D') G., Description statistique des données ordinales: analyse multidimensionnelle, Thesis, Joseph Fourier University, Grenoble, France, 1975.

[AUB 79] AUBIGNY (D') G., "Analyse factorielle et visualisation des préférences", BATTEAU P., JACQUET-LAGRÈZE E., MONJARDET B., Eds., *Analyse et Agrégation des Préférences dans les Sciences Sociales, Économiques et de Gestion*, p. 177–212, Economica, Paris, 1979.

[AUB 84] AUBIGNY (D') G., "Un modèle algébrique pour l'analyse des données de dissimilarité", *Actes des troisièmes journées franco-soviétiques de statistique et analyse des données*, Paris, 1984.

[AUB 88] AUBIGNY (D') G., "The additive decomposition of some entropy functions and (constrained-) ordination methods", *Proceedings of the XIV-th IBC Conference*, Namur, Belgium, 1988.

[AUB 89] AUBIGNY (D') G., L'Analyse multidimensionnelle des données de dissimilarité, PhD Thesis, Joseph Fourier University, Grenoble, France, 1989.

[AUB 93] AUBIGNY (D') G., "Analyse des proximités et programmes de codage multidimensionnel", *Revue Modulad*, vol. 11, num. 3, p. 1–34, 1993.

[AUB 98] AUBIGNY (D') G., "Vers un renouveau des méthodes de positionnement multidimensionnel", *Journées Modulad-Renault*, Renault, Guyancourt, France, Modulad, Paris, November 1998.

[BAR 72] BARLOW R. E., BARTHOLOMEW D. J., BREMNER J. M., BRUNK H. D., *Statistical Inference under Order Restrictions: The Theory of Isotonic Regression.*, John Wiley & Sons, New York, 1972.

[BAR 95] BARVINOK A. I., "Problems of distance geometry and convex properties of quadratic maps", *Discrete and Computational Geometry*, vol. 13, p. 189–202, 1995.

[BAS 01] BASALAJ W., Proximity Visualisation of Abstract Data,   External research studentship, Trinity College, Cambridge, UK, January 2001.

[BEN 73] BENZÉCRI J.-P., *L'Analyse des Données*, Dunod, Paris, 1973.

[BEZ 95] BEZDEK J. C., PAL N. R., "An index of topological preservation for feature extraction", *Pattern Recognition*, vol. 28, num. 3, p. 381–391, 1995.

[BIS 96] BISHOP C. M., SVENSÈN M., WILLIAMS C. K. I., GTM: A principled alternative to the self organizing map, Technical report, Neural Computing Research Group, Aston University, Birmingham, 1996.

[BIS 98] BISHOP C. M., SVENSÈN M., WILLIAMS C. K. I., "GTM: the generative topographic mapping", *Neural Computation*, vol. 10, num. 1, p. 215–234, 1998.

[BOR 81] BORG I., *Multidimensional Data Representationsn: When and Why?*, Mathesis Press, Ann Arbor, USA, 1981.

[BOR 97] BORG I., GROENEN P., *Modern Multidimensional Scaling. Theory and Applications*, Springer Series in Statistics, Springer, New York, 1997.

[BRO 06] BRONSTEIN M. M., BRONSTEIN A. M., KIMMEL R., YAVNEH I., "Multigrid multidimensional scaling", *Numerical Linear Algebra with Applications*, vol. 13, p. 149–171, 2006.

[CAI 76] CAILLIEZ F., PAGES J.-P., *Introduction à l'Analyse des Données*, SMASH, Paris, 1976.

[CAI 83] CAILLIEZ F., "The analytical solution of the additive constant problem", *Psychometrika*, vol. 48, p. 305–310, 1983.

[CAR 70] CARROLL J. D., CHANG J. J., "Analysis of individual differences in multidimensional scaling via an N-way generalization of Eckart-Young decompositio", *Psychometrika*, vol. 35, p. 238–319, 1970.

[CAR 80] CARROLL J. D., ARABIE P., "Multidimensional scaling", *Annual Review of Psychology*, vol. 31, p. 607–649, 1980.

[CHA 81] CHANDON J. L., PINSON S., *Analyse Typologique, Théorie et Applications*, Masson, Paris, 1981.

[CLA 76] CLARK A. K., "Re-evaluation of Monte Carlo studies in non-metric multidimensional scaling", *Psychometrika*, vol. 41, p. 401–404, Springer-Verlag, 1976.

[COO 72] COOPER L. G., "A new solution to the additive constant problem in metric multidimensional scaling", *Psychometrika*, vol. 37, p. 311–322, 1972.

[COX 94] COX T. F., COX M. A., *Multidimensional Scaling*, Monographs on Statistics and Applied Probability 59, Chapman & Hall, London, 1994.

[CRI 88] CRIPPEN G. M., HAVEL T. F., *Distance Geometry and Molecular Conformation*, John Wiley & Sons, New York, 1988.

[DAV 83] DAVISON M. L., *Multidimensional Scaling*, Wiley series in probability and mathematical statistics, John Wiley & Sons, New York, 1983.

[DEG 70] DEGERMAN R. L., "Multidimensional analysis of complex structure mixtures of class and quantitative variation", *Psychometrika*, vol. 35, p. 475–471, Springer-Verlag, 1970.

[DEM 94] DEMARTINES P., Analyse des données par réseau de neurones auto-organisés, Thesis, Institut national polytechnique de Grenoble, Grenoble, France, November 1994.

[DEM 97] DEMARTINES P., HÉRAULT J., "Curvilinear component analysis: a self-organizing neural network for non-linear mapping of data sets", *IEEE Transaction on Neural Networks*, vol. 8, num. 1, p. 148–154, January 1997.

[EKM 54] EKMAN G., "Dimensions of color vision", *Journal of Psychology*, vol. 38, p. 467–474, 1954.

[ELH 88] EL HADRI K., Lissage monotone et optimisation non différentiable pour la résolution d'un problème de codage multidimensionnel, PhD Thesis, Pierre Mendès University, France, Grenoble, 1988.

[ELH 98] EL HAOUZI EL KARI M. F., Contribution à la résolution numérique de problèmes de codage multidimensionnel: approche séquentielle, Report , Grenoble, France, December 1998.

[EVE 97] EVERITT B. S., RABE-HESKETH S., *The Analysis of Proximity Data*, Kendall's Library Statistics 4, Arnold, London, 1997.

[EVE 01] EVERETT J. E., "Algorithms for multidimensional scaling", CHAMBERS L. D., Ed., *The Practical Handbook of Genetic Algorithms*, p. 203–233, Chapman & Hall, CRC, 2nd edition, 2001.

[GLU 93] GLUNT W., HAYDEN T. L., RAYDAN M., "Molecular conformation from distance matrices", *Journal of Computational Chemistry*, vol. 14, p. 114–120, 1993.

[GOO 95] GOODHILL G. J., FINCH S., SEJNOWSKI T. J., "A unifying measure for neigbourhood preservation in topographic mapping", *Proceedings of the 2nd Joint Symposium on Neural Computation*, vol. 5, California Institute of Technology, p. 191–202, 1995.

[GOO 96] GOODHILL G. J., FINCH S., SEJNOWSKI T. J., "Quantifying neighbourhood preservation in topographic mappings", *Proceedings of the 3rd Joint Symposium on Neural Computation*, vol. 6, California Institute of Technology, p. 61–82, 1996.

[GOW 66] GOWER J. C., "Some distance properties of latent roots and vectors used in multivariate analysis", *Biometrika*, vol. 53, p. 325–338, 1966.

[GOW 68] GOWER J. C., "Adding a point to vector diagrams in multivariate analysis", *Biometrika*, vol. 55, p. 582–585, 1968.

[GOW 96] GOWER J. C., HAND D. J., *Biplots*, Monographs on Statistics and Applied Probability 54, Chapman & Hall, London, 1996.

[GRO 93] GROENEN P. J. F., *The Majorization Approach to Multidimensional Scaling*, DSWO Press, Leiden, 1993.

[GRO 95] GROENEN P. J. F., MATHAR R., HEISER W. J., "The majorization approach to multidimensional scaling for Minkowski distances", *Journal of Classification*, vol. 12, num. 1, p. 3–19, March 1995.

[GUT 68] GUTTMAN L., "A general non-metric technique for finding the smallest coordinate space for a configuration of points", *Psychometrika*, vol. 33, p. 469–504, 1968.

[HAV 91] HAVEL T., "An evaluation of computational strategies for use in the determination of protein structure from distance constraints obtained by nuclear magnetic resonance", *Proceedings of Biophys. Molec. Biol.*, vol. 56, p. 43–78, 1991.

[HEI 81] HEISER W. J., Unfolding Analysis of Proximity Data, PhD Thesis, University of Leiden, Netherlands, 1981.

[HEN 95] HENDRICKSON B. A., "The molecular problem: exploring structure in global optimization", *SIAM Journal on Optimization*, vol. 5, p. 835–857, 1995.

[HOL 72] HOLMAN E. W., "The relation between hierarchical and Euclidean models for psychological distances", *Psychometrika*, vol. 37, p. 417–423, 1972.

[KEA 94] KEARSLEY A., TAPIA R. A., TROSSET M. W., The solution of the metric Stress and Sstress problems in Multidimensional Scaling using Newtons method, Report num. TR94-44, Department of computational and applied mathematics, Rice University, Houston, USA, 1994.

[KLA 69] KLAHR D. A., "Monte Carlo investigation of the statistical significance of Kruskal's scaling procedure", *Psychometrika*, vol. 34, p. 319–330, 1969.

[KOH 82] KOHONEN T., "Self-organization of topologically correct feature maps", *Biological Cybernetics*, vol. 43, p. 59–69, 1982.

[KRU 64a] KRUSKAL J. B., "Multidimensional scaling by optimizing goodness of fit to a non-metric hypothesis", *Psychometrika*, vol. 29, p. 1–27, 1964.

[KRU 64b] KRUSKAL J. B., "Nonmetric multidimensional scaling: a numerical method", *Psychometrika*, vol. 29, p. 115–129, 1964.

[KRU 78] KRUSKAL J. B., WISH M., *Multidimensional Scaling*, SAGE publ., Beverly Hills, 1978.

[LAF 06] LAFON S., LEE A. B., "Diffusion maps and coarse-graining, a unified framework for dimensionality reduction, graph partitioning, and data set parametrization", *IEEE Transactions on Pattern Analysis and Machine Intelligence*, vol. 28, num. 9, p. 1393–1403, 2006.

[LEE 77] DE LEEUW J., HEISER W., "Convergence of correction matrix algorithm for multidimensional scaling with restrictions on the configuration", LINGOES J., Ed., *Geometric Representation of Relational Data*, p. 735–752, Mathesis Press, Ann Arbor, 1977.

[LEE 80] DE LEEUW J., HEISER W., "Multidimensional scaling", KRISHNAAH P. R., Ed., *Multivariate Analysis*, p. 501–522, North Holland, Amsterdam, 1980.

[LEE 82] DE LEEUW J., HEISER W., "Theory of multidimensional scaling", KRISHNAIAH P. R., KANAL L., Eds., *Handbook of Statistics*, vol. 2, p. 285–316, North Holland, Amsterdam, 1982.

[LEE 88] DE LEEUW J., "Convergence of the majorization method for multidimensionnel scaling", *Journal of Classification*, vol. 5, p. 163–180, 1988.

[LEE 01] LEE M. D., "Determining the dimensionality of multidimensional scaling models for cognitive modeling", *Journal of Mathematical Psychology*, vol. 45, p. 187–205, 2001.

[LEE 07] LEE J. A., VERLEYSEN M., *Nonlinear Dimensionality Reduction*, Springer Series in Information Science and Statistics, Springer-Verlag, New York, 2007.

[LER 95] LERMAN I. C., NGOUENET R. F., Algorithmes génétiques séquentiels et parallèles pour une représentation affine des proximités, Report num. 2570, INRIA, Rennes, France, 1995.

[LET 98] LE THI H. A., PHAM D. T., Large scale molecular optimization from distance matrices by a DC optimization approach, Report, 26 pages, INSA, University of Rouen, France, 1998.

[LEV 85] LEVY S., GUTTMAN L., "A faceted cross-cultural analysis of some core social values", CANTER D., Ed., *Facet Theory: Approach to Social Research*, Springer Verlag, Heidelberg, 1985.

[LI 95] LI S., DE VEL O., COOMANS D., "Comparative performance of non-linear dimensionality reduction methods", *5th International Workshop on Artificial Intelligence and Statistics*, Fort Lauderdale, USA, 1995.

[MAC 81] MACKAY D. B., ZINNES J. L., "Probabilistic scaling of spatial judgments", *Geographical Analysis*, vol. 19, num. 1, p. 21–37, 1981.

[MAC 89] MACKAY D., "Probabilistic multidimensional scaling: an anisotropic model for distance judgments", *Journal of Mathematical Psychology*, vol. 33, p. 187–205, 1989.

[MAC 92] MACHMOUCHI M., Contributions à la mise en oeuvre des méthodes d'analyse des données de dissimilarité, Thesis, Pierre Mendès University, Grenoble, France, 1992.

[MAC 94] MACHMOUCHI M., D'AUBIGNY G., "Simulated annealing scaling (SASCAL) algorithm: application for solving the multidimensionnal scaling (MDS) problem", *Proceedings of CompStat*, Physica-Verlag, Vienna, August 1994.

[MAI 79] MAILLES J.-P., Analyse Factorielle des Tableaux de dissimilarités, Thesis, Pierre and Marie Curie University, Paris, 1979.

[MAR 78] MARDIA K. V., "Some properties of classical multidimensional scaling", *Communications in Statistics A: Theory and Method*, vol. A7, p. 1233–1241, 1978.

[MES 56] MESSICK S. J., ABELSON R. P., "The additive constant problem in multidimensional scaling", *Psychometrika*, vol. 21, p. 1–15, 1956.

[MOR 96] MORE J. J., WU Z., Distance Geometry for Protein Structures,   Preprint num. MCS-P628-1296, Argone National Laboratory, Argone, Illinois, 1996.

[NAU 01] NAUD A., Neural and statistical methods for the visualization of multidimensional data, PhD Thesis, University Mikolaja Kopernika, Toruniu, 2001.

[N'G 95] N'GOUENET R. F., Analyse géométrique des données de dissimilarité par la multidimensional scaling: une approche parallèle basée sur les algorithmes génétiques. Application aux séquences biologiques, PhD thesis, University of Rennes 1, 1995.

[NIE 80] NIEMANN, "Linear and non-linear mappings of patterns",   *Pattern Recognition*, vol. 12, p. 83–87, 1980.

[OH 01] OH M.-S., RAFTERY A. E., "Bayesian multidimensional scaling and choice of dimension", *Journal of American Statistical Association*, vol. 96, p. 1031–1044, 2001.

[PEK 99] PEKALSKA E., DE RIDDER D., DUIN R. P. W., KRAAIJVELD M. A., "A new method for generalizing Sammon mapping with application to algorithm speed-up", *ASCI'99 Proceedings of 5th Annual Conference of the Advanced School for Computing and Image*, Heijen, Netherlands, p. 221–228, 1999.

[PEK 05] PEKALSKA E., DUIN R. P. W., *The Dissimilarity Representation for Pattern Recognition: Foundations and Applications*, Machine Perception and Artificial Intelligence, World Scientific Publishing Company, New York, 2005.

[PHA 97] PHAM D., LE THI H. A., "Convex analysis approach to DC programming: theory, algorithms and applications", *Acta Mathematica Vietnamica*, vol. 22, num. 1, p. 289–355, 1997.

[RAB 75] RABINOWITZ G. B., "An introduction to non-metric multidimensional scaling", *American Journal of Political Science*, vol. 19, p. 343–390, 1975.

[RAM 77] RAMSAY J. O., "Maximum likelihood estimation in multidimensional scaling", *Psychometrika*, vol. 42, p. 241–266, 1977.

[RAM 78a] RAMSAY J. O., "Confidence regions for multidimensional scaling analysis", *Psychometrika*, vol. 43, p. 145–160, 1978.

[RAM 78b] RAMSAY J. O., *MULTISCALE: Four programs for multidimensional scaling by the method of maximum likelihood*, International Education Service, Chicago, 1978.

[RAM 80] RAMSAY J. O., "Some small sample results for Maximum Likelihood Estimation in Multidimensional Scaling", *Psychometrika*, vol. 45, p. 141–146, 1980.

[RAM 82] RAMSAY J. O., "Some Statistical Approaches to Multidimensional Scaling data", *Journal of the Royal Statistical Society, Series A*, vol. 145, p. 285–311, 1982.

[RIP 96] RIPLEY B. D., *Pattern Recognition and Neural Networks*, Cambridge University Press, Cambridge, 1996.

[ROM 84] ROMESBURG H. C., *Cluster Analysis for Researcher*, Lifetime Learning Publications, Belmont, USA, 1984.

[SAI 78] SAITO T., "The problem of the additive constant and eigenvalues in metric multidimensional scaling", *Psychometrika*, vol. 43, p. 193–201, 1978.

[SAM 69] SAMMON J. W., "A nonlinear mapping for data structure analysis", *IEEE Transactions on Computers*, vol. 18, p. 401–409, 1969.

[SAM 92] SAMPSON P. D., GUTTORP P., "Non parametric estimation of stationary spatial covariance structure", *Journal of the American Statistical Association*, vol. 87, p. 108–119, 1992.

[SAM 94] SAMPSON P. D., GUTTORP P., MEIRING W., "Spatio-temporal analysis of regional ozone data for operational evaluation of an air quality model", *Proceedings of the Section of Statistics and the Environment*, American Statistical Association, 1994.

[SAR 90] DE SARBO W. S., MANRAI A. K., BURK R. R., "A nonspatial methodology for the analysis of two-way proximity data incorporating the distance-density hypothesis", *Psychometrika*, vol. 55, p. 229–253, Springer-Verlag, 1990.

[SAX 79] SAXE J. B., "Embeddability of weighted graphs in k-space is strongly NP-hard", *Proceedings of 17th Allerton Conference in Communications, Control, and computing*, p. 480–489, 1979.

[SCH 35] SCHOENBERG I. J., "Remarks to Maurice Frechet's article: Sur la définition axiomatique d'une classe d'espaces distanciés vectoriellement applicables sur l'espace de Hilbert", *Annals of Mathematics*, vol. 36, p. 724–732, 1935.

[SCH 81] SCHIFFMAN S. S., REYNOLDS M. L., YOUNG F. W., *Introduction to Multidimensional Scaling: Theory, Methods and Applications*, Academic Press, New York, 1981.

[SHE 62a] SHEPARD R. N., "The analysis of proximities: multidimensional scaling with unknown distance function, I", *Psychometrika*, vol. 27, p. 125–140, 1962.

[SHE 62b] SHEPARD R. N., "The analysis of proximities: multidimensional scaling with unknown distance function, II", *Psychometrika*, vol. 27, p. 219–246, 1962.

[SHE 72a] SHEPARD R. N., ROMNEY A. K., NERLOVE S. B., *Multidimensional Scaling: Theory and Applications in the Behavioral Sciences*, vol. 1–2, Seminar Press, New York, 1972.

[SHE 72b] SHERMAN C. R., "Nonmetric multidimensional scaling: a Monte Carlo study of the basic parameters", *Psychometrika*, vol. 37, p. 323–355, 1972.

[SPE 72] SPENCE I., "A Monte Carlo evaluation of three non-metric multidimensional scaling algorithms", *Psychometrika*, vol. 37, p. 461–486, 1972.

[SPE 74a] SPENCE I., "On random rankings studies in non-metric scaling", *Psychometrika*, vol. 39, p. 267–268, 1974.

[SPE 74b] SPENCE I., GRAEF J., "The determination of the underlying dimensionality of an empirically obtained matrix of proximities", *Multivariate Behavioral Research*, vol. 9, p. 331–342, 1974.

[SPE 78] SPENCE I., YOUNG F. W., "Monte Carlo Studies in non-metric scaling", *Psychometrika*, vol. 43, p. 115–117, 1978.

[STE 69] STENSON H. H., KNOLL R. L., "Goodness of fit for random rankings in Kruskal's non-metric scaling procedure", *Psychological Bulletin*, vol. 72, p. 122–126, 1969.

[TAK 77a] TAKANE Y., Statistical Procedures for Nonmetric Multidimensional Scaling, PhD thesis, University of North Carolina, Chapel Hill, 1977.

[TAK 77b] TAKANE Y., YOUNG F. W., DE LEEUW J., "Nonmetric individual differences multidimensional scaling: an alternating least squares method with optimal scaling features", *Psychometrika*, vol. 42, p. 7–67, 1977.

[TAK 78] TAKANE Y., "A maximum likelihood method of multidimensional scaling I: the case in which all pairwise orderings are independent - theory", *Japanese Psychological Research*, vol. 20, num. 1, p. 7–17, 1978.

[TIP 96] TIPPING M. E., Topographic Mappings and Feedforward Neural Networks, PhD Thesis, ASTON University, Birmingham, 1996.

[TOR 52] TORGERSON W. S., "Multidimensional scaling: I. theory and methods", *Psychometrika*, vol. 17, p. 401–419, 1952.

[TOR 58] TORGERSON W. S., *Theory and Methods of Scaling*, John Wiley, New York, 1958.

[TOU 93] TOURNOIS J., DICKES P., *Pratique de l'Échelonnement Multidimensionnel: De l'Observation à l'Interprétation*, De Boeck-Wesmael, Brussels, 1993.

[TRO 97] TROSSET M. W., "Numerical algorithms for multidimensional scaling", KLAR R., OPITZ D., Eds., *Classification and Knowledge Organization*, p. 80–92, Springer-Verlag, Berlin, 1997.

[TRO 06] TROSSET M. W., PRIEBE C. E., The out-of sample problem for classical multidimensional scaling, Report, Department of Statistics, Indiana University, Bloomington, IN, 2006.

[TSU 96] TSUCHIYA T., "A probabilistic multidimensional scaling with unique axes", *Japanese Psychological research*, vol. 38, num. 4, p. 204–211, 1996.

[TUC 51] TUCKER L. R., A method for the synthesis of factor analysis studies, Technical report 984, Department of the Army, Washington, 1951.

[VAR 06] VARONECKAS A., ZILINSKAS A., ZILINSKAS J., "Multidimensional scaling using parallel genetic algorithm", BOGLE I. D. L., ZILINSKAS J., Eds., *Computer-aided Methods in Optimal Design and Operations*, p. 129–136, World Scientific Publishing Company, 2006.

[WEZ 01] VAN WEZEL M. C., KOSTERS W. A., "Nonmetric multidimensional scaling with neural networks", *Lecture Notes in Computer Sciences*, vol. 2189, p. 145–156, Springer-Verlag, 2001.

[WEZ 04] VAN WEZEL M. C., KOSTERS W. A., "Nonmetric multidimensional scaling: Neural networks versus traditional techniques", *Intelligent Data Analysis*, vol. 8, p. 601–613, 2004.

[WIL 70] WILKINSON C., "Adding a point to principal coordinate analysis", *Systematic Zoology*, vol. 19, p. 258–263, 1970.

[YOU 38] YOUNG G., HOUSEHOLDER A. S., "Discussion of a set of points in terms of their mutual distances", *Psychometrika*, vol. 3, p. 19–22, 1938.

[YOU 87] YOUNG F., *Multidimensional Scaling: History, Theory and Applications*, Lawrence Erlbaum Associates, London, 1987.

[YOU 94] YOUNG F. W., HAMER R. M., *Theory and Applications of Multidimensional Scaling*, Erlbaum, Hillsdale, NJ, 1994.

[ZIL 07] ZILINSKAS A., ZILINSKAS J., "Parallel genetic algorithm: assessment of performance in Multidimensional scaling", *Gecco'07*, London, England, UK, p. 1492–1499, July 7–11 2007.

Chapter 5

Statistical Modeling of Functional Data

5.1. Introduction

This chapter is devoted to statistical modeling for functional data, i.e. when we have a sample of n observations that cannot be considered as vectors belonging to $\mathbb{R}^p$ but can be thought of as curves or more generally functions. These functions depend on an index that is most often time t and belongs to a closed interval T. We suppose from now on, without loss of generality, that this interval is $T = [a, b]$ with $a < b$. In practice, such functional data are observed at some discretization or design points that may be equidistant and that may be the same for all curves. We present in Figure 5.1 an example of monthly cumulative precipitation curves.

With the improvement of storage capacities and the performances of computers, this field of statistics has developed greatly during the last 20 years and there is now a huge number of papers published on this topic in statistics but also in chemometrics, climatology, image processing, pattern recognition and biometry.

From a historical point of view, the first studies dealing with functional data were certainly made by climatologists or chemists, the first scientists confronted with such types of data. The involved tools were based on signal processing, principal components analysis (PCA) and discrete Karhunen–Loeve expansion. In France, Deville [DEV 74] introduced the notion of principal components analysis of curves whereas Dauxois and Pousse [DAU 76] proposed a general mathematical framework for such statistical objects, generalizing multivariate statistics to Hilbert-valued random variables.

Chapter written by Philippe BESSE and Hervé CARDOT.

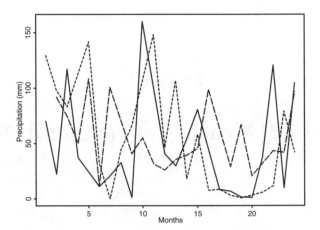

Figure 5.1. *Three examples of curves representing monthly cumulative precipitation over two years*

Indeed, most multivariate statistical methods can be extended to deal with curves and more generally with functional objects: canonical analysis [LEU 93], discriminant analysis [DO 99, HAS 95], linear regression models [CAR 99, RAM 91], generalized linear models [MAR 99], sliced inverse regression [YAO 03], k nearest neighbors [ABR 06] and Support Vector Machine [ROS 06] for supervised classification. An exhaustive review of all the proposed extensions is almost impossible now and the reader is referred to the books by Ramsay and Silverman [FER 06, RAM 02, RAM 05] for more references. Let us note that the statistical study of curves may necessitate, in a first step, transformations by non-linear functions of the time index with warping techniques (curve registration) in order to obtain curves that are well synchronized. Such preparation of the data is needed e.g. in order to attenuate occurrence of events such as puberty which have deep impact on the shape of individual growth curves. This aspect of functional data analysis is not developed in this chapter (see Ramsay and Li [RAM 98] or Kneip *et al.* [KNE 00]).

The adaptation of the usual multivariate statistical techniques to functional data requires the introduction of sophisticated mathematical tools that can be unattractive to applied statisticians. These tools are (only) necessary to study the asymptotic properties of the estimators in order to assess the relevance and the stability of a particular approach and provide, when possible, its symptotic distribution in order to build confidence intervals. They generally assume that the sample size grows to infinity and the discretization scheme becomes finer and finer. Nevertheless, with real datasets, we always have at hand discretized data and estimators are built by considering matrices and vectors in a finite dimension. Tools borrowed from numerical analysis are then needed to transform discretized curves into functions. The main techniques are interpolation, smoothing (splines) and basis expansions (Fourier, wavelets) of the discretized trajectories. The main questions of interest are then:

– in which situation is a functional approach more effective than a multivariate approach based on the usual Euclidean framework?

– what are, in practice, the complementary mathematical tools needed to perform such studies?

We can identify frequent situations which illustrate or give answers to the first question:

– When the *curves* are *smooth* functions, discretization leads to highly correlated vectors for time points close to each other. Very badly conditioned matrices then occur in the smoothing step and this very strong structure often masks some more interesting features of the data; derivatives of the curves must be taken into account. For PCA, Besse and Ramsay [BES 86] gave a first approach that has been developed by Ramsay [RAM 96, RAM 00]. Ferraty and Vieu [FER 06] propose adapting non-parametric regression based on kernel smoothers with distance between objects measured by semi-norms which generally rely on derivatives, whereas [MAS 07b] propose linear models depending on derivatives. It can also be important to improve supervised classification [ROS 08].

– When the data are noisy observations of a smooth underlying phenomenon, it is important to jointly introduce a denoising feature within the multivariate procedure. This generally gives rise to a double tuning problem for both optimizing the smoothing parameter and, for example, the number of components in the case of PCA.

– When studying singular functions such as spectrometric data obtained by high performance liquid chromatography (HPLC), both levels and positions of peaks can be noisy. In this case, it seems efficient to associate a wavelet decomposition and then a thresholding strategy with the multivariate method [BES 03, MAL 00].

To answer the second question, we have chosen in this chapter to give at the end of each section only a brief description and bibliography for the mathematical tools necessary for the asymptotic studies. The next section introduces the functional framework and Hilbertian tools which are required. They can be passed over by the reader interested in applications. Note that this is always a very active area of research in mathematical statistics and many theoretical properties of functional approaches still need to be developed. We prefer to insist in this chapter on the practical and matrix implementation of three of the most popular methods, also providing the description of the associated numerical tools. We will only consider the case in which all the curves are discretized at the same design points with noisy measurements and emphasize approximation tools based on smoothing splines. When the design is not the same from one curve to another, Besse *et al.* [BES 97] propose dealing with hybrid splines merging B-splines [DEB 78, DIE 93] and smoothing splines. This also allows us to manage with missing values (see section 5.3.3).

After a short introduction of the functional framework, three examples of functional data analysis are developed. They are illustrated on real life datasets.

PCA. The statistical description of a set of curves observed with noise at the same design points leads to the proposal of an adaptation of principal components analysis. This approach, in such a context, can be connected with the *simultaneous* nonparametric regressions of a sample of curves.

Modelization. The aim is now to consider models in which the covariates are functions. We first present the functional linear regression model which is a direct extension of the linear model as well as the generalized linear model, based on the notion of link function.

Forecasting. A continuous time process with a natural period can be modeled as a functional process, considering discrete time processes whose values are functions and are built with the whole sample paths within each time period of the original continuous time process. Bosq [BOS 91, BOS 00] has developed a theoretical framework defining a functional autoregressive model (FAR(1)) for that kind of situation, whereas the last section presents a solution for the forecasting problem by means of a regression based on the spline approximation under dimensionality constraints.

5.2. Functional framework

This section presents a general framework for functional data analysis. It first aims to give notation useful for dealing with random functions. The second objective is to define smoothness criterion or equivalently roughness penalties involved in the optimization procedures described below. These penalties are defined by semi-norms that can be written matricially in finite dimension Euclidean spaces.

5.2.1. *Functional random variable*

We consider a random function Z taking values in a Hilbert space H, assumed to be separable. The inner product in H is denoted $\langle \cdot, \cdot \rangle_H$ and the norm of an element x of H is denoted $||x||_H$. We also assume that Z is of second order, i.e. $\mathbb{E}||Z||_H^2 < \infty$. When this condition is fulfilled, we can define the expectation of Z, say $\mathbb{E}(Z) \in H$, and its covariance operator, say Γ, which is compact and non-negative. Identifying H and its topological dual, we can write with the help of the Riesz theorem the first two moments of Z as:

$$\forall f \in H, \quad \langle \mathbb{E}(Z), f \rangle_H = \mathbb{E}[\langle f, Z \rangle_H],$$

$$\forall (u, v) \in H \times H, \quad \langle \Gamma u, v \rangle_H = \mathbb{E}[\langle Z - \mathbb{E}(Z), u \rangle_H \langle Z - \mathbb{E}(Z), v \rangle_H].$$

Introducing now a tensorial product notation, the covariance operator can be written

$$\Gamma = \mathbb{E}\left[(Z - \mathbb{E}(Z)) \otimes (Z - \mathbb{E}(Z))\right],$$

where $a \otimes b(u) = \langle a, u \rangle_H \, b$, $\forall(a, b) \in H \times H$ and $u \in H$.

In the particular case $H = L^2(T)$, the space of square integrable functions defined on T, it can easily be checked that the covariance operator is an integral operator

$$\forall f \in L^2(T), \ \forall t \in T, \quad \Gamma f(t) = \int_T \gamma(s, t) f(s) \, ds, \tag{5.1}$$

where $\gamma(s, t)$ is the covariance function of the underlying continuous time process $Z(t)$ with expectation $a(t)$:

$$\gamma(s, t) = \mathbb{E}\left[(Z(t) - a(t))(Z(s) - a(s))\right], \quad \forall(s, t) \in T \times T. \tag{5.2}$$

We will also consider in the following a discrete time process $(Z_i)_{i \in \mathbb{Z}}$, with finite second moment and taking values in H. Let us denote the covariance operator by Γ and the cross-covariance operator by Δ. If the process Z is stationary, these operators do not depend on i and are defined by

$$\Gamma = \mathbb{E}\left[(Z_0 - a) \otimes (Z_0 - a)\right],$$
$$\Delta = \mathbb{E}\left[(Z_0 - a) \otimes (Z_1 - a)\right],$$
$$\Delta^* = \mathbb{E}\left[(Z_1 - a) \otimes (Z_0 - a)\right],$$

where function $a = \mathbb{E}(Z_i)$ is the expectation of the process. Moreover, if Z_i is an autoregressive process of order 1 (see the definition below), they satisfy:

$$\Delta^*(Z_i - a) = \Gamma(E(Z_{i+1}|Z_i, Z_{i-1}, \ldots) - a),$$
$$\Delta = \rho\Gamma.$$

5.2.2. Smoothness assumption

The estimation procedures developed in this chapter consist of looking for smooth solutions of an optimization problem, where smoothness is controlled by the norm of derivatives of the function of interest. Such criterion dealing with a sequence of derivatives is commonly used to define spline functions. Considering the Sobolev space defined by

$$W^m(T) = \{z : z, z', \ldots, z^{(m-1)} \text{ absolutely continuous}, z^{(m)} \in L^2(T)\},$$

the regularity of any function z belonging to W^m can be measured or controlled by the semi-norm:

$$\|z\|_m^2 = \|D^m z\|_{L^2(T)}^2 = \int_T (z^{(m)}(t))^2 dt. \tag{5.3}$$

Such a criterion can easily be generalized to other equivalent semi-norms [WAH 90], replacing the differential operator D^m by any differential operator containing at least a differential term of order m. This generalization leads to the definition of more general spline functions called Tchebicheff splines.

5.2.3. *Smoothing splines*

Smoothing splines are widely used for performing non-parametric regression and the reader is referred to [GRE 94, WAH 90] for more details on this topic. Let us consider the usual non-parametric framework:

$$x_j = z(t_j) + \varepsilon_j \; ; \quad \mathbb{E}(\varepsilon_j) = 0, \; \mathbb{E}(\varepsilon_j \varepsilon_k) = \sigma^2 \delta_{kj}, \; j,k = 1,\ldots,p \tag{5.4}$$

$$a \le t_1 < t_2 < \ldots < t_p \le b$$

where x_j is a noisy observation at point t_j of function z which is assumed to be a smooth function, $z \in W^m(T)$.

The smoothing spline estimator, say $\widehat{z}$ of function z, is by definition the solution of the following optimization problem subject to a smoothness constraint:

$$\min_{z \in W^2} \left\{ \frac{1}{p} \sum_{j=1}^{p} (z(t_j) - x(t_j))^2; \; \|z\|_m^2 < c \; (c \in \mathbb{R}_+) \right\}. \tag{5.5}$$

Introducing a Lagrange multiplier, the solution is found by considering

$$\min_{z \in W^2} \left\{ \frac{1}{p} \sum_{j=1}^{p} (z(t_j) - x(t_j))^2 + \ell \|D^m z\|_{L^2}^2 \right\}. \tag{5.6}$$

The smoothing parameter ℓ, which is positive and depends on c, allows us to make a tradeoff between the fit to the data ($\ell \to 0$) and the regularity of the solution ($\ell \to +\infty$). Its value is generally selected by minimizing a prediction error estimated by the generalized cross-validation criterion (GCV) [WAH 90].

The solution $\widehat{z}$ of this optimization problem is a piecewise polynomial, constructed of polynomials of degree $(2m-1)$ between the knots t_j and t_{j+1}, and polynomials of degree $(m-1)$ at the boundary intervals $[a, t_1]$ and $[t_p, b]$. The limit of the solution when $\ell = 0$ is the interpolation spline that exactly fits all the observed values. At the opposite extreme, when ℓ tends to infinity, the solution is given by the polynomial regression of degree $(m - 1)$ so that the penalty term equals zero.

The explicit construction of the estimator depends on the basis chosen to define the subspace S_p of spline functions. The most general and simple basis is, from a theoretical point of view, the basis built with reproducing kernels [WAH 90]. This approach, adopted for instance by Besse and Ramsay [BES 86], is difficult to implement in practice because of numerical problems of instability, particularly when the number of design points p is large. We restrict ourself in this chapter to natural cubic splines, which are piecewise polynomials of degree 3 $(m = 2)$ whose second and third order derivatives are null at the boundary points a and b. The Reisch algorithm [GRE 94], which is known to have good numerical properties, is then useful to build a basis of such a vectorial space. It relies on the resolution of a tridiagonal system due to the Cholesky decomposition. The number of operations required to solve the system is of order p.

Let us consider the band matrix $\mathbf{Q}$ of size $(p \times (p - 2))$ with non-zero generic elements

$$q_{j-1,j} = \frac{1}{t_j - t_{j-1}}, q_{j,j} = -\frac{1}{t_j - t_{j-1}} - \frac{1}{t_{j+1} - t_j}, q_{j+1,j} = -\frac{1}{t_{j+1} - t_j},$$

and the $(p - 2) \times (p - 2)$ symmetric band matrix $\mathbf{R}$ with generic elements

$$r_{j,j} = \frac{1}{3}(t_{j+1} - t_{j-1}), r_{j+1,j} = r_{j,j+1} = \frac{1}{6}(t_{j+1} - t_j).$$

Let us denote by $\mathbf{M}$ the matrix associated with the semi-inner product in $W^2(T)$ restricted to S_p: if z is a function of $W^2(T)$ and $\mathbf{z}$ the p dimensional vector containing the values of z at the knots, then

$$\int_T z''(t)^2 dt = \|z\|_2^2 = \mathbf{z}'\mathbf{M}\mathbf{z} = \|\mathbf{z}\|_{\mathbf{M}}^2,$$

and matrix $\mathbf{M}$ satisfies $\mathbf{M} = \mathbf{Q}\mathbf{R}^{-1}\mathbf{Q}$.

Then, if vector $\mathbf{x}$ is composed of the p observations made at knots t_j, the value of the smoothing spline at the same design points is evaluated as follows:

$$\hat{\mathbf{z}} = \mathbf{A}_\ell \mathbf{x} \quad \text{with} \quad \mathbf{A}_\ell = (\mathbf{I} + \ell\mathbf{M})^{-1} \tag{5.7}$$

where matrix $\mathbf{A}_\ell$ is referred to as the *hat matrix* or *smoothing matrix*. Finally, the function $\hat{z}$ is obtained by interpolating vector $\hat{\mathbf{z}}$ with interpolation splines. Note that the solution does not depend on the boundary points a and b of interval T.

Let us now introduce another matrix useful to derive the inner product between two spline functions. Considering two vectors $\mathbf{y}_1$ and $\mathbf{y}_2$ associated with the same design points, the inner product in $L^2(T)$ between their interpolating splines $\widehat{y}_1$ and $\widehat{y}_2$ is given by:

$$\mathbf{y}_1' \mathbf{N} \mathbf{y}_2 = \int_T \widehat{y}_1(t)\widehat{y}_2(t)\, dt. \tag{5.8}$$

More generally, this matrix can be derived due to the basis generated by reproducing kernels associated with spline functions [BES 86, RAM 91]. It is also possible to obtain an approximation using a quadrature rule by considering, for instance, $\mathbf{N} = \mathrm{diag}(w_1, \ldots, w_p)$, where $w_1 = (t_2 - t_1)/2$, $w_j = (t_{j+1} - t_{j-1})/2$, $j = 2, \ldots, p-1$ and $w_p = (t_p - t_{p-1})/2$. Evaluation of the inner product is then very fast and generally effective.

5.3. Principal components analysis

We are now interested in describing and estimating trajectories z_i for $i = 1, \ldots, n$ of a random function Z. With notation defined in the previous section, we assume that the random variable X is a noisy observation of the smooth random function Z at some design points. An illustration of such functional data is given in Figure 5.1.

Each trajectory z_i is assumed to be observed at p different instants of time $t_1 < \ldots < t_p$ in interval T, and we assume these instants are the same for all the trajectories. We also assume that the discrete measurements are made with errors that are independent and identically distributed, with mean zero and variance σ^2. We therefore have n independent realizations of model (5.4) with the assumption that the correlation between the noise and the smooth underlying random function is null, $\mathbb{E} < (\epsilon_i \mathbf{z}_{i'}') = 0$ for all i' and i.

At this level, it would be possible to consider the question of estimating the n trajectories of Z as n independent non-parametric regression problems. Nevertheless, it seems important to take into account that the different trajectories are drawn from the same distribution and, in particular, have the same temporal covariance structure. We propose in the following to add the constraint that the random function Z takes values in a finite dimensional subspace, with dimension q, of $W^2(T)$. This leads to the expansion of the estimated trajectories as linear combinations of q functions, which are smooth by definition.

5.3.1. *Model and estimation*

For $i = 1, \ldots, n$, we denote the ith observed discrete trajectory by $\mathbf{x}_i = (x_i(t_1), \ldots, x_i(t_p))' \in \mathbb{R}^p$. We denote a q-dimensional affine subspace of $\mathbb{R}^p$ by

A_q, with $q < p$. Regularity assumptions of the trajectories as well as dimensionality constraints imply the model is expressed as follows:

$$
\mathbf{x}_i = \mathbf{z}_i + \boldsymbol{\varepsilon}_i; \quad i = 1, \ldots, n \quad \text{with} \quad
\begin{cases}
\mathbb{E}(\boldsymbol{\varepsilon}_i) = 0 \text{ and } \mathbb{E}(\boldsymbol{\varepsilon}_i \boldsymbol{\varepsilon}_i') = \sigma^2 \mathbf{I}, \\
\sigma \text{ unknown}, (\sigma > 0) \\
\mathbf{z}_i \text{ independent of } \boldsymbol{\varepsilon}_{i'}, \ i' = 1, \ldots, n, \\
\mathbf{z}_i \in A_q \text{ a.s.} \quad \text{and} \quad \|\mathbf{z}_i\|_m^2 \leq c \text{ a.s.}
\end{cases}
\tag{5.9}
$$

(where a.s. means 'almost surely'.) The particular feature of this model is that it combines two types of constraints: the first, based on the dimension, leads to a definition of PCA proposed e.g. by [CAU 86], whereas the second is based on smoothness. Estimation in such a model by a weighted least squares approach leads to the solution of an optimization problem in which the regularity constraint that depends on c has been replaced by a Lagrange multiplier ℓ.

With previous notation and introducing weights w_i that allow different importance for different individuals to be given, we aim to minimize

$$
\min_{\mathbf{z}_i, A_q} \left\{ \sum_{i=1}^n w_i \left(\|\mathbf{z}_i - \mathbf{x}_i\|_I^2 + \ell \|\mathbf{z}_i\|_M^2 \right) ; \mathbf{z} \in A_q, \ \dim A_q = q \right\}.
\tag{5.10}
$$

Let us define $\overline{\mathbf{x}} = \sum_{i=1}^n w_i \mathbf{x}_i$ the weighted mean of the coordinates and $\overline{\mathbf{x}}$ the $n \times p$ centered matrix of observations. In the context of climatic trajectories described below, this corresponds to the matrix of anomalies $(\mathbf{x}_i - \overline{\mathbf{x}})$, i.e. the matrix of deviations from the mean trajectory. Denoting the diagonal matrix with diagonal elements w_i by $\mathbf{D}$, the empirical covariance matrix can be written $\mathbf{S} = \overline{\mathbf{X}}' \mathbf{D} \overline{\mathbf{X}}$.

The solution of optimization problem (5.10) (see [BES 97] for a proof) is given by

$$
\widehat{\mathbf{z}}_i = \mathbf{A}_\ell^{1/2} \widehat{\mathbf{P}}_q \mathbf{A}_\ell^{1/2} \mathbf{x}_i + \mathbf{A}_\ell \overline{\mathbf{x}}, \ i = 1, \ldots, n.
$$

The matrix $\widehat{\mathbf{P}}_q = \mathbf{V}_q \mathbf{V}_q'$ is the orthogonal projection onto the subspace $\widehat{E}_q$ generated by the q eigenvectors of matrix

$$
\mathbf{A}_\ell^{1/2} \mathbf{m} S \mathbf{A}_\ell^{1/2},
$$

associated with the q largest eigenvalues.

Smooth estimations of the trajectories are then given by interpolation splines of the vectors $\widehat{\mathbf{z}}_i$.

Denoting the eigenvectors of the matrix $\mathbf{A}_\ell^{1/2} \mathbf{m} S \mathbf{A}_\ell^{1/2}$ associated with the largest eigenvalues by $\mathbf{v}_j, j = 1, \ldots, q$, we determine that the estimated discrete trajectories are projected onto the subspace generated by the vectors

$$
\widetilde{\mathbf{v}}_j = \mathbf{A}_\ell^{1/2} \mathbf{v}_j, \quad j = 1, \ldots, q.
$$

An equivalent decomposition is given by considering the orthogonal projection with metric $\mathbf{A}_\ell^{-1}$ of the transformed data $\mathbf{A}_\ell \mathbf{x}_i$ onto the space generated by the orthonormal (according to this particular metric) basis $\widetilde{\mathbf{m}v}_j, j = 1, \ldots, q$:

$$\hat{\mathbf{z}}_i = \mathbf{A}_\ell \bar{\mathbf{x}} + \sum_{j=1}^{q} \langle \tilde{\mathbf{v}}_j, \mathbf{A}_\ell \mathbf{x}_i \rangle_{\mathbf{A}_\ell^{-1}} \tilde{\mathbf{v}}_j. \tag{5.11}$$

5.3.2. *Dimension and smoothing parameter selection*

In this procedure, there are two important parameters which play the role of smoothing parameters: the dimension q of the subspace and the smoothing parameter ℓ which controls the smoothness of the solutions. Choosing values for these parameters has to be carried out simultaneously since, in practice, the parameters have dependent roles. Indeed, the dimension parameter q also acts as a smoothing parameter since, in practice, higher values of q are associated with high frequency components of the signal. For instance, when the continuous time stochastic process $Z(t)$ is stationary, it is well known that the trigonometric functions are the eigenfunctions of the covariance operator; PCA can then be seen as a Fourier decomposition of the signal. Note that Deville [DEV 74], in his seminal work on functional principal components analysis, introduced the term harmonic analysis of a random function. Reducing the dimension q therefore allows smaller values for the smoothing parameter ℓ to be chosen, which is why the functional PCA leads to better estimations of the true signal compared with independent non-parametric regressions where each smoothing parameter is chosen independently by cross-validation. In a different context, some mathematical arguments validating the use of undersmoothing procedures can be found (e.g. [KNE 01]) when the trajectories are densities estimated with kernels.

The following criterion, proposed by Besse [BES 92], can be useful for choosing both the dimension and the smoothing parameter value. It is based on a quadratic mean error criterion that measures the distance between the true projection and its estimation:

$$R_q = \mathbb{E}\frac{1}{2} \left\| \mathbf{P}_q - \widehat{\mathbf{P}_q} \right\|^2 = q - \mathrm{tr}\, \mathbf{P}_q \widehat{\mathbf{P}_q}.$$

A *jackknife* approximation of this criterion can be derived with the perturbation theory of linear operators and leads us to consider the following risk:

$$\widehat{R_{Pq}} = \frac{1}{n-1} \sum_{k=1}^{q} \sum_{j=k+1}^{p} \frac{\frac{1}{n}\sum_{i=1}^{n} c_{ik}^2 c_{ij}^2}{(\lambda_j - \lambda_k)^2} \tag{5.12}$$

where c_{ij} is the generic element of matrix $\mathbf{x}\mathbf{A}_\ell^{1/2}\mathbf{V}_q$.

Besse *et al.* [BES 97] have shown on simulated data the interest of considering simultaneously dimension reduction and roughness penalties, with better estimation of the true signal compared to individual non-parametric regression performed independently. Taking into account the common covariance structure as well as the smoothness of the signals allows better estimations to be made. Some theoretical arguments can be found in [CAR 00].

Note also that when the noise variance is relatively important, i.e. larger than the small eigenvalues of the covariance operator, it is generally more effective to reduce the dimension of the estimated signal. This work has also shown that criterion $\widehat{R_{Pq}}$ is an operational and effective selection criterion.

5.3.3. *Some comments on discretization effects*

Many procedures have been developed in the literature to deal with discretization of the observed trajectories (see [RAM 05] for a review). When the design points are the same for all the trajectories, the simplest rely on quadrature rules which allow approximations of inner products to be made. When the number of design time points is sufficiently large, approximation of the trajectories are generally performed with regression splines, smoothing splines, wavelets or kernel smoother. It can be proven [BEN 09, CAR 00] that discretization effects are negligible compared to sampling effects, even if the observations are corrupted by noise, provided the true trajectories are smooth enough.

We can also have sparse measurements of the trajectories, i.e. only a few (from 2 to 10) measurements for each curves. This is often the case e.g. with medical studies. In this context, James *et al.* [JAM 00] and Rice and Wu [RIC 01] propose performing PCA by expanding the curves in B-splines basis and treating sparse curves as vectors with missing values. The expectation–maximization (EM) algorithm is then employed to obtain estimations of the mean and covariance functions.

More recently, Hall *et al.* [HAL 06] propose a different approach, based on a direct non-parametric approach for estimating the mean and covariance functions. It consists of considering these objects as univariate and bivariate functions of time. These authors prove that the discretization effect may become the leading term in the asymptotic error. A result of particular interest is that we still get consistent estimators even if the number of discretization points does not tend to infinity, provided that their location is random, at the expense of lower rates of convergence compared to the parametric rates.

5.3.4. PCA of climatic time series

We have chosen to illustrate the estimation procedure previously described on a
set of discrete functions which are particularly noisy. We consider the square root of
monthly cumulative precipitation in 26 cities of France, observed during a period of 10
consecutive years [ECO 91]. Data have been transformed with the square root function
in order to stabilize the variance, as usually done for Poisson processes, and we will
study the sample made by the 26×5 trajectories observed during 2 consecutive years.
Note that the study of annual curves gives approximately the same results.

A classical multivariate PCA was first performed and the first three eigenvectors
(or eigenfunctions) are drawn in Figure 5.2. These functions are very noisy and
difficult to interpret.

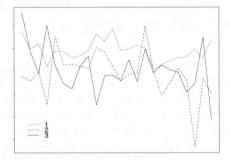

Figure 5.2. *First three eigenfunctions when performing a classical PCA (without roughness
penalty) of precipitation data*

The functional PCA which included a roughness penalty controlled by a smoothing
parameter ℓ was then performed. The simultaneous choice of the dimension and
the smoothing parameter value were made using Figure 5.3, which represents the
evolution of the stability of the subspace in which the trajectories are expanded
according to the smoothing parameter value. The index $\widehat{R_{Pq}}$ depends on the gap
between consecutive eigenvalues, and appears to be unstable and consequently rather
difficult to interpret. Nevertheless, we can note that for small values of parameter ℓ
($\log(\ell) < -5$), the only stable component is the first one. As ℓ get larger ($\log(\ell) > 6$),
the data are clearly oversmoothed and many components disappear. The behavior of
R_{P5}, which has a single minimum, leads us to retain the following values: $q = 5$ and
$\ell \approx 1$.

The eigenfunctions obtained by incorporating a roughness penalty in the PCA are
much easier to interpret and reveal e.g. different periodic components (see Figure 5.4).

Another example of PCA of climatic time series (temperature) is presented in
Antoniadou *et al.* [ANT 01].

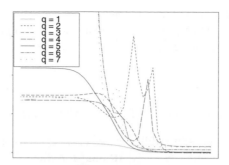

Figure 5.3. *Jackknife approximation $\widehat{R_{Pq}}$ of the stability of projection for different dimensions as a function of $\log(\rho)$*

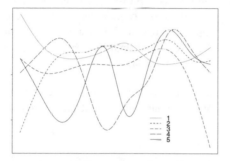

Figure 5.4. *Representation of the $q = 5$ eigenfunctions associated with the largest eigenvalues, generating the estimated projection $\widehat{P}_q$*

5.4. Linear regression models and extensions

We are now interested in estimating the link between a scalar response and a functional variable. The first model which has been extended from the multivariate framework to the functional frame was the linear model [HAS 93, RAM 91]. In chemometrics, for example [FRA 93, GOU 98], the purpose was to explain and predict the concentration of certain types of chemical components with the help of spectrometric responses measured at more than 100 different frequencies.

The number of explanatory variables (the different frequencies) can be larger than the sample size; we need to introduce a constraint in order to obtain a unique and stable solution to the least squares criterion. Mainly two strategies have been developed in the literature to handle this issue. The first strategy relies on performing a functional principal components analysis and expanding the regression function in the basis of eigenfunctions. The second approach is based on the introduction of a roughness penalty in the least squares criterion, that allows us to control the smoothness of the estimator. Note that partial least squares have also been adapted to that functional context by Preda and Saporta [PRE 05].

The regression problem can presented as follows: let (Z, Y) be a pair of random variables taking values in $H \times \mathbb{R}$, where Z is the functional covariate (the spectrometric curve, for instance) and Y is the response variable (the chemical concentration). We can assume

$$\mathbb{E}[Y|Z = z] = \int_0^1 \psi(t)z(t)\, dt, \qquad z \in H, \tag{5.13}$$

where function $\psi \in H$ is the regression function to be estimated.

An adaptation of the generalized linear model was proposed by Eilers and Marx [MAR 96, MAR 99] or Cardot *et al.* [CAR 03a] for when the response variable is discrete or bounded. In this latter work, a multilogit model is built to predict land use with the help of the temporal evolution of coarse resolution remote sensing data. We will describe this example with more details at the end of the section. More recently, non-parametric models of the regression function were proposed by Ferraty and Vieu [FER 06]. This topic, of growing interest in the statistical community, relies on building kernel smoothers for functional data-based semi-metrics, that allow the distance between two curves to be measured.

Note that all these models readily extend to the case of more than one functional covariate considering, for instance, new functional spaces that are themselves products of functional spaces.

5.4.1. *Functional linear models*

Denoting the covariance operator of Z by Γ and the cross-covariance function between Z and Y by Δ,

$$\Delta(t) = \mathbb{E}[Z(t)Y], \quad t \in T.$$

It is easy to check, using equation (5.13), that

$$\Delta = \Gamma\psi. \tag{5.14}$$

It is therefore necessary to invert operator Γ to build an estimator of ψ. The problem is that Γ is a compact operator and, consequently, it does not have a bounded generalized inverse. To deal with that difficulty, we have to look for a regularized solution in the same spirit as in inverse problems. We now present the two main types of regularization estimators studied in the literature. The first, based on a dimension reduction, is the principal components regression based on smoothed PCA. It consists of inverting equation (5.14) in the finite dimensional space generated by the first q eigenfunctions of the (smoothed) empirical covariance operator. The second type of estimators is based on penalized least squares.

We now assume we have a sample $(\mathbf{z}_i, y_i)$, $i = 1, \ldots, n$ of independent and identically distributed (iid) observations, where $\mathbf{z}_i = (z_i(t_j), j = 1, \ldots, p) \in \mathbb{R}^p$.

5.4.2. *Principal components regression*

Principal components regression is a classic technique in multivariate regression when there are numerous and/or highly correlated covariates. It consists of estimating the eigenfunctions of the empirical covariance operator, i.e. performing the FPCA of the curves z_i and expanding the regression function in the basis of the empirical eigenfunctions. As a matter of fact, this is equivalent to inverting equation (5.14) in the subspace generated by the eigenfunctions of the empirical covariance operator associated with the largest eigenvalues. This decomposition is widely used because it is optimal according to the variance of the approximated signals z_i under a dimensionality constraint.

The principal components regression estimator can be written

$$\widehat{\psi}(t) = \sum_{j=1}^{q} \beta_j \widehat{v}_j(t), \tag{5.15}$$

where $\{\widehat{v}_j\}_{j=1,\ldots,q}$ are the q eigenfunctions of the empirical covariance operator associated with the largest eigenvalues.

Vector β of coordinates in this new basis is determined by considering a least squares criterion in which the principal components $\langle \widehat{z}_i, \widehat{v}_j \rangle_{L^2} = \mathbf{z}_i' \mathbf{N} \mathbf{v}_j$ now play the role of explanatory variables:

$$\min_{\beta \in \mathbb{R}^q} \sum_{i=1}^{n} w_i \left(Y_i - \sum_{j=1}^{q} \beta_j \mathbf{z}_i' \mathbf{N} \mathbf{v}_j \right)^2.$$

Assuming the trajectories z_i are smooth, we can add a roughness penalty in the dimension reduction process as described in the previous section. This approach is described with more detail in section 5.5.

5.4.3. *Roughness penalty approach*

We now present the second approach which relies upon spline approximation and minimization of a penalized least squares criterion [GOU 98, HAS 93, MAR 96], similar to the well-known *ridge regression*. Regularization no longer holds on the dimension as was the case for principal components regression, but acts directly on the smoothness of the solution. More precisely, we consider the following criterion:

$$\min_{\psi} \sum_{i=1}^{n} w_i \left(Y_i - \psi' \mathbf{N} \mathbf{z}_i \right)^2 + \ell ||\psi||_2^2, \tag{5.16}$$

where $\boldsymbol{\psi} = (\psi(t_1), \ldots, \psi(t_p))'$. The smoothing parameter (or regularization parameter) ℓ allows the stability of the solution as well as its regularity to be controlled. Matricially, solving equation (5.16) is equivalent to

$$\min_{\psi} \ \boldsymbol{\psi}'\mathbf{NSN}\boldsymbol{\psi} - 2\ \boldsymbol{\psi}'\mathbf{NZ}'\mathbf{DY} + \ell\ \boldsymbol{\psi}'\mathbf{M}\boldsymbol{\psi}. \tag{5.17}$$

The solution can be written explicitly and is given by

$$\widehat{\boldsymbol{\psi}} = (\mathbf{NSN} + \ell\mathbf{M})^{-1}\mathbf{NZ}'\mathbf{DY}. \tag{5.18}$$

We recognize in this formulation of the estimator ψ the inverse of the covariance operator plus a roughness penalty $\mathbf{NSN} + \ell\mathbf{M}$ as well as the transpose of the cross-covariance function $\mathbf{NZ}'\mathbf{DY}$.

5.4.4. *Smoothing parameters selection*

Selecting good values for the dimension q and for the smoothing parameter ℓ is of major importance in practice. These tuning parameters control the tradeoff between the bias of the estimator and its variability. Too small a value for the smoothing parameter (or too large a dimension) leads to an estimator that fits the data too well and has bad prediction qualities. On the other hand, if the smoothing parameter value is too large (or if the dimension is too small) we obtain a stable but biased estimator.

In practice, different techniques inspired by model selection approaches are used to select values for the smoothing parameters. Generalized cross-validation, initially proposed for smoothing splines [WAH 90], can be adapted to the context of functional linear models:

$$\mathrm{GCV} = \frac{\sum_{i=1}^n \left(y_i - \widehat{\boldsymbol{\psi}}'\mathbf{N}\mathbf{z}_i\right)^2}{\mathrm{tr}\left(\mathbf{I} - \mathbf{A}\right)^2}$$

where matrix $\mathbf{A}$ is the *hat matrix*. It depends on the regularization parameter (ℓ or q) and, by definition, satisfies $\widehat{\mathbf{y}} = \mathbf{A}\mathbf{y}$ where $\mathbf{y}$ is the vector of observations and $\widehat{\mathbf{y}}$ is the vector of fitted values. For the penalized spline approach, it can be written as:

$$\mathbf{A} = \mathbf{Z}\left(\mathbf{NSN} + \ell\mathbf{M}\right)^{-1}\mathbf{NZ}'\mathbf{D}.$$

As far as principal components regression is concerned, the GCV criterion is simple since $\mathrm{tr}(\mathbf{A}) = q$.

Note that other selection criteria such as prediction error on a test sample or penalized mean square error of prediction by the number of components (Akaïke, BIC, etc.) have been proposed [MAR 99, RAM 05].

5.4.5. *Some notes on asymptotics*

The first asymptotic studies [CAR 99, CAR 03c] have shown that, to obtain consistent estimators, the smoothing parameter ℓ (respectively, q) should not tend to zero (respectively, to infinity) too rapidly. Recently, Crambes *et al.* [CRA 09] gave the asymptotic rates of convergence of the penalized splines approach according to a mean square prediction error criterion, and proved that it is optimal. This rate depends on many ingredients such as the regularity of the function Z, the regularity of the function ψ and the behavior of its Fourier decomposition in the basis of the eigenfunctions of the covariance operator. Considering a different loss function based on the mean squared error of estimation of the function ψ itself, [HAL 07] have also proved (under rather similar assumptions) the consistency of the principal components regression estimator and the optimality of the rates of convergence.

Crambes *et al.* [CRA 09] have also proved that choosing the smoothing parameter value by minimizing the generalized cross-validation criterion previously defined leads to a consistent estimator of the regression function, with optimal asymptotic rate of convergence according to a prediction error criterion.

The asymptotic distribution of predictions made with principal components regression has been derived by [CAR 07], allowing prediction confidence intervals to be built. Tests of nullity of the regression function or of equality to a deterministic function are described [CAR 03b]. They are based on the asymptotic behavior of the empirical cross-covariance function.

5.4.6. *Generalized linear models and extensions*

Various extensions of the linear model have been proposed in the literature. The most direct approach is to consider a generalized linear model in order to describe statistically the dependence between a categorical, discrete or bounded real variable and a random function. Marx and Eilers [MAR 99] study a penalized maximum likelihood approach, similar to the least squares method given in section 5.4.3, to predict the survival probability two years after a cancer with the help of the density function of DNA in malignant cells. The model can be written in its general form as:

$$\mathbb{E}[Y|Z = z] = g^{-1}\left(< \psi, z >\right), \qquad z \in H, \tag{5.19}$$

where function g is the link function (logit, probit, log, etc.). When g is the identity function, then model (5.19) is simply the functional linear model described previously.

The estimator of the regression function ψ cannot generally be written explicitly and is obtained by maximizing a penalized likelihood criterion,

$$\max_{\psi} \ \sum_{i=1}^{n} \log \mathcal{L}(y_i, \mathbf{z}_i, \boldsymbol{\psi}) - \ell \, \|\psi\|_2^2 \tag{5.20}$$

where $\mathcal{L}(.)$ is the likelihood function. Marx and Eilers [MAR 99] use a procedure relying on iterated weighted least squares to find the solution (see [FAH 94] for the description of such a method in the multivariate framework) whereas [JAM 02] adapts the EM algorithm in a sparse discretization context. The smoothing parameter value is then chosen by minimizing an information criterion such as Bic or Akaïke.

The second approach, which is similar to principal components regression, consists of reducing the dimension of the data by projection of ψ onto the space generated by the eigenfunctions associated with the largest eigenvalues of the empirical covariance operator. We adopt this point of view in the remote sensing example described below.

A few consistency results are given in the literature. Cardot and Sarda [CAR 05] prove the consistency of the penalized maximum likelihood approach, whereas some properties of the principal components regression estimator in this context have been studied [MUE 05].

5.4.7. *Land use estimation with the temporal evolution of remote sensing data*

We present a brief illustration of the interest of considering generalized linear models for functional data. This study was made at INRA (French National Institute for Agricultural Research) and is part of the project UIC-Végétation; for more details see [CAR 03a, CAR 08].

The VEGETATION sensor of satellite SPOT4 acquires images of Europe (2,000 km × 2,000 km) with a low spatial resolution (each pixel represents an area of 1 km × 1 km) at a high temporal frequency (daily measurements). Information collected by this sensor at different frequencies (red, blue, near infrared, shortwave infrared) allow the development of cultures and crops to be characterized at a regional scale. The observed response of each pixel is actually the aggregated responses of the different crops (wheat, forest, etc.) in the underlying area; these pixels are generally called *mixed pixels*.

The aim is to use the temporal evolution of such mixed pixels to discriminate the different types of cultures and estimate the land use, i.e. the proportion of each crop inside a mixed pixel. We know precisely the land use in a small region of France and therefore it is possible to build a prediction model of the land use using VEGETATION remote sensing data. One of the main tasks of this project is to predict, using such a model, the land use of neighboring areas and, using meteorological data, predict yield for different crops such as maize or wheat.

The statistical question is to estimate proportions (of land use) with longitudinal or varying time covariates; a multilogit model with functional covariates has been constructed to answer this question. The interest of adopting a functional point of view is that measurement times may vary from one pixel to another due, for example, to the presence of clouds in part of the image.

We denote the vector of proportions associated with land use in pixel i by $\pi_i = [\pi_{ik}]_{k=1,\ldots,g}$ where π_{ik} is the proportion of area in pixel i devoted to crop or theme k for $k = 1,\ldots,g = 10$. By definition, we have $\pi_{ik} \geq 0$ and $\sum_k \pi_{ik} = 1$ for $i = 1,\ldots,n = 1{,}554$. In the area under study (40 km × 40 km), we also observe coarse resolution remote sensing data (VEGETATION sensor) at a high temporal frequency from March 1998 to August 1998 for four spectral bands of interest: red (R), near infrared (NIR), blue (B) and shortwave infrared (SWIR). Two indexes are widely used in bioclimatoloy to characterize vegetation development and are constructed: $NDVI = (NIR - R)/(NIR + R)$ and $PVI = (NIR - 1.2R)/\sqrt{2.2}$. They are combinations of the responses in the red and near infrared wavelengths.

The discrete evolution of these indexes, observed for each pixel i at time $t_1,\ldots,t_k,\ldots,t_p$, are denoted by $\mathbf{z}_i = [z_i(t_1),\ldots,z_i(t_p)]'$. Having suppressed dates in which there were too many clouds, we finally obtain $p = 32$ time points for each trajectory.

We now assume the proportion π_{ik} of crop k in pixel i is drawn from a multinomial distribution whose parameters satisfy

$$\mathbb{E}(\pi_{ik}|z_i) = \frac{\exp\left(\delta_k + \int_T \psi_k(t)\left(z_i(t) - \mathbb{E}(z_i(t))\right) dt\right)}{\sum_{l=1}^{g} \exp\left(\delta_l + \int_T \psi_l(t)\left(z_i(t) - \mathbb{E}(z_i(t))\right) dt\right)}. \tag{5.21}$$

For identification, we fix $\psi_g = 0$ and $\delta_g = 0$. Then, each regression function ψ_k for $k = 1,\ldots,g-1$ has to be interpreted by comparing with the class of reference g.

The number of parameters in the model, which depends on the number of different classes of land use and on the number of explanatory variables, is very important $(32 \times (10 - 1))$ and allows us to reduce the dimension of the data.

In a first step, we perform an FPCA of the discrete trajectories of the pixels $\mathbf{z}_i, i = 1,\ldots,n$ in order to reduce the dimension in a optimal way according to the minimum variance criterion. We can then expand the regression functions ψ_k on the basis of the eigenfunctions of the empirical covariance operator, such as in equation (5.15), and estimate their coordinates by maximum likelihood. The class of reference was chosen to be *urban* because its response does not vary much with time.

In a second step, significant directions (i.e. significant coordinates) in the eigenfunction basis are selected using a likelihood ratio test with an ascendant procedure.

The $n = 1,554$ pixels of the initial sample are split into a learning sample of size 1,055 pixels and a test sample of 499 pixels. These samples are used to evaluate the prediction error and determine the 'best' model. The basic model which predicts land use by its empirical mean in the learning sample is used as a benchmark, and denoted M_0.

Estimations of the regression function for the NDVI index are depicted in Figure 5.5. We can remark upon the similarity, up to a constant term, between the regression functions associated with the labels *forest* and *permanent crops*. The reader is referred to [CAR 05] for more details on this application.

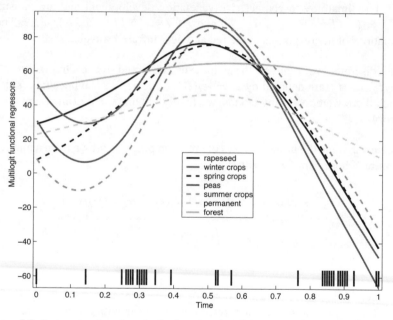

Figure 5.5. *Regression functions in the functional multilogit models with the NDVI index*

We also note that prediction errors of the multilogit models, according to the criterion

$$R_{ik} = \frac{|\pi_{ik} - \widehat{\pi_{ik}}|}{\frac{1}{n}\sum_{i=1}^{n} \pi_{ik}},$$

are systematically better than that of M_0 (see Table 5.1).

As a conclusion to this brief example, we can say that the temporal information provided by the coarse resolution remote sensing data is useful in predicting land use. The PVI index, which is a linear combination of R and NIR responses, seems to be the most effective index among those tested here to predict land use, giving the best results for 5 labels out of 10.

Labels	NDVI	PVI	Blue	Red	NIR	SWIR	M_0
Urban	0.49	**0.36**	0.47	0.54	0.41	0.51	0.86
Water	0.43	**0.29**	0.78	0.62	0.61	0.31	1.30
Rapeseed	0.48	0.46	**0.45**	0.50	0.47	0.47	0.59
Winter crops	0.20	0.21	**0.19**	0.20	0.22	**0.19**	0.30
Spring crops	0.58	**0.56**	0.60	0.61	0.65	0.61	0.69
Peas	0.50	**0.43**	0.45	**0.43**	0.48	0.46	0.63
Summer crops	0.61	0.68	0.61	0.60	0.76	**0.53**	0.88
Permanent crops	0.47	**0.46**	0.52	0.49	**0.46**	0.50	0.61
Forest	0.34	0.36	0.34	**0.31**	0.45	0.35	0.98
Potatoes	0.90	0.93	0.94	0.90	1.06	**0.85**	1.31

Table 5.1. *Median risks on the test sample for the multilogit models and the model M_0; numbers in bold indicate the best prediction error for each label*

5.5. Forecasting

In this section, we propose models that can be useful to predict a continuous time stochastic process over a whole interval of time. Of particular interest are processes having a naturally periodic component such as those studied in section 5.5.6 where the aim is to predict *El Niño*. Rather than employing classical time series techniques such as SARIMA models, the proposed approach is based on a *global prediction* of the phenomenon over the considered period, i.e. 1 year. Models and examples presented below are essentially drawn from Bosq [BOS 00], Besse and Cardot [BES 96] and Besse *et al.* [BES 00]. Such models have also been useful for ozone prediction [DAM 02].

The aim is to forecast a continuous time stochastic process $(X_t)_{t \in \mathbb{R}}$, not necessarily assumed to be stationary, over a time interval of length δ where $\delta = 1$ is assumed to be equal to the length of the main period, without loss of generality. We can then define a discrete time stochastic process taking values in a function space as:

$$Z_n = \{X_{t+(n-1)}, t \in [0,1]\},$$

which is assumed to be stationary and of second order i.e. $\mathbb{E}\|Z_n\|^2 < \infty$.

5.5.1. *Functional autoregressive process*

The mathematical framework for such processes taking values in function space is described in detail in [BOS 00]. We are interested here in functional autoregressive processes of order 1, referred to as ARH(1) and defined as follows. Process $(Z_i)_{i \in \mathbb{Z}}$ is an ARH(1) process, with mean $a \in H$ and correlation operator ρ if it satisfies

$$\forall i \in \mathbb{Z}, \quad Z_i - a = \rho(Z_{i-1} - a) + \epsilon_i. \tag{5.22}$$

Operator ρ is assumed to be compact and we assume that $\sum_{n \geq 1} \|\rho^n\| < +\infty$. The innovation sequence $\{\epsilon_i\}$ is a sequence of iid functional random variables, centered with finite variance $\mathbb{E}\|\epsilon_i\|_H^2 = \sigma^2 < +\infty$. The conditional expectation $\mathbb{E}(Z_{i+1}|Z_i, Z_{i-1}, \ldots)$, which is the key tool for prediction, is given by

$$\mathbb{E}(Z_{i+1}|Z_i, Z_{i-1}, \ldots) - a = \rho(Z_i - a) , \quad i \in \mathbb{Z}. \tag{5.23}$$

In $L^2[0,1]$, this operator admits an integral operator representation with kernel $\rho(s,t)$; equation (5.23) can therefore be written

$$\mathbb{E}(Z_{i+1}(t)|Z_i, Z_{i-1}, \ldots) - a(t) = \int_0^1 \rho(s,t)(Z_i - a)(s)\, ds, \quad t \in [0,1], \ i \in \mathbb{Z}. \tag{5.24}$$

Covariance operators $\Gamma = \mathbb{E}((Z_i-a)\otimes(Z_i-a))$ and $\Delta = \mathbb{E}((Z_i-a)\otimes(Z_{i+1}-a))$ satisfy an equation similar to equation (5.14) in the functional linear model:

$$\Delta = \rho\Gamma. \tag{5.25}$$

Operator Δ also has an integral operator representation:

$$\Delta f(t) = \int_0^1 \gamma(s,t+1)f(s)\, ds, \quad t \in [0,1], \ f \in L^2[0,1]. \tag{5.26}$$

As in the functional linear model, we need to invert the covariance operator Γ to build an estimator of operator ρ. Regarding the functional linear model, Γ has a bounded inverse and the following steps are used to build an estimator of ρ:

1) eigenanalysis of operator Γ, which consists of looking for the solution of

$$\int_{[0,1]} \gamma(s,t)f(s)ds = \lambda f(t) , \ t \in [0,1];$$

2) projection of the trajectories onto the functional subspace H_q of dimension q generated by the eigenfunctions $\phi_1, \ldots, \phi_q$ of Γ associated with the q largest eigenvalues $\lambda_1 \geq \ldots \geq \lambda_q > 0$;

3) inversion of the restriction of Γ in H_q; and

4) finite rank approximation of ρ using equation (5.25).

Regarding principal components regression, estimator $\widehat{\rho}$ of ρ is expanded in the basis of the eigenfunctions of the empirical covariance operator:

$$\widehat{\rho}(s,t) = \sum_{j=1}^q \sum_{j'=1}^q \widehat{\beta}_{jj'}\widehat{v}_j(t)\widehat{v}_{j'}(s). \tag{5.27}$$

In practice, trajectories are discretized and the procedures developed by Bosq [BOS 91] and Pumo [PUM 92] rely on the two important points:

1) a linear interpolation of the sample paths of the stochastic process between measurement points; and

2) a finite rank estimator of operator ρ.

As seen before, many strategies can be employed to deal with the discretization. We present in the following a procedure which combines dimension reduction and smooth functional approximations to the discretized observations, similar to the approach presented in section 5.3.

5.5.2. *Smooth ARH(1)*

It has been proposed in [BES 96] to anticipate the necessary dimension reduction in the approximation step of the discretized trajectories. This approach proved to be more effective than the linear interpolation procedure developed by Pumo [PUM 92] on simulations and real datasets. It consists of considering the optimization problem:

$$\min_{\widehat{z}_i \in H_q} \frac{1}{n} \sum_{i=1}^{n} \left(\frac{1}{p} \sum_{j=1}^{p} (z_i(t_j) - \widehat{z}_i(t_j))^2 + \ell \left\| D^2 \widehat{z}_i \right\|_{L^2}^2 \right) \tag{5.28}$$

where H_q is a subspace of H with dimension q. As already seen in section 5.3, the mean trajectory of the processes Z_i is given by interpolating splines, at time points $(t_1, \ldots, t_j)$, of vector $\widehat{\mathbf{a}} = \mathbf{A}_\ell \bar{\mathbf{z}}$. Denoting the solution of problem (5.28) by $\widehat{z}_i = \widehat{\mathbf{a}} + \widehat{\mathbf{x}}_i$, we get the matrix representation of the covariance operators:

$$\widehat{\boldsymbol{\Gamma}}_{q,\ell} = \frac{1}{n} \sum_{i=1}^{n} \widehat{\mathbf{x}}_i \widehat{\mathbf{x}}_i' \mathbf{N}, \tag{5.29}$$

and

$$\widehat{\boldsymbol{\Delta}}_{q,\ell} = \frac{1}{n-1} \sum_{i=1}^{n-1} \widehat{\mathbf{x}}_{i+1} \widehat{\mathbf{x}}_i' \mathbf{N}. \tag{5.30}$$

Then, considering the generalized inverse $\widehat{\boldsymbol{\Gamma}}_{q,\ell}^{-1}$ of matrix $\widehat{\boldsymbol{\Gamma}}_{q,\ell}$, we can deduce the matrix representation of operator ρ:

$$\widehat{\rho}_{q,\ell} = \widehat{\boldsymbol{\Gamma}}_{q,\ell}^{-1} \widehat{\boldsymbol{\Delta}}_{q,\ell}. \tag{5.31}$$

Finally, prediction of a new curve $\mathbf{z}_{n+1}$ is given by

$$\widehat{\mathbf{z}}_{n+1} = \widehat{\rho}_{q,\ell} \widehat{\mathbf{x}}_n + \mathbf{A}_\ell \bar{\mathbf{z}}. \tag{5.32}$$

In practice, choosing good values for the regularization parameters ℓ and q is important to obtain effective predictions. Selection procedures are discussed in section 5.5.4.

5.5.3. *Locally ARH(1) processes*

In some situations, the stationarity assumption of process $(Z_i)_{i \in \mathbb{Z}}$ is too strong a hypothesis which is not satisfied. This is, for example, probably the case for *El Niño* where exceptional events can occur. Besse *et al.* [BES 00] proposed building local approximations to the covariance operators which can avoid such hypotheses. This approach leads to the proposition of local linear approximations to the autocorrelation operator, which are built by kernel smoothers in the same spirit as in [FER 06]. Let us define the real Gaussian kernel K_h as:

$$K_h(x) = \frac{1}{\sqrt{2\pi}} \exp\left(\frac{-x^2}{2h}\right) \tag{5.33}$$

and the local estimators of the operators of covariance:

$$\widehat{\Delta}_h(\mathbf{x}) = \frac{\sum_{i=1}^{n-1} K_h\left(\|\mathbf{x}_i - \mathbf{x}\|_{\mathbf{N}}\right) \mathbf{x}_i \mathbf{x}_{i+1}' \mathbf{N}}{\sum_{i=1}^{n-1} K\left(\|\mathbf{x}_i - \mathbf{x}\|_{\mathbf{N}}\right)}, \tag{5.34}$$

$$\widehat{\Gamma}_h(\mathbf{x}) = \frac{\sum_{i=1}^{n} K_h\left(\|\mathbf{x}_i - \mathbf{x}\|_{\mathbf{N}}\right) \mathbf{x}_i \mathbf{x}_i' \mathbf{N}}{\sum_{i=1}^{n-1} K_h\left(\|\mathbf{x}_i - \mathbf{x}\|_{\mathbf{N}}\right)}. \tag{5.35}$$

Introducing the diagonal weight matrix $\mathbf{W}_h(\mathbf{x})$ with diagonal generic elements:

$$w_{i,h}(\mathbf{x}) = \frac{K_h\left((\mathbf{x}_i - \mathbf{x})'\mathbf{N}(\mathbf{x}_i - \mathbf{x})\right)}{\sum_{\ell=1}^{n} K\left((\mathbf{x}_\ell - \mathbf{x})'\mathbf{N}(\mathbf{x}_\ell - \mathbf{x})\right)}, \quad i = 1, \ldots, n, \tag{5.36}$$

the estimators of the covariance operators can be written

$$\widehat{\Delta}_h(\mathbf{x}) = \sum_{i=1}^{n} w_{i,h}(\mathbf{x}) \, \mathbf{x}_i \mathbf{x}_{i+1}' \mathbf{N} \quad \text{and} \quad \widehat{\Gamma}_h(\mathbf{x}) = \mathbf{X}'\mathbf{W}_h(\mathbf{x})\mathbf{X}\mathbf{N}.$$

The local estimator of operator ρ is then simply defined by performing the eigen-analysis of operator $\widehat{\Gamma}_h$ with $\mathbf{x} = \mathbf{z}_n - \widehat{\mathbf{a}}$. The rank q inverse of $\widehat{\Gamma}_h$ can be written

$$\left(\widehat{\mathbf{P}}_q \widehat{\Gamma}_h \widehat{\mathbf{P}}_q\right)^{-1} = \sum_{\ell=1}^{q} \frac{1}{\widehat{\lambda}_\ell} \widehat{\mathbf{v}}_\ell \widehat{\mathbf{v}}_\ell' \mathbf{N}$$

where $(\widehat{\mathbf{v}}_\ell, \ell = 1, \ldots, q)$ are the eigenvectors of $\widehat{\Gamma}_h$ with metric $\mathbf{N}$ associated with the largest eigenvalues $(\widehat{\lambda}_\ell, \ell = 1, \ldots, q)$. Inverting the equation that links operators ρ, Δ and Γ, evaluated at point $\mathbf{z}_n - \widehat{\mathbf{a}}$, allows a local estimator of operator ρ to be defined by

$$\widehat{\rho}_{q,h} = \left(\sum_{\ell=1}^{q} \widehat{\mathbf{v}}_\ell \widehat{\mathbf{v}}_\ell'\right) \mathbf{N} \widehat{\Delta}_h \left(\sum_{\ell=1}^{q} \frac{1}{\widehat{\lambda}_\ell} \widehat{\mathbf{v}}_\ell \widehat{\mathbf{v}}_\ell'\right) \mathbf{N}. \tag{5.37}$$

This estimator depends on smoothing parameters q and h whose values are determined by minimizing the prediction error in the test sample (see section 5.5.4 for more details). To make a prediction of $\mathbf{z}_{n+1}$, we simply need to apply formula (5.32).

This approach can also be adapted with a presmoothing step, with smoothing splines, of the trajectories. This is particularly useful when they are observed with noise at the design points. The drawback is that the estimator depends on three smoothing parameters.

5.5.4. *Selecting smoothing parameters*

The procedures described above involve smoothing parameters whose role is of major importance for the quality of prediction. For instance, too large a value of the smoothing parameter ℓ for smooth ARH(1) implies an oversmoothing of the trajectories and consequently a bad estimation of operator ρ. It is therefore important to propose effective data-driven procedures for selecting 'good' smoothing parameter values. Techniques proposed in the framework of functional linear models may no longer be adequate due to the non-independence between the trajectories Z_i. These parameters are now chosen by splitting the initial sample into a learning sample which is constituted by the first trajectories of the stochastic process and a test sample. The estimators are built using the learning sample and their prediction accuracies are compared with the test sample. The smoothing parameters are selected in order to give the best predictions in the test sample.

5.5.5. *Some asymptotic results*

The literature on the asymptotic properties of estimators for functional processes is not abundant and the reader is referred to [BOS 00] for the main references.

The first results were given by Bosq in the 1990s, assuming the trajectories were observed over the whole interval without discretization. Conditions ensuring the consistency of the estimators are similar to those considered in the functional linear model, and assume the dimension q does not tend to infinity with a rate that depends on the eigenvalues of the covariance operator. Interpolation and smoothing spline procedures have been proposed to deal with discretized trajectories. Note that recently, Mas [MAS 07a] has given such results on the asymptotic distribution for predicted values in order to build confidence intervals.

5.5.6. *Prediction of climatic time series*

Besse and Cardot [BES 96] considered functional approaches for prediction on real datasets (highway traffic) and simulated data. They have noticed that considering estimators that are based on simultaneous dimension and smoothness constraints could lead to better performances. We present in this section a brief example drawn from [BES 00] which provides an illustration of the ability of functional approaches to outperform more classical time series techniques.

ENSO (*El Niño Southern Oscillation*) is a climatic phenomenon of interaction between the Pacific ocean and the atmosphere, of major importance. *El Niño* (EN) is characterized by important interannual variations of temperatures, particularly in December, and a high correlation with oscillations of the atmospheric pressure. EN is therefore observed through the evolution of monthly mean values of the temperature measured at the surface of the ocean close to Chili, whereas the Southern Oscillation (SO) is measured by the atmospheric pressure at the sea level near Tahiti. These climatic series begin in 1950 and are used in this study. We give a graphic representation in Figure 5.6.

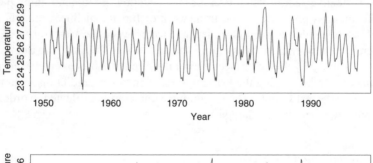

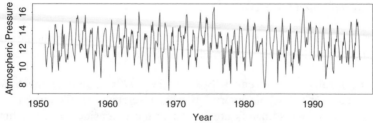

Figure 5.6. *Monthly mean values of the sea surface temperature (Niño-3) and of the atmospheric pressure at Tahiti*

Different predictors are compared from this data using the selection procedure described above for the different smoothing parameter values. The data from 1950 to 1986 (36 years) are used to build the estimators whose prediction ability are compared on ten predictions, one year ahead, for the period 1987–1996. More precisely, the first 36 years are split into a learning sample made with the 31 first years and a test sample (years 32–36).

The different approaches are compared with the most simple model, referred to as climatology, which consists of predicting the evolution during one year by the mean annual trajectory over the past periods.

The best parametric models where obtained by considering a SARIMA model $(0, 1, 1) \times (1, 0, 1)_{12}$ for the temperature index (EN) and a SARIMA $(1, 1, 1) \times (0, 1, 1)_{12}$ for the pressure index (SO). The 'portemanteau' test concluded at the absence of correlation for the innovation.

The other considered approaches are non-parametric:

Kernel is a non-parametric prediction made with classical kernel smoothers considering horizons varying from 1 to 12 months. The optimal values for the bandwidths are 0.9 and 1.5 for EN and SO, respectively.

Functional kernel is also a non-parametric regression approach with functional response and functional explanatory variables. The optimal values of the smoothing parameters are 0.3 and 0.6.

Smooth ARH(1) requires 4 (respectively, 3) dimensions associated with a smoothing parameter with value 1.6×10^{-5} (respectively, 8×10^{-5}).

Local ARH(1) requires 3 (respectively, 2) dimensions associated with a smoothing parameter with value 0.9×10^{-5} (respectively, 5×10^{-5}).

The pressure series are noticeably less smooth than the temperature series and need more significant smoothing stages. Figure 5.7 presents a comparison of predictions of El Niño made for year 1986. Scalar predictions are linearly interpolated and functional predictions are built by spline interpolation. The prediction made by the SARIMA model for this year is clearly poor, whereas functional and non-parametric approaches have similar good performances. The same type of conclusions are drawn with the SO index. In Table 5.2 we give a synthetic comparison of the performances of the different approaches during a 10 year period according to the mean squared error (MSE) and mean relative absolute error (MRAE) of prediction.

Method	El Niño index		S. Osc. index	
	MSE	MRAE	MSE	MRAE
Climatology	0.73	2.5 %	0.91	6.3 %
SARIMA	1.45	3.7 %	0.95	6.2 %
Kernel	0.60	2.3%	0.87	6.1%
Functional kernel	0.58	2.2%	0.82	6.0%
Smooth ARH(1)	0.55	2.3%	**0.78**	**5.8%**
Local ARH(1)	**0.53**	**2.2%**	0.82	5.8%

Table 5.2. *Comparison of mean squared and mean relative absolute errors for predictions one year ahead made during the period 1987–1996*

It seems that functional approaches perform better than classical parametric time series techniques (SARIMA), which give even worse results than the climatic benchmark prediction.

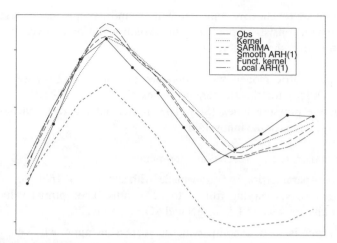

Figure 5.7. *Comparison of El Niño during year 1986 and different predicted values*

5.6. Concluding remarks

Examples on real or simulated datasets presented in this chapter, as well as many others described in the quoted literature, clearly show that it is worth introducing a functional framework even if it needs to consider more abstract spaces as well as more complex estimation algorithms. This last practical aspect can require the use of sophisticated numerical tools which are not very common in a statistical framework. Nevertheless, as many software programs such as R or MatLab include both statistical and numerical tools, many authors now give access to their packages and these approaches are becoming easier to implement. This technical effort allows most statistical methods to be adapted to functional frameworks in order to extract the most relevant part of the data. This requires that five levels of examination of the data and of the statistical modeling be considered:

1) Is a curve registration or a warping procedure of the timescale required to better synchronize the observed trajectories?

2) If the discretization grid is dense enough, then which functional basis is best to interpolate or smooth the discretized data from Fourier, splines and wavelets?

3) Is a smoothing step required to remove some noise within data? This can be achieved by wavelet thresholding, spline smoothing or high frequencies filtering.

4) What kind of metric is well suited to measure distances between curves and between statistical units? For example, a Sobolev norm, which takes derivatives into account, is well adapted to spline basis and regular curves.

5) How should the values of tuning parameters such as PCA dimension, smoothing parameters, number of neighbors in kNN, etc. be jointly optimized?

In order to answer the above questions, much theory still needs to be developed and many experiments on real data need to be conducted.

5.7. Bibliography

[ABR 06] ABRAHAM C., BIAU G., CADRE B., "On the kernel rule for function classification", *Ann. Inst. Statist. Math.*, vol. 58, p. 619–633, 2006.

[ANT 01] ANTONIADOU T., BESSE P., FOUGÈRES A., LE GALL C., STEPHENSON D., "L'Oscillation Atlantique Nord (NAO) et son influence sur le climat européen", *Revue de Statistique Appliquée*, vol. 49, p. 39–60, 2001.

[BEN 09] BENKO M., HARDLE W., KNEIP A., "Common functional principal components", *Annals of Statistics*, vol. 37, num. 1, p. 1–24, 2009.

[BES 86] BESSE P., RAMSAY J., "Principal component analysis of sampled curves", *Psychometrika*, vol. 51, p. 285–311, 1986.

[BES 92] BESSE P., "PCA stability and choice of dimensionality", *Statistics & Probability Letters*, vol. 13, p. 405–410, 1992.

[BES 96] BESSE P., CARDOT H., "Approximation spline de la prévision d'un processus fonctionnel autorégressif d'ordre 1", *Canadian Journal of Statistics*, vol. 24, p. 467–487, 1996.

[BES 97] BESSE P., CARDOT H., FERRATY F., "Simultaneous nonparametric regressions of unbalanced longitudinal data", *Computational Statistics and Data Analysis*, vol. 24, p. 255–270, 1997.

[BES 00] BESSE P., CARDOT H., STEPHENSON D., "Autoregressive forecasting of some functional climatic variations", *Scandinavian Journal of Statistics*, vol. 27, p. 673–688, 2000.

[BES 03] BESSE P., FARKAS E., "Discrimination par forêt aléatoire de chromatogrammes décomposés en ondelettes", ZIGHED A. D., Ed., *Journées de Statistique*, Lyon, 2003.

[BOS 91] BOSQ D., "Modelization, non-parametric estimation and prediction continuous time processes", ROUSSAS G., Ed., *Nonparametric Functional Estimation and Related Topics*, Nato, Asi series, p. 509–529, 1991.

[BOS 00] BOSQ D., *Linear Processes in Function Spaces*, Springer, 2000.

[CAR 99] CARDOT H., FERRATY F., SARDA P., "Functional linear model", *Statist. & Prob. Letters*, vol. 45, p. 11–22, 1999.

[CAR 00] CARDOT H., "Nonparametric estimation of the smoothed principal components analysis of sampled noisy functions", *Journal of Nonparametric Statistics*, vol. 12, p. 503–538, 2000.

[CAR 03a] CARDOT H., FAIVRE R., GOULARD M., "Functional approaches for predicting land use with the temporal evolution of coarse resolution remote sensing data", *Journal of Applied Statistics*, vol. 30, p. 1185–1199, 2003.

[CAR 03b] CARDOT H., FERRATY F., MAS A., SARDA P., "Testing ypotheses in the functional linear model", *Scandinavian Journal of Statistics*, vol. 30, p. 241–255, 2003.

[CAR 03c] CARDOT H., FERRATY F., SARDA P., "Spline estimators for the functional linear model", *Statistica Sinica*, vol. 13, p. 571–591, 2003.

[CAR 05] CARDOT H., SARDA P., "Estimation in generalized linear models for functional data via penalized likelihood", *Journal of Multivariate Analysis*, vol. 92, p. 24–41, 2005.

[CAR 07] CARDOT H., MAS A., SARDA P., "CLT in functional linear regression models", *Prob. Theory and Rel. Fields*, vol. 18, p. 325–361, 2007.

[CAR 08] CARDOT H., MAISONGRANDE P., FAIVRE R., "Varying-time random effects models for longitudinal data: unmixing and temporal interpolation of remote sensing data", *Journal of Applied Statistics*, vol. 35, p. 827–846, 2008.

[CAU 86] CAUSSINUS H., "Models and uses of principal component analysis", DE LEEUW J. et al., Eds., *Multidimensional Data Analysis*, DSWO Press, p. 149–170, 1986.

[CRA 09] CRAMBES C., KNEIP A., SARDA P., "Smoothing splines estimators for functional linear regression", *Annals of Statistics*, vol. 37, num. 1, p. 35–72, 2009.

[DAM 02] DAMON J., GUILLAS S., "The inclusion of exogenous variables in functional autoregressive ozone forecasting", *Environmetrics*, vol. 13, p. 759–774, 2002.

[DAU 76] DAUXOIS J., POUSSE A., Les analyses factorielles en calcul des Probabilités et en Statistique: Essai d'étude synthètique, PhD thesis, University of Toulouse, 1976.

[DEB 78] DE BOOR C., *A Practical Guide to Splines*, Springer Verlag, 1978.

[DEV 74] DEVILLE J., *Méthodes statistiques et numériques de l'analyse harmonique*, Ann. Insee, 15, 1974.

[DIE 93] DIERCKX P., *Curve and Surface Fitting with Splines*, Clarendon Press, Oxford, 1993.

[DO 99] DO K., KIRK K., "Discriminant analysis of event-related potential curves using smoothed principal components", *Biometrics*, vol. 55, p. 174–181, 1999.

[ECO 91] ECOSTAT, "http://www.math.univ-toulouse.fr/besse/pub/data/precipdon.dat", 1991.

[FAH 94] FAHRMEIR L., TUTZ G., *Multivariate Statistical Modelling Based on Generalized Linear Models*, Springer Verlag, 1994.

[FER 06] FERRATY F., VIEU P., *Nonparametric Functional Data Analysis: Methods, Theory, Applications and Implementations*, Springer-Verlag, London, 2006.

[FRA 93] FRANK I., FRIEDMAN J., "A statistical view of some chemometrics regression tools", *Technometrics*, vol. 35, p. 109–148, 1993.

[GOU 98] GOUTIS C., "Second-derivative functional regression with applications to near infra-red spectroscopy", *Journal of the Royal Statistical Society, B*, vol. 60, p. 103–114, 1998.

[GRE 94] GREEN P., SILVERMAN B., *Nonparametric Regression and Generalized Linear Model: A Roughness Penalty Approach*, Chapman and Hall, 1994.

[HAL 06] HALL P., MUELLER H.-G., WANG J.-L., "Properties of principal component methods for functional and longitudinal data analysis", *Annals of Statistics*, vol. 34, p. 1493–1517, 2006.

[HAL 07] HALL P., HOROWITZ J. L., "Methodology and convergence rates for functional linear regression", *Annals of Statistics*, vol. 35, num. 1, p. 70–91, 2007.

[HAS 93] HASTIE T., MALLOWS C., "A discussion of 'A statistical view of some chemometrics regression tools' by I.E. Frank and J.H. Friedman", *Technometrics*, vol. 35, p. 140–143, 1993.

[HAS 95] HASTIE T., BUJA A., TIBSHIRANI R., "Penalized discriminant analysis", *Annals of Statistics*, vol. 23, p. 73–102, 1995.

[JAM 00] JAMES G., HASTIE T., SUGAR C., "Principal component models for sparse functional data", *Biometrika*, vol. 87, p. 587–602, 2000.

[JAM 02] JAMES G., "Generalized linear models with functional predictor variables", *Journal of the Royal Statistical Society, B*, vol. 64, p. 411–432, 2002.

[KNE 00] KNEIP A., LI X., MACGIBBON B., RAMSAY J. O., "Curve registration by local regression", *Canadian Journal of Statistics*, vol. 28, p. 19–30, 2000.

[KNE 01] KNEIP A., UTIKAL K., "Inference for density families using functional principal component analysis", *Journal of the American Statistical Society*, vol. 96, p. 519–542, 2001.

[LEU 93] LEURGANS S., MOYEED R., SILVERMAN B., "Canonical correlation analysis when the data are curves", *Journal of the Royal Statistical Society, B*, vol. 55, p. 725–740, 1993.

[MAL 00] MALLET Y., COOMANS D., DE VEL O., "Application of adaptative wavelets in classification and regression", WALCZAK B., Ed., *Wavelets in Chemistry. Series data handling in Science and Technology*, p. 437–456, 2000.

[MAR 96] MARX B., EILERS P., "Generalized linear regression on sampled signals with penalized likelihood", FORCINA A., MARCHETTI G., R. H., GALMACCI G., Eds., *11th International workshop on Statistical Modelling*, 1996.

[MAR 99] MARX B., EILERS P., "Generalized linear regression on sampled signals and curves: a *P*-spline approach", *Technometrics*, vol. 41, p. 1–13, 1999.

[MAS 07a] MAS A., "Weak convergence in the functional autoregressive model,", *Journal of Multivariate Analysis*, vol. 98, p. 1231–1261, 2007.

[MAS 07b] MAS A., PUMO B., "The ARHD model", *Journal of Statistical Planning and Inference*, vol. 137, p. 538–553, 2007.

[MUE 05] MUELLER H.-G., STADTMUELLER U., "Generalized functional linear models", *Annals of Statistics*, vol. 33, p. 774–805, 2005.

[PRE 05] PREDA C., SAPORTA G., "PLS regression on a stochastic process", *Computational Statistics & Data Analysis*, vol. 48, p. 149–158, 2005.

[PUM 92] PUMO B., Estimation et prévision de processus fonctionnels autorégressifs: application aux processus à temps continu, PhD thesis, University of Paris VI, 1992.

[RAM 91] RAMSAY J., DALZELL C., "Some tools for functional data analysis, with discussion", *Journal of the Royal Statistical Society, B*, vol. 53, p. 539–572, 1991.

[RAM 96] RAMSAY J., "Principal differential analysis: data reduction by differential operators", *Journal of the Royal Statistical Society Ser. B*, vol. 58, p. 495–508, 1996.

[RAM 98] RAMSAY J. O., LI X., "Curve registration", *Journal of the Royal Statistical Society, Series B,*, vol. 60, p. 351–363, 1998.

[RAM 00] RAMSAY J., "Differential equation models for statistical functions", *Canadian Journal of Statistics*, vol. 28, num. 1, p. 225–240, 2000.

[RAM 02] RAMSAY J., SILVERMAN B., *Applied Functional Data Analysis*, Springer, New York, 2002.

[RAM 05] RAMSAY J., SILVERMAN B., *Functional Data Analysis*, Springer, New York, 2nd edition, 2005.

[RIC 01] RICE J., WU C., "Nonparametric mixed effects models for unequally sampled noisy curves", *Biometrics*, vol. 57, p. 253–259, 2001.

[ROS 06] ROSSI F., VILLA N., "Support vector machine for functional data classification", *Neurocomputiong*, vol. 69, p. 730–742, 2006.

[ROS 08] ROSSI F., VILLA N., "Recent advances in the use of SVM for functional data classification", DABO-NIANG S., FERRATY F., Eds., *Functional and Operatorial Statistics*, Physica Verlag, Heidelberg, p. 273–280, 2008.

[WAH 90] WAHBA G., *Spline Models for Observational Data*, SIAM, 1990.

[YAO 03] YAO A. F., FERRÉ L., "Functional sliced inverse regression analysis", *Statistics*, vol. 37, p. 475–488, 2003.

Chapter 6

Discriminant Analysis

6.1. Introduction

We consider a population of statistical units where the output is a qualitative variable Z with g modalities and the vector of inputs is $X^1, \ldots, X^p$. The dependent variable Y defines g classes $C_1, \ldots, C_g$, often denoted as *a priori* classes. The discriminant analysis problem, also known as *classification* or *supervised learning statistical pattern recognition*, is to predict and sometimes to explain the output from the d inputs, often called the predictors. Thus, discriminant analysis is the set of statistical methods aiming to predict the output of an individual based on its values on the d predictors.

In fact, the precise purpose of discriminant analysis depends on the application context. In most cases, a good prediction is of primary interest and the aim is to minimize the misclassification rate for unlabeled observations. In some cases, however, providing a relevant *interpretation* of the classes from the predictors could be important since prediction is not really the point. In addition, the prediction problem is embedded in a complex decision process for which statistical learning is a simple element to control to help a human diagnosis.

Thus, discriminant analysis methods may be classified according to the relative importance they gave to the prediction or interpretation focus of the analysis. It is noteworthy that these two focuses can be somewhat contradictory.

Roughly speaking, three large families of methods exist:

Chapter written by Gilles CELEUX.

– Statistical methods are often older and consider both aspects of discriminant analysis. They are based on a probabilistic model in the statistical decision theory framework.

– Non-parametric methods focus on the prediction problem. They are seldom based on a probabilistic model. They lead to a 'black box' decision rule aiming to minimize the misclassification error rate. Neural networks, or support vector machines (SVM), are examples of this kind of approach.

– Decision tree methods favor a clear interpretation and lead to simple and readable decision rules.

In this chapter, we essentially consider the decisional face of discriminant analysis. It is actually the most important and difficult problem to deal with. In this framework, after providing a detailed description of particular methods, section 6.2 covers the many steps to be considered when facing the problem of designing a classification rule: understanding the basic concepts; defining the datasets; choosing a method; selecting the predictive variables; and assessing the classification error rate. In section 6.3, five standard classification methods are presented. All of them are representative of a particular point of view to deal with a classification problem. In section 6.4, some special questions on recent developments in classification methodology are considered.

6.2. Main steps in supervised classification

6.2.1. *The probabilistic framework*

To predict to which of g classes $G_1, \ldots, G_g$ an object described with p predictive variables belongs, a *learning* set

$$A = \{(\boldsymbol{x}_i, z_i), \ldots, (\boldsymbol{x}_n, z_n), \boldsymbol{x}_i \in \mathbb{R}^p \text{ and } z_i \in \{1, \ldots, g\}\},$$

is employed where vector $\boldsymbol{x}_i$ gives the values taken by object i for the p predictive variables and z_i the label to which it belongs. For this statistical decision problem, the learning sample A is to be used to design a classification function $\delta(\mathbf{x})$, relating every $\mathbb{R}^p$ vector to one of the g classes in order to minimize the error risk in a sense which is now specified.

We define the prior probabilities of the classes $p(G_1), \ldots, p(G_g)$, the class-conditional densities of the predictive variables $p(\boldsymbol{x}|G_1), \ldots, p(\boldsymbol{x}|G_g)$ and the mis-classification costs of an object from class G_k in class G_ℓ $c(\ell|k)$ for $\ell, k = 1, \ldots, g$, with $c(k|k) = 0$. Any decision function δ defines a partition of $\mathbb{R}^p, (D_1, \ldots, D_g)$ with $D_k = \{\boldsymbol{x} \in \mathbb{R}^p / \delta(\boldsymbol{x}) = k\}$.

The mean risk of the decision function δ can be written

$$R(\delta) = \sum_{k=1}^{g} p(G_k) \sum_{\ell=1}^{g} c(\ell|k) \int_{D_\ell} p(\boldsymbol{x}|G_k) d\boldsymbol{x}.$$

The optimal decision function δ^*, referred to as *Bayes decision rule*, is minimizing $R(\delta)$. For instance, for a two-class problem $(g = 2)$, it takes the form

$$\delta^*(\boldsymbol{x}) \;=\; 1 \iff c(2|1)p(G_1)p(\boldsymbol{x}|G_1) \geq c(1|2)p(G_2)p(\boldsymbol{x}|G_2)$$
$$\delta^*(\boldsymbol{x}) \;=\; 2 \iff c(2|1)p(G_1)p(\boldsymbol{x}|G_1) < c(1|2)p(G_2)p(\boldsymbol{x}|G_2).$$

With equal misclassification costs, this leads to a vector $\boldsymbol{x}$ being assigned to the class providing the greatest conditional probability $p(G_k|\boldsymbol{x}) \propto p(G_k)p(\boldsymbol{x}|G_k)$.

Choosing the misclassification cost is not concerned with the statistical analysis of the learning sample. According to the sampling scheme, it may be possible to estimate the prior probabilities $p(G_k)$ from A. (We are concerned with this question in the next section.)

The main statistical task is to estimate the class-conditional densities $p(\boldsymbol{x}|G_k)$ or the conditional probabilities $p(G_k|\boldsymbol{x})$. The classification methods and models differ in their assumptions regarding the class-conditional densities or the conditional probabilities of the classes and on the procedures to estimate the features of those quantities.

6.2.2. *Sampling schemes*

First, it is important not to neglect the sampling scheme leading to the learning sample A. The two most employed sampling schemes are (1) the *mixture* sampling where A is sampled at random from the whole population under study and (2) the *retrospective* sampling where A is the concatenation of g independent sub-samples with fixed size n_k and with distribution $p(\boldsymbol{x}|G_k)$ for $k = 1, \ldots, g$.

With the mixture sampling, the x_is are the realizations of a mixture distribution

$$p(\boldsymbol{x}) = \sum_{k=1}^{g} p(G_k)p(\boldsymbol{x}|G_k).$$

This sampling scheme is the most natural and allows us to estimate the prior probabilities by the proportions n_k/n for $k = 1, \ldots, g$, where n_k denotes the number of objects arising from class G_k in sample A. However, it cannot be used in the common case where some classes, corresponding to rare and disastrous events (serious disease, material failure, cataclysm, etc.) have a very low prior probability.

Retrospective sampling is regularly used in medicine and is well suited for properly taking rare classes into account. However, it requires that class prior probabilities be known since they cannot be estimated from the fixed proportions n_k. More details on sampling schemes in discriminant analysis can be found in [CEL 94].

6.2.3. *Decision function estimation strategies*

The main statistical task to obtain a good approximation of the Bayes decision rule is to propose an estimation of the class-conditional densities $p(x|G_k)$ or a direct estimation of the conditional probabilities of the classes $p(G_k|x)$.

Methods concerned with the estimation of the class-conditional densities are included in the so-called *generative* approach in supervised classification. They differ by the assumptions they make on those densities. Parametric and non-parametric methods can be distinguished. Parametric methods assume that the class-conditional densities belong to a parametric family of densities. A standard choice is the Gaussian family of distributions, as for linear discriminant analysis (LDA) detailed in section 6.3.1. Non-parametric methods do not make such assumptions on the class-conditional densities and use local methods to estimate those distributions. The k nearest-neighbor method (section 6.3.3) is an example of a non-parametric generative method.

Methods concerned with the direct estimation of the conditional probabilities of the classes are included in the so-called *predictive* approach in supervised classification. Logistic regression (section 6.3.2) is a typical example of the predictive approach in discriminant analysis. It takes advantage of a linear logit property verified by many class-conditional densities (section 6.3.2). From this point of view, it can be regarded as a *semi-parametric* model. Other methods such as classification trees (CART) (section 6.3.4) work directly on the conditional probabilities of the classes without making any assumption about the class-conditional densities and are fully non-parametric.

Some other methods which model the class-conditional densities or the conditional probabilities of the classes will be presented. Those methods are aimed at finding directly a decision function with a fixed form. Examples of such methods are neural networks (e.g. [BIS 95]) and support vector machine (SVM) [VAP 96]. For instance, the Perceptron method provides a linear decision function without any reference to the Gaussian model. This method and the single hidden layer back-propagation network are described in 6.3.5. This latter method is the most widely used neural network providing non-linear decision functions.

6.2.4. *Variables selection*

Each of the p predictive variables brings information (its own ability to separate the classes) and noise (its sampling variance). It is important to select the variables bringing more discriminant information than noise. In practice, there are three types of variables:

1) useful variable: e.g. to separate men and women, the size is clearly a useful variable;

2) redundant variable: discriminant information is essentially provided by other variables e.g. weight does not give information about the sex of the person; and

3) irrelevant or noisy variables: eye color does not contain any information about the sex of a person.

Variable selection is generally an important step to obtain a reliable decision function in supervised classification. Some methods such as decision trees intrinsically achieve such a selection, but most classification methods need to be used with an additional variable selection procedure. Variable selection procedures make use of a criterion of selection and of an algorithm of selection.

6.2.4.1. *Selection criteria*

In order to define the most standard variable selection criteria, which are variance analysis (ANOVA) criteria, we need to define inertia matrices related to the classes $G_1, \ldots, G_g$. First we have the *total inertia matrix* $\mathbf{S}_T$:

$$\mathbf{S}_T = \frac{1}{n} \sum_{i=1}^{n} (\boldsymbol{x}_i - \overline{\boldsymbol{x}})(\boldsymbol{x}_i - \overline{\boldsymbol{x}})'$$

where $\overline{\boldsymbol{x}} = \frac{1}{n} \sum_{i=1}^{n} \boldsymbol{x}_i$ is the empirical mean of the whole sample. We then consider the *within-classes inertia matrix* $\mathbf{S}_W$ and the *between-classes inertia matrix* $\mathbf{S}_B$

$$\mathbf{S}_W = \frac{1}{n} \sum_{k=1}^{g} \sum_{i/z_i=k} (\boldsymbol{x}_i - \overline{\boldsymbol{x}}_k)(\boldsymbol{x}_i - \overline{\boldsymbol{x}}_k)',$$

where $\overline{\boldsymbol{x}}_k = \frac{1}{n_k} \sum_{i/z_i=k} \boldsymbol{x}_i$ is the empirical mean of class G_k with size n_k, and

$$\mathbf{S}_B = \sum_{k=1}^{g} \frac{n_k}{n} (\overline{\boldsymbol{x}}_k - \overline{\boldsymbol{x}})(\overline{\boldsymbol{x}}_k - \overline{\boldsymbol{x}})'.$$

It is worth noting that, from Huygen's Theorem, we obtain the important relation

$$\mathbf{S}_T = \mathbf{S}_B + \mathbf{S}_W. \tag{6.1}$$

Using these definitions, the most standard criteria are now described. When $g = 2$, the Mahalanobis distance D is used. The criterion to be maximized is the classes separation:

$$D^2 = ||\overline{x}_1 - \overline{x}_2||^2_{\mathbf{S}_W^{-1}} = (\overline{x}_1 - \overline{x}_2)\mathbf{S}_W^{-1}(\overline{x}_1 - \overline{x}_2)'.$$

When $g > 2$, an equivalent of the Mahalanobis distance, due to relation (6.1), is the Wilks' lambda:

$$\lambda = \frac{|\mathbf{S}_W|}{|\mathbf{S}_T|}.$$

The criterion to be minimized is the mean homogenity of a class.

As will be seen in section 6.3.1, those criteria are essentially well suited to LDA despite the fact they can be used in a more general context. Other criteria are equivalent to the Mahalanobis distance under the LDA assumption. An example of such criterion is the Kullback–Leibler distance defined:

$$\int \ln\left(\frac{p(x|G_2)}{p(x|G_1)}\right) p(x|G_2)dx + \int \ln\left(\frac{p(x|G_1)}{p(x|G_2)}\right) p(x|G_1)dx.$$

We will also see, in the logistic regression framework, that model-based classification methods allow us to select variables using specific statistical tests measuring the variable information with respect to the model at hand.

6.2.4.2. Selection procedures

The *filter* approach consists of selecting the variables in a preliminary step without taking account of the classification method at hand. Despite its simplicity, it is not generally recommended.

Conversely, *wrapper* methods which consist of using algorithms aiming to select variables minimizing the classification error rate, or a criterion related to a particular classification method, are the most widely used methods. In this setting, an exhaustive selection considering all possible ways of combining variables is prohibitive as soon as there are more than 10 variables. However, for two-class problems, it is possible to use an optimal *leaps and bounds* or *branch and bounds* algorithm with about 30 variables in input [FUR 74].

The most popular wrapper method makes use of a *forward* selection algorithm. At the first step, the variable optimizing the variable selection criterion is chosen, then the best couple of variables containing the first chosen variable is selected, etc.

This method is often sub-optimal, and a *backward* variable selection algorithm can be preferred despite the fact it is more costly in CPU time and memory. It starts from the whole set of variables. At the first step to eliminate the variable providing the largest deterioration of the criterion, the variable providing the largest deterioration of the criterion (knowing that the first variable has been discarded), is eliminated and so on.

Whatever the variable selection algorithm is, the procedure must be stopped in a proper way. This difficult task is clearly quite important since it aims to restrict the selection to informative variables and to eliminate redundant or noisy variables. First, it should be noted that this question cannot be treated independently of the classification method at hand. A widely used and highly recommended stopping rule consists of minimizing the cross-validated error rate of the associated decision function (section 6.2.5). The main drawback of this method is that it is painfully slow for classification methods and highly CPU time-consuming.

For generative (respectively, predictive) parametric methods, the likelihood of the model $\prod_{i=1}^{n} p(\boldsymbol{x}_i|G_{z_i})$ (respectively, $\prod_{i=1}^{n} p(G_{z_i}|\boldsymbol{x}_i)$) is available. It is possible to use penalized likelihood criteria such as the Akaike criterion (AIC) [AKA 74]:

$$AIC(J) = -2\ln(L(m(J))) + 2M(J)$$

where $L(m(J))$ is the model with J variables likelihood and $M(J)$ the number of independent parameters of this model. This criterion is asymptotically equivalent to cross-validation [STO 77] but it has a practical tendency to select models which are too complex. Thus, the *Bayesian information criterion* (BIC) of Schwarz [SCH 78]:

$$BIC(J) = -2\ln(L(m(J))) + M(J)\ln(n)$$

is preferred.

This criterion has a Bayesian background and is equivalent to the *minimum description length* of Rissanen [RIS 83] based on coding theory. Finally, it is worth highlighting that in the LDA context, it is possible to regard the variable selection problem as an analysis of a variance problem. It allows the definition of a stepwise forward selection algorithm, detailed in [CEL 91] or [MCL 92], based on the F statistic of Fisher which is equivalent to minimizing the Wilks lambda criterion. This algorithm is available in most statistical packages including discriminant analysis programs. Its great advantage is that it allows selected variables which lose their discriminative efficiency to be eliminated, with the introduction of new variables.

6.2.5. *Assessing the misclassification error rate*

Supervised classification aims to provide a decision function (a classifier) to predict the unknown labels of future observations. It is therefore important to assess the expected misclassification error rate when this classifier, based on a finite learning sample A, is used on a population of potentially infinite size. In this section, techniques to estimate this expected misclassification rate are reviewed.

First, it is important to state precisely the misclassification error rate to be optimized. In many situations, the classification problem involves separating a high prior probability class of normal objects and a low prior probability risk class of abnormal objects. In such cases, minimizing the unconditional error rate is of little interest. The main interest lies in maximizing the *sensibility*, namely the probability of predicting risk class, given that true class is risk class. In other words, the error rates to be minimized are conditional error rates. However, in what follows, the misclassification costs of the classes are assumed to be equal. This simplification allows us to restrict attention to the estimation of the unconditional error rate.

We first point out two wrong methods of assessing this misclassification error rate:

– Estimation by the *apparent error rate*: this consists of estimating the misclassification error rate of the decision function δ by the error rate calculated for the learning set A. Acting in such a way, we obtain too optimistic an error rate since the objects are used twice to build the decision function and to assess its performance. This optimistic bias is larger for smaller A and for more complex classification methods (e.g. non-parametric or parametric with many parameters).

– Estimation by a *theoretical* computation: under the assumptions of the model used to design the decision function δ, it is often easy to derive a closed form estimation of the error rate by inserting the parameter estimation in the model. An example of this way of estimating the misclassification error rate is presented in 6.3.1. Not only are such computations rarely possible, but they also involve an optimistic bias of the error rate since they do not take the sampling fluctuation into account or question the model relevance.

One of the simplest methods to assess the misclassification error rate of a classifier consists of estimating it with a *test sample*. The misclassification error rate is computed on a test sample T independent of the learning sample A. The sample T has to be of reasonable size (at least 25% of the size of A) and has to be randomly obtained from the initial sample. This method involves no optimistic bias but cannot be recommended. It requires a large sample size, since the objects in T are not used in building the classifier δ. Moreover, the test sample method does not allow assessment of the variance of the misclassification error rate.

The most widely used methods to assess the misclassification error rate of a classifier are resampling techniques such as *bootstrap* and *cross-validation*.

Cross-validation in its most common version is known as *leave-one-out*. For $i = 1, \ldots, n$, the classifier is designed from the learning sample A except the ith; the ith element is assigned to one of the classes with this classifier. The error rate is then estimated by the frequency of misclassified points using this procedure. The resulting estimated error rate is almost unbiased. However, its variance is larger for larger n since in these cases, the different classifiers designed with $n-2$ common observations have a strong tendency to be the same. Moreover, leave-one-out is costly for large n even if it is often possible to derive the $(n-1)$th decision function from the nth decision function by simplified calculations.

Thus v-fold cross-validation is often preferred to leave-one-out. It consists of dividing at random the learning sample A into v parts of approximately equal sizes. For $\ell = 1, \ldots, v$, a classifier is designed from A with the ℓ part of the data removed. The ℓ part of the data is assigned to one of the classes with this classifier. If $v = n$, v-fold cross-validation is *leave-one-out*. Conversely, if $v = 2$, v-fold cross-validation is known as *half sampling* where the whole sample is separated into two equal parts used as learning and test samples alternatively.

The choice of v is sensitive. The larger v is, the smaller the estimation bias but the greater the variance. Empirical studies show that when n is small, $v = n$ could be preferred. When n is large, $L = 2$ could be preferred. In many cases where A is of reasonable size, choosing v around 5–10 is sensible. Moreover, for small values of v, it is possible to use a Monte Carlo version of the v-fold procedure.

The *bootstrap* technique replaces the unknown probability distribution of the data with the empirical distribution of the learning sample A. It builds a B (typically $B = 100$) bootstrap sample of size n from A. For every bootstrap sample b, two error rates are obtained: the apparent error rate on b and an estimation of the error rate on the whole sample A. The difference between these two error rates is an estimation of the optimism bias of the apparent error rate on b. The mean O_B of this optimism bias on the B bootstrap samples is the estimator of the optimism bias of the apparent error rate E_A computed on A. Finally, the bootstrap estimate of the misclassification error rate is

$$E_B = E_A + O_B.$$

The bootstrap method can be computationally expensive and provide a biased estimation of the error rate since bootstrap samples and A have common observations. Better bootstrap estimates using the so-called '0.632 bootstrap' have been proposed to remove this bias [EFR 97]. Those procedures are mimicking cross-validation and we see no reason to prefer bootstrap to cross-validation. In practice, the two techniques give similar results and the conception of cross-validation is simpler.

6.2.6. *Model selection and resampling techniques*

There are larger numbers of classification methods using tuning parameters: width window, number of neighbors, regularization parameters, etc. Moreover, the performances of methods depend on the number of chosen variables, number of nodes of a tree-classifier, complexity of a neural network, etc. Some examples of these different situations are provided in the following sections. It is quite important to choose those independent tuning parameters in the correct way in order to minimize the expected misclassification rate. Again, the problem here is to solve the bias–variance dilemma.

For this purpose, it would be ideal to obtain a learning sample A and a *validation* sample R to choose the tuning parameters, and also to select a model in a large family of models and a test sample to assess the misclassification error rate. If the initial dataset is quite large, it is possible for instance to use 50% of the whole dataset for A and 25% each for R and T.

In general, it is necessary to use resampling techniques (especially cross-validation) to properly achieve the validation step (model selection) and test step (assessing the misclassification error rate of the selected model).

The general procedure is as follows. Among the models competing to design the classifier and which differ by some parameter linked to their complexity, we choose the model which provides the smallest estimated error rate using a resampling technique as cross-validation. It is important to remember that the expected error rate of the selected model remains to be assessed. The available error rates are too optimistic. To assess the actual error rate, there is the need to use cross-validation twice. It is used the first time to select a model or to choose a tuning parameter. The second time, it is used to assess the misclassification error rate.

This embedded way to use cross-validation is costly but necessary to assess the final classifier in a reliable way. We insist on the fact that using a single cross-validation procedure to select a model and assess its error rate in the same exercise would produce an over-optimistic estimation of the actual error rate of the classifier.

6.3. Standard methods in supervised classification

Obviously, there are numerous methods in discriminant analysis with different points of view. Moreover, the research in classification is very active because there are a lot of potential applications in many domains (e.g. medicine, biology, pattern recognition, marketing and engineering).

It would be impossible to review all the techniques of discriminant analysis in this chapter. The reader is referred to the following books among others for a detailed description of classification methods: [MCL 92] proposes a complete presentation of statistical classification methods; [HAN 96] and [RIP 95] present the statistical methods and neural networks in a statistical perspective; [BIS 95] focuses on neural networks and [BIS 06] proposes a comprehensive panorama of machine learning techniques; [HAS 01] proposes a statistical perspective (data mining); the book of [SCH 02] can be recommended for an extensive presentation of kernel methods in machine learning including SVM; [DEV 96] proposes a presentation of classification emphasizing its mathematical aspects; [CEL 94] focuses on discriminant analysis for discrete data; and [HAN 01] gives a presentation of classification in a data mining perspective.

In this chapter, we simply give a synthetic presentation of five of the most widely used classification methods, each representing a particular point of view of discriminant analysis:

 – *linear discriminant analysis* (LDA) is a generative parametric method;

 – *logistic regression* is a predictive semi-parametric method;

 – *k nearest neighbors methods* is a generative non-parametric method;

 – *classification tree* is a predictive non-parametric method providing readable decision functions; and

 – *single hidden layer back-propagation network* is the most widely used neural network method.

6.3.1. *Linear discriminant analysis*

This method assumes that the class-conditional densities $p(\boldsymbol{x}|G_k), k = 1, \ldots, g$ are Gaussian distributions $\mathcal{N}(\boldsymbol{\mu}_k, \boldsymbol{\Sigma})$ with means $\boldsymbol{\mu}_k$ and a common covariance matrix $\boldsymbol{\Sigma}$:

$$p(\boldsymbol{x}|G_k) = \frac{1}{(2\pi)^{d/2}|\boldsymbol{\Sigma}|^{1/2}} \exp\left(-\frac{1}{2}(\boldsymbol{x} - \boldsymbol{\mu}_k)'\boldsymbol{\Sigma}^{-1}(\boldsymbol{x} - \boldsymbol{\mu}_k)\right).$$

The Bayes decision function therefore leads to linear separation between classes. For instance, the Bayes classifier for $g = 2$ and equal misclassification costs is

$$\delta(\boldsymbol{x}) = 1 \iff (\boldsymbol{x} - \frac{\boldsymbol{\mu}_1 + \boldsymbol{\mu}_2}{2})'\boldsymbol{\Sigma}^{-1}(\boldsymbol{\mu}_1 - \boldsymbol{\mu}_2) \geq 0.$$

The parameters of the Gaussian distributions are estimated by their empirical values

$$\hat{\boldsymbol{\mu}}_k = \overline{\boldsymbol{x}}_k, \quad k = 1, \ldots, g$$

and

$$\hat{\boldsymbol{\Sigma}} = \frac{\sum_{k=1}^{g} \sum_{i/z_i=k} (\boldsymbol{x}_i - \overline{\boldsymbol{x}}_k)(\boldsymbol{x}_i - \overline{\boldsymbol{x}}_k)'}{n - g}.$$

Note that in order to obtain an unbiased estimator of $\boldsymbol{\Sigma}$, $\hat{\boldsymbol{\Sigma}}$ is not exactly equal to $\mathbf{S}_W$.

It is worth noting that, assuming that the class covariance matrices are free, we have *quadratic discriminant analysis* (QDA) whose name comes from the fact that this method provides quadratic separation equations between classes. In this case, the class covariance matrices $\Sigma_k, k = 1, \ldots, g$ are estimated by

$$\hat{\Sigma}_k = \frac{\sum_{i/z_i=k}(\boldsymbol{x}_i - \overline{\boldsymbol{x}}_k)(\boldsymbol{x}_i - \overline{\boldsymbol{x}}_k)'}{n_k - 1}.$$

LDA is available in much statistical and machine learning software programs and is one of the most widely used classification methods. Reasons for its success include the following:

– LDA provides a good compromise between relevance and complexity; in other words, it often allows us to solve the bias–variance dilemma in a satisfactory way. LDA often outperforms QDA which is dependent upon a much larger number of parameters. For instance, for $p = 10$ and $g = 4$, LDA requires 95 parameters to be estimated while QDA requires 260 parameters.

– In the LDA framework, variable selection can be achieved in an almost optimal way by using F Fisher statistics [CEL 91, MCL 92]. As a matter of fact, standard variable selection criteria are theoretically justified under the LDA assumptions (Gaussian distributions and common class covariance matrices). Thus, when $g = 2$, it can be shown under the LDA assumptions that the optimal misclassification probability is $\Phi(-\Delta/2)$, Φ being the distribution function of a standard Gaussian distribution with mean 0 and variance 1 and where Δ is the Mahalanobis distance between the two classes:

$$\Delta^2 = ||\boldsymbol{\mu}_1 - \boldsymbol{\mu}_2||_{\Sigma^{-1}} = (\boldsymbol{\mu}_1 - \boldsymbol{\mu}_2)'\Sigma^{-1}(\boldsymbol{\mu}_1 - \boldsymbol{\mu}_2).$$

– LDA provides *stable* (small sampling variances) and *robust* (i.e. not too sensitive to reasonable deviations from the normality and homoscedastic assumptions) performances.

For these reasons, LDA should be considered as a reference classification method.

6.3.2. *Logistic regression*

Instead of modeling class-conditional densities, logistic regression models the conditional probabilities of the classes. This method is described in the $g = 2$ classes case and its generalization to the general case is sketched. Moreover, this presentation of logistic regression is restricted to linear logistic regression which is the most widely logistic regression method.

Setting $\Pi(\boldsymbol{x}) = p(G_1|\boldsymbol{x})$, the logistic regression equation is

$$\text{logit}(\Pi(\boldsymbol{x})) = \ln\left(\frac{\Pi(\boldsymbol{x})}{1 - \Pi(\boldsymbol{x})}\right) = \beta_0 + \sum_{j=1}^{p}\beta_j\mathbf{x}^j.$$

In this section, the label indicator variable z is often replaced by the binary variable y defined by $y = 1$ if $z = 1$ and $y = 0$ if $z = 2$. The model, described by an equation that we refer to as the *linear logit equation*, involves lighter assumptions than parametric assumptions on the class-conditional densities. Many parametric models of discriminant analysis fit the linear logit equation. A sufficient condition for this is that the joint density $p(\boldsymbol{x}, y)$ has the form [VEN 98]:

$$p(\boldsymbol{x}, y) \propto \exp\left(\sum_{j=1}^{p} \beta_j x^j y + \beta_0 y + \sum_{r=1}^{R} \gamma_r \phi_r(\boldsymbol{x})\right)$$

for integer R, scalars γ_r and functions ϕ_r. This form is verified by e.g. the multivariate Gaussian distribution (for quantitative predictors) and for the multivariate multinomial distribution (for qualitative predictors). This is why logistic regression is considered as a semi-parametric model.

Linear logistic regression is part of the family of generalized linear models [MCC 89] where the linear dependence between the dependent variable ($\Pi(\boldsymbol{x})$ here) and the predictors is described with a *link function*, namely the logit function. Note that other link functions are also possible, such as the *probit function* which is the inverse Φ^{-1} of the distribution function of a standard Gaussian distribution with mean 0 and variance 1. The resulting probit model leads, in practice, to the same results as the logistic regression.

Using the convention $x^0 = 1$ in the rest of this section, it is straightforward to derive from the linear logit equation that

$$\Pi(\boldsymbol{x}|\boldsymbol{\beta}) = \frac{\exp(\sum_{j=0}^{p} \beta_j x^j)}{1 + \exp(\sum_{j=0}^{p} \beta_j x^j)}.$$

The model parameters are $\boldsymbol{\beta} = (\beta_0, \beta_1, \ldots, \beta_p)$. In general, they are estimated by maximizing the conditional loglikelihood for sample A, defined by

$$\ell(\boldsymbol{\beta}) = \sum_{i=1}^{n} \ln p(G_{z_i}|\boldsymbol{x}_i) = \sum_{i=1}^{n} \ln p(y_i|\boldsymbol{x}_i).$$

From this, it follows that

$$\ell(\boldsymbol{\beta}) = \sum_{i=1}^{n} \left(y_i \boldsymbol{\beta}' \boldsymbol{x}_i - \ln(1 + e^{\boldsymbol{\beta}' \boldsymbol{x}_i}) \right),$$

where $y_i = 1$ if $z_i = 1$ and $y_i = 0$ if $z_i = 2$. The corresponding likelihood equations are $p + 1$ non-linear equations

$$\sum_{i=1}^{n} (y_i - \Pi(\boldsymbol{x}_i|\boldsymbol{\beta}))\boldsymbol{x}_i = 0,$$

that can be solved with a Newton–Raphson algorithm. It is important to note that this maximization of the conditional likelihood leading to simple equations is valid, despite a maximization of the joint likelihood related to a retrospective or a mixture sampling which appears more natural. This is due to a useful property of logistic regression: the three likelihoods are equivalent except when estimating the constant term f β_0 (see [PRE 79] or [CEL 94, sections 3.4, 3.7]). More precisely, denoting the constant term for the retrospective and mixture sampling by β_0^R and β_0^M, respectively, we have the relations

$$\beta_0^R = \beta_0 + \ln \frac{\pi_1}{\pi_2}$$

and

$$\beta_0^M = \beta_0 + \ln \frac{n_1}{n_2}.$$

Logistic regression takes advantage of the generalized linear model tools. In particular, it allows tests of the form $\beta_j = 0$ to be used to select variables. It also allows goodness of fit tests (such as likelihood ratio tests, Wald tests, etc.) to be used. Several well-documented tests (e.g. [CEL 94, MCC 89]) are not described here. The reader seeking a detailed presentation of logistic regression, including its practical and numerical aspects, is referred to the book [HOS 89].

Generalizing logistic regression to the case where $(g > 2)$ is carried out as follows. A reference class is considered, say G_1, and $g - 1$ logistic regression G_1 versus G_k, $k = 2, \ldots, g$ are performed. Denoting these logistic regression equations $(1 < k \leq g)$ by $e_k(\boldsymbol{x})$, we have

$$e_k(\boldsymbol{x}) = \beta_{k0} + \beta_{k1} x^1 \ldots + \beta_{kd} x^p.$$

The logistic regression equation of G_r versus G_s is obtained by the formula

$$e_{rs}(\boldsymbol{x}) = e_r(\boldsymbol{x}) - e_s(\boldsymbol{x}).$$

Hence, this procedure is not dependent upon the reference class and it can be shown easily that

$$p(G_k|\boldsymbol{x}) = \frac{\exp[\beta_{k0} + \ldots + \beta_{kp} x^p]}{1 + \sum_{\ell=2}^{g} \exp[\beta_{\ell 0} + \ldots + \beta_{\ell p} x^p]}, \quad k = 2, \ldots, g$$

and

$$p(G_1|\boldsymbol{x}) = \frac{1}{1 + \sum_{\ell=2}^{g} \exp[\beta_{\ell 0} + \ldots + \beta_{\ell p} x^p]}.$$

Logistic regression has many advantages. As stated above, it provides a linear classifier without the need for strong parametric assumptions. Since it directly estimates the conditional probabilities of the classes, it allows quantitative and qualitative predictors to be dealt with in the same exercise. Logistic regression is widely used in medicine because it allows the *odds-ratio* for binary predictors to be considered. These odds-ratios can be interpreted as a multiplicative risk augmentation. (This characteristic is also interesting in domains such as marketing and credit-scoring.) Assume that class G_1 is the risk class. Its prior probability is therefore small and, for all $\boldsymbol{x}$, $\Pi(\boldsymbol{x})$ is small. Setting

$$\Pi_j^\ell(\boldsymbol{x}) = \Pi(\boldsymbol{x}), \text{ if } x^j = \ell,$$

the odds-ratio for variable j is

$$ODD(j) = \exp(\beta_j) = \frac{\Pi_j^1(\boldsymbol{x})}{\Pi_j^0(\boldsymbol{x})} \frac{1 - \Pi_j^0(\boldsymbol{x})}{1 - \Pi_j^1(\boldsymbol{x})} \approx \frac{\Pi_j^1(\boldsymbol{x})}{\Pi_j^0(\boldsymbol{x})}$$

for $\Pi_j^1(\boldsymbol{x})$ and $\Pi_j^0(\boldsymbol{x})$ small. In other words, individuals with variable $j = 1$ have a probability of belonging to risk class $ODD(j)$ which is greater than for the individuals for which variable $j = 0$.

Linear logistic regression therefore has an important place among classification methods leading to linear classifiers. As for LDA, linear logistic regression is expected to provide stable and robust decision functions. From a theoretical point of view, linear discriminant analysis and linear logistic regression are different. From a practical point of view, however, these methods have a strong tendency to provide analogous performances.

Finally, we want to stress that, as Gaussian classifiers, logistic regression can lead to non-linear classifiers. It is possible to consider a quadratic logit $\Pi(x)$, a higher degree logit or even a non-parametric logit using a basis of spline functions [KOO 97]. The possibility to extend logistic regression to non-linear situations could be of great interest for taking non-linear relations between qualitative predictors into account.

6.3.3. *The K nearest neighbors method*

We now consider non-parametric methods which do not assume any particular form of the class-conditional densities. In this framework, the well-documented kernel method, one of the most widely used methods for non-parametric estimation of a density function, can be considered [SIL 86]. In this method, the class-conditional densities are estimated by smoothing:

$$p(\boldsymbol{x}|G_k) = \frac{1}{n_k h^p} \sum_{i/z_i=k} \text{Ker}(\frac{\boldsymbol{x} - \boldsymbol{x}_i}{h}),$$

where Ker is a probability density (Gaussian, uniform, etc.) and h, the window width, is a sensitive smoothing parameter. If h is too small, the estimated density will have too many modes. If h is too large, the density will be too smooth. The method of choosing the window width h has received much attention and many strategies have been proposed [SIL 86]. Among them, a good strategy consists of choosing the value of h to maximize the cross-validated likelihood. As for all non-parametric methods, this method needs large sample sizes. It is rarely employed in the discriminant analysis context. A reason for this lack of practical success of kernel density estimation for classification is that it involves numerical difficulties in high dimension contexts (large p) which are common in classification.

In the classification framework, the old K nearest neighbors method [FIX 51] is more widely used. This method has the same local point of view than the kernel method. In its simplest version, it can be summarized as follows. For any vector x to be classified, its K nearest neighbors in the learning sample A are determined and x is assigned to the majority class among those K neighbors.

This strategy is equivalent to computing a local estimation of the class-conditional densities around a small neighborhood with volume V of x. Indeed, when the learning sample A has been obtained using the mixture sampling, we have

$$p(G_k|x) = \frac{p(x|G_k)p(G_k)}{p(x)} \approx \frac{(a_k/(n_k V))(n_k/n)}{a/(nV)} = \frac{a_k}{a},$$

where a is the number of points belonging to A in the small neighborhood and a_k is the number of points belonging to A and G_k in this small neighborhood.

When the learning sample A has been obtained using the retrospective sampling, we have

$$p(G_k|x) \approx \frac{a_k}{a} \frac{n}{n_k} \pi_k.$$

This formula leads to a modification of the majority rule of the K nearest neighbors method.

Despite the K nearest neighbors estimate of a class-conditional density $p(x|G_\ell)$, where $1 \leq \ell \leq g$ is not a density, it can be shown that it is consistent if K tends to infinity with n_ℓ and if k/n_ℓ tends to 0 [DEV 96].

Clearly, choosing K is a sensitive task. A good and simple strategy is to choose K to minimize the leave-one-out error rate. On the other hand, the choice of the distance could also be sensitive. It is difficult to give a simple guideline for this choice, however. In many cases, the chosen distance is the standard Euclidean distance.

In order to be efficient and provide stable classifiers, the K nearest neighbors method needs large sample sizes. For this very reason, the basic K nearest neighbors method can appear to be painfully slow. Many authors have therefore proposed algorithms to accelerate the K nearest neighbors research.

In practice, this method is essentially efficient when the boundaries between the classes are highly non-linear. The reader interested by non-parametric methods in classification is referred to the book [DEV 96] which contains a detailed presentation of asymptotic properties of non-parametric estimators of class-conditional densities.

6.3.4. *Classification trees*

We now consider non-parametric classification methods which emphasize building simple and readable classifiers. Despite the aim to obtain interpretable decision functions, those methods are quite different from descriptive discriminant analysis methods which are essentially exploratory analysis tools. Factor discriminant analysis is essentially a principal component analysis of the class centers in the space $\mathbb{R}^p$ with the Mahalanobis distance $\mathbf{S}_W^{-1}$ [HAS 01, SAP 90].

Favoring readability of a classifier, the classification trees methods are quite attractive. Introduced in the 1950s, they had almost disappeared from statistical software programs when the book of Breiman *et al.* was published [BRE 84]. This book answered many limitations of previous classification methods and, in particular, they proposed a powerful pruning strategy in their software CART (classification and regression trees) using cross-validation. Today, CART (or related software) has become a standard *data mining* tool. Figure 6.1 is a fictional illustration of the classifier that can be obtained with a classification tree to separate, for instance, 'good' and 'bad' customers to repay a bank loan.

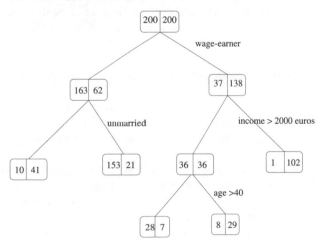

Figure 6.1. *Classification tree example*

To build such a classification tree, a set of binary questions has to be considered. These questions are formed directly from the p predictors and are characterized with a cut c. They are of the form '$x > c$?' if predictor x is quantitative or 'is the characteristic c present?' if predictor x is qualitative. The classification power of a cut c at the node t of the tree is measured with a *classification criterion*. Moreover, a *stopping rule* for building the tree is to be defined as a *classification rule* applied to one of the classes for the terminal nodes. This latter classification rule simply consists of assigning a terminal node to the class providing the largest empirical conditional probability $p(k|t), 1 \leq k \leq g$. Typically, it is a majority rule which can be weighted for unequal misclassification costs for the classes.

The classification criterion of a cut is based on the impurity of a node. The most widely used impurity measures are the entropy

$$i(t) = -\sum_{k=1}^{g} p(k|t) \log p(k|t)$$

and Gini criterion

$$i(t) = -\sum_{k \neq \ell} p(k|t)p(\ell|t).$$

The classification criterion of a cut is then the impurity reduction of the node

$$\Delta i(t) = i(t) - p_l i(t_l) - p_r i(t_r),$$

where p_l and p_r are the empirical probability (weight) of falling to the left and right node, respectively. It is worthwhile mentioning that, in practice, the choice of the classification criterion is not very sensitive. The important point is that the impurity measure is a concave function to ensure a positive decrease of the impurity for each cut.

For $g > 2$, two strategies are possible to design the classifier: either the impurity is decreased for the classes at each node or a 'one-against-one' rule is used. For each node, the g classes are grouped into two clusters in order to maximize the decrease in impurity. There is no sensitive difference in the practical performances of each strategy, but the one-against-one strategy is somewhat easier to interpret.

The main point when designing a classification tree consists of pruning the big tree built in the first step of the algorithm, without any stopping rule, to resolve the bias variance dilemma of the final decision tree. The bad solution, which was the only available solution for a long time, consists of using a threshold defined *a priori* to stop a branch of the tree. The good strategy consists of building the tree without any stopping rule and to prune this big tree with a test sample or by using a cross-validation procedure. The much more efficient procedure proposed in [BRE 84] is as follows. They define the cost-complexity of a tree T by

$$C_\alpha(T) = R(T) + \alpha|T|$$

where $R(T)$ is the apparent misclassification cost of T computed on the learning sample A, $|T|$ is the number of terminal nodes of T and α is a non-negative scalar. For any α, there exists a tree with minimal cost-complexity.

If, as described in section 6.2.6, a test sample independent of the learning sample A is available, it is possible to determine from $\alpha = 0$ a sequence of J nested trees with minimal cost-complexity for an increasing sequence $\alpha_1, \ldots, \alpha_J$. One tree among the J trees is then selected by minimizing the misclassification error rate of the test sample.

If such a test sample is not available (when n is relatively small), the pruning procedure is based on cross-validation. The learning sample A is divided into v parts of approximately equal sizes. v decision trees are then built on those v parts. A difficulty arises in the fact that those trees are not nested and a trick is used to circumvent it. The sequence $\alpha_1, \ldots, \alpha_J$ is computed from the big tree built on the whole learning sample A. The sequence $(\alpha'_j = \sqrt{\alpha_j \alpha_{j+1}})$ is then defined and the selected pruned tree is the tree whose complexity, chosen in the sequence (α'_j), minimizes the cross-validated error rate computed on the v trees. Details and variations of this pruning procedure can be found in [BRE 84] or [CEL 94, Chapter 7].

In practice, decision trees can be seen as a reference model if the focus is on deriving an easily interpretable classifier. They have many advantages; they provide simple and compact classifiers. The associated decision function is readable, non-parametric and uses conditional information in a powerful way. Decision trees are available for any kind of data and allow quantitative and qualitative predictors to be mixed. They propose intrinsically a variable selection. Finally, the classifiers they propose are insensitive to outliers and invariant by monotone transformation of the predictors. However, their limits are also well documented. They require a large sample size to produce classifiers with good performances.

Building a decision tree can be painfully slow because of the pruning step. Error rates of decision trees are rarely optimal because they do not take into account the linear relations between predictor. Their main limitation, however, is that they produce unstable decision functions since the choice of the first nodes is very important to determine the final structure of a tree. Recent techniques to dramatically reduce the instability of decision trees will be presented in section 6.4.3, where the price to be paid is the loss of readability.

6.3.5. *Single hidden layer back-propagation network*

Neural networks are a family of methods for regression and classification which has been developed in the machine learning and artificial intelligence communities rather than the statistical community. The starting point of view of neural networks is actually not probabilistic. These methods are not aiming to estimate the class-conditional densities or the conditional probabilities of the classes, but to directly design boundaries between the classes by minimizing a least squares criterion.

For instance, the Perceptron method, which is probably the oldest neural network technique, aims to find a linear separation between the classes. In the neural network framework [BIS 95] for the two class case, we can summarize as follows. Setting $x = (x^0, x^1, \ldots, x^p)$ with $x^0 = 1$, Perceptron aims to compute weights $\mathbf{w} = (w_0, \ldots, w_p)$ to design a classifier of the form:

$$\delta(x) = 1 \quad \text{if} \quad h\left(\sum_{j=0}^{p} w_j f_j(x^j)\right) = 1$$

$$\delta(x) = 2 \quad \text{if} \quad h\left(\sum_{j=0}^{p} w_j f_j(x^j)\right) = 0$$

where the function h, named activation function of the network, is the *Heaviside* function, defined

$$h(a) = \begin{cases} 0 & \text{if } a < 0 \\ 1 & \text{if } a > 0, \end{cases}$$

where the f_js have been fixed in advance to obtain linearly separable classes. In general, $f_j(x^j) = x^j$ is considered and a standard transformation such as the log transformation is used to make the classes linearly separable.

The loss function to be minimized is

$$E(\mathbf{w}) = \sum_{i \in C_A} \mathbf{w}' \mathbf{f}_i y_i$$

where C_A denotes the set of misclassified points in the learning sample A, $\mathbf{f}_i = (1, f_1(x_i^1), \ldots, f_p(x_i^p))$ and $y_i = 1$ if i belongs to G_1 and $y_i = -1$ if i belongs to G_2. The Perceptron algorithm is a sequential gradient algorithm:

$$\mathbf{w}^{r+1} = \mathbf{w}^r + \alpha(z_r - \delta_r)x,$$

where r denotes the iteration index, $z_r = 1, 2$ is the label of the observation presented to the neural network and $\delta_r = 1, 2$ is the class label to which this observation is assigned using the weights $\mathbf{w}^r$.

In general, $\alpha = 1$ and the Perceptron algorithm consists of doing nothing if a vector is well classified and to add (respectively, remove) this vector to the weight vector if it is wrongly assigned to class G_1 (respectively, G_2). If the classes are linearly separable, the Perceptron algorithm converges in a finite number of iterations and assigns all the vectors of A to their actual class. This optimality for the classification of the learning sample does not guarantee an optimal or even a good classification error rate for future vectors. If the classes are not linearly separable, the Perceptron algorithm does not converge. Several *ad hoc* modifications of the Perceptron algorithm have been proposed to ensure convergence in this case, e.g. allowing $\alpha \propto 1/r$.

The Perceptron is therefore able to derive an optimal hyperplane to separate the classes by using geometric arguments on A even when the assumptions of LDA described in section 6.3.1 are not verified, as soon as the classes are well separated. Conversely, the Perceptron is not guaranteed to find the optimal hyperplane of LDA when the classes are not well separated (no linear separation between classes).

Finally, the Perceptron does not appear as a versatile method for classification. By using an hidden layer and choosing the weights in a softer activation function than the Heaviside function, is is possible to obtain a neural network, greatly improving the classification performances of the Perceptron.

In the learning step, the introduction of a hidden layer allows us to obtain a vector $\mathbf{f} = (f_0, f_1, \ldots, f_p)$ which is data-dependent and to modify it along the iterations. Thus, the single hidden layer back-propagation network is a neural network of the form displayed in Figure 6.2, which allows us to obtain non-linear classifiers.

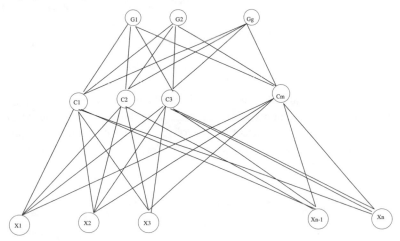

Figure 6.2. *Single hidden layer back-propagation network structure*

More precisely, it is based on the computation of g functions

$$y_k(\boldsymbol{x}) = \tilde{h}\left[\sum_{\ell=0}^{m}\beta_{k\ell}h\left(\sum_{j=0}^{p}\alpha_{\ell j}x^j\right)\right], \ k = 1,\ldots,g, \qquad (6.2)$$

where m is the number of nodes in the hidden layer. This leads to the decision function $\delta(\boldsymbol{x}) = k$ iff $k = \arg\max(y_k(\boldsymbol{x}))$. This neural network makes use of two vectors of adaptive weights and two activation functions. The output activation function $\tilde{h}$ could be the identity function as in regression, but in the classification context the *softmax* activation function is most widely used

$$\tilde{h}(a_k) = \frac{\exp(a_k)}{\sum_{\ell=1}^{g}\exp(a_\ell)}.$$

For the activation function h of the hidden layer, the single hidden layer back-propagation network uses a function which transforms a conditional probability distribution in a conditional probability distribution. It therefore replaces the Heaviside activation function by a *sigmoid* function

$$h(a) = \frac{1}{1 + \exp(-a)}$$

or by an analogous function $h(a) = \tanh(a)$. With these activation functions, a single hidden layer back-propagation network can produce complex non-linear separation between the classes. (The approximation error of such hidden layer neural networks decreases is $O(1/m)$.)

The weights $\mathbf{w} = (\beta_{k0},\ldots,\beta_{km}, k = 1,\ldots,g; \alpha_{\ell 0},\ldots,\alpha_{\ell p}, \ell = 1,\ldots,m)$ of this neural network have to be computed in order to minimize the quadratic loss function

$$R(\mathbf{w}) = \sum_{k=1}^{g}\sum_{i=1}^{n}\left(z_i^k - y_k(\boldsymbol{x}_i)\right)^2, \qquad (6.3)$$

where $z_i^k = 1$ if $i \in G_k$ and 0 otherwise. This can be done with the back-propagation gradient algorithm that we now describe. First, notice that an additive decomposition of $R(\mathbf{w}) = \sum_{i=1}^{n}R_i$ with respect to the observations of the learning sample is possible. Setting $\boldsymbol{\alpha}_k = (\alpha_{k0},\ldots,\alpha_{kp})$, $\boldsymbol{\beta}_k = (\beta_{k0},\ldots,\beta_{km})$, $c_{\ell i} = h(\boldsymbol{\alpha}_\ell'\boldsymbol{x}_i)$ and $\mathbf{c}_i = (c_{0i},\ldots,c_{mi})$, we have

$$\frac{\partial R_i}{\partial \beta_{k\ell}} = -2(z_i^k - y_k(\boldsymbol{x}_i))\tilde{h}'(\boldsymbol{\beta}_k'\mathbf{c}_i)c_{\ell i} = t_{ki}c_{\ell i}$$

and

$$\frac{\partial R_i}{\partial \alpha_{\ell j}} = -2\sum_{k=1}^{g}(z_i^k - y_k(\boldsymbol{x}_i))\tilde{h}'(\boldsymbol{\beta}_k'\mathbf{c}_i)h'(\boldsymbol{\alpha}_\ell'\boldsymbol{x}_i)x_i^j = s_{\ell i}x_i^j.$$

An iteration of the gradient algorithm is

$$\beta_{k\ell}^{(r+1)} = \beta_{k\ell}^{(r)} - \gamma_r \sum_{i=1}^{n} \frac{\partial R_i}{\partial \beta_{k\ell}^{(r)}}$$

$$\alpha_{\ell j}^{(r+1)} = \alpha_{\ell j}^{(r)} - \gamma_r \sum_{i=1}^{n} \frac{\partial R_i}{\partial \alpha_{\ell j}^{(r)}} \qquad (6.4)$$

where γ_r is a non-increasing sequence, which can be constant, but rather is proportional to $1/r$ to verify standard sufficient conditions of convergence of this kind of stochastic approximation algorithm. The t_{ki}s and $s_{\ell i}$s now verify the back-propagation equations:

$$s_{\ell i} = h'(\alpha_\ell' x_i) \sum_{k-1}^{g} t_{ki}. \qquad (6.5)$$

These equations lead to the following back-propagation algorithm. For fixed weights, the $y_k(x_i)$ are computed using equation (6.2), the t_{ki}s are then computed and the $s_{\ell i}$s are derived from back-propagation equations (6.5). Finally, the t_{ki}s and $s_{\ell i}$s are used to update the weights by formula (6.4).

Practical implementation of the single hidden layer back-propagation network is delicate, and requires heuristics to overcome many important difficulties. These difficulties and remedies are as follows:

– Practically, the gradient algorithm is very slow and other optimization algorithms such as the conjugate gradient algorithm is to be preferred [BIS 95].

– An efficient use of neural networks requires that we proceed to elementary statistical treatments (as standardization of the predictors, selection and often transformation of the predictors, etc.). Such previous statistical work is needed for a sensible use of neural networks and is helpful in choosing the initial weights in an efficient way.

– The choice of the initial weights is crucial for good performances. Choosing weights near zero could be a good choice to avoid degenerate solutions.

– According to its starting position, the single hidden layer back-propagation network can lead to quite different weights since the function to be minimized $R(\mathbf{w})$ is not convex. A simple way to limit the influence of starting conditions is to start from different positions and to linearly combine the different obtained classifiers [BIS 95]. In section 6.4.3, more sophisticated strategies to combine classifiers will be presented.

– The number of nodes m in the hidden layer is a sensitive tuning parameter. Typically, the larger the value of m, the more flexible the network becomes. However, a network which is too complex would imply a larger risk of *overfitting* on the learning sample. The ideal solution would probably consist of choosing m via cross-validation as described in section 6.2.6. Such an approach is not used in general because a cross-validation procedure would be too expensive for most datasets. Finally, m is often determined on a subjective ground.

– The main problem with the single hidden layer back-propagation network and, in general, with neural networks, lies in the fact that neural networks are, by their very nature, tools to approximate functions and not prediction tools in statistical decision problems. Consequently, they can often lead to overfitting classifiers. As stated above, overfitting is difficult to control with resampling tools such as cross-validation or bootstrap. Thus, regularization methods are used to avoid large variations in the neural network weights $w_0, \ldots, w_L$. Such a regularization is achieved by using a penalized loss criterion

$$\tilde{R}(\mathbf{w}) = R(\mathbf{w}) + \lambda \sum_{\ell=0}^{L} w_\ell^2$$

where λ is a parameter tuning the smoothing strength. The larger λ is, the smaller are the weights. This tuning parameter is generally chosen on subjective ground.

6.4. Recent advances

Discriminant analysis is a very active domain in many scientific communities. Each year, many new methods are proposed. It would not be possible to give an overview of research activities in this 'hot' domain. The interested reader is referred to [BIS 06, DEV 96, HAN 01, HAS 01, SCH 02] for solid elements on that issue from different points of view. However, it is interesting to give, in this section, a synthetic view of recent efforts to improve the toolkit of classification methods. In the presentation of recent advances in discriminant analysis, we opt for a plan related to that of section 6.3.

6.4.1. *Parametric methods*

We present here methods which assume that the class-conditional densities are Gaussian. As stated above, linear discriminant analysis is a reference method providing good, stable and robust classifiers. However, LDA classifiers could be jeopardized with small sizes of learning samples and the situation is clearly worst for QDA. The new methods that are now presented aim for good performances for small learning sample sizes.

6.4.1.1. *Predictive discriminant analysis*

This method, introduced by Geisser [GEI 66], considers discriminant analysis in a Bayesian framework and bases the decision function on the class-conditional densities integrated with respect to the parameters. This approach is sketched in the LDA framework for which the class-conditional density of an observation is a Gaussian distribution with mean $\boldsymbol{\mu}_k$ and covariance matrix $\boldsymbol{\Sigma}$ for class G_k. Setting $\boldsymbol{\theta}_k = (\boldsymbol{\mu}_k, \boldsymbol{\Sigma})$, the predictive approach consists of considering a prior distribution $p(\boldsymbol{\theta}_k)$ on $\boldsymbol{\theta}_k$ and deriving the classifier from the integrated density $p_k(\boldsymbol{x}) = \int p(\boldsymbol{x}|\boldsymbol{\theta}_k, G_k)p(\boldsymbol{\theta}_k)d\boldsymbol{\theta}_k$. In general, a non-informative prior distribution is assumed for $\boldsymbol{\theta}_k$. In the LDA framework, this non-informative prior distribution is an improper prior distribution (namely a prior distribution which is not a probability distribution) $p(\boldsymbol{\theta}_k) \propto |\boldsymbol{\Sigma}|^{-a/2}$, and leads to the following predictive density:

$$p_k(\boldsymbol{x}) = (n\pi)^{-p/2}\left(\frac{n_k}{n_k+1}\right)^{p/2}\frac{\Gamma(\frac{1}{2}(n-g+p-a+2))}{\Gamma(\frac{1}{2}(n-g+-a+2))}|\hat{\boldsymbol{\Sigma}}|^{-1/2}$$
$$\times \left[1+\frac{n_k}{n(n_k+1)}(\boldsymbol{x}-\hat{\boldsymbol{\mu}}_k)'\hat{\boldsymbol{\Sigma}}^{-1}(\boldsymbol{x}-\hat{\boldsymbol{\mu}}_k)\right]^{\frac{1}{2}(n-g+p-a+2)}.$$

This distribution is the t multivariate distribution of Student with $n - g - a + 2$ degrees of freedom. Most of the authors choose $a = p + 1$. This distribution is interesting because it has heavier tails than the Gaussian distribution. In practice, predictive LDA and LDA demonstrate analogous performances except for small samples. Predictive LDA is of particular interest when the size of the classes n_k are unbalanced with some very small values. Moreover, predictive QDA can be defined in the same way and this method is expected to provide good classifiers in situations where QDA cannot be performed for numerical reasons (small n_k values). The interested reader will find a clear and precise presentation of predictive discriminant analysis in [RIP 95] and the article of [JOH 96] for additional theoretical details.

6.4.1.2. *Regularized discriminant analysis*

An often efficient way to provide decision functions with diminished variance in the small sample setting consists of regularizing the class covariance matrices estimates. One of the most convincing methods, in this context, is the regularized discriminant analysis (RDA) method of Friedman [FRI 89]. Friedman proposed introducing two tuning parameters λ and γ to the class covariance matrix estimate $\hat{\boldsymbol{\Sigma}}_k$:

$$\hat{\boldsymbol{\Sigma}}_k(\lambda, \gamma) = (1-\gamma)\hat{\boldsymbol{\Sigma}}_k(\lambda) + \gamma(\frac{\text{tr}(\hat{\boldsymbol{\Sigma}}_k(\lambda))}{p})\mathbf{I},$$

where

$$\hat{\boldsymbol{\Sigma}}_k(\lambda) = \frac{(1-\lambda)(n_k-1)\hat{\boldsymbol{\Sigma}}_k + \lambda(n-g)\hat{\boldsymbol{\Sigma}}}{(1-\lambda)(n_k-1) + \lambda(n-g)}.$$

The complexity parameter λ ($0 \leq \lambda \leq 1$) controls the contribution of the empirical estimators $\hat{\boldsymbol{\Sigma}}_k$ and $\hat{\boldsymbol{\Sigma}}$ (defined in section 6.3.1), while regularization parameter γ ($0 \leq$

$\gamma \leq 1$) controls the shrinking of the eigenvalues towards equality since $\text{tr}(\hat{\boldsymbol{\Sigma}}_k(\lambda))/p$ is the mean of eigenvalues of $\hat{\boldsymbol{\Sigma}}_k(\lambda)$.

With fixed $\gamma = 0$, varying λ therefore leads to an intermediate method between LDA and QDA. With fixed $\lambda = 0$, increasing γ leads to a less biased estimate of the eigenvalues of the covariance matrices. For $\lambda = 1$ and $\gamma = 1$, we obtain a simple classifier which consists of assigning any vector to the class whose center is the nearest. Parameters λ and γ are computed by minimizing the cross-validated error rate. In practice, this method performs well compared to LDA and QDA for small sample sizes. The drawbacks of RDA are that it provides classifiers which are generally difficult to interpret, that the parameters λ and γ are not very sensitive and it provides the same performances for large ranges of values.

An alternative method avoids using these kinds of tuning parameters and leads to decision functions with simple geometric interpretation, while performing in a manner analogous to RDA. Eigenvalue decomposition discriminant analysis (EDDA) makes use of the eigenvector decomposition of the class covariance matrices

$$\boldsymbol{\Sigma}_k = \lambda_k \mathbf{D}_k \mathbf{A}_k \mathbf{D}'_k \qquad (6.6)$$

where $\lambda_k = |\boldsymbol{\Sigma}_k|^{1/p}$, $\mathbf{D}_k$ is the eigenvector matrix of $\boldsymbol{\Sigma}_k$ and $\mathbf{A}_k$ is a diagonal matrix with determinant 1, whose diagonal contains the normalized eigenvalues of $\boldsymbol{\Sigma}_k$ in decreasing order [BEN 96]. Parameter λ_k determines the volume of class G_k, $\mathbf{D}_k$ its orientation and $\mathbf{A}_k$ its shape.

By allowing some of these elements to vary between classes, we get different (more or less parsimonious) and easy to interpret models. Moreover, by considering particular situations (diagonal or proportional to identity matrices), we obtain other interesting models. Finally, 14 models are in competition and EDDA chooses the model minimizing cross-validated error rate, the parameters being estimates with maximum likelihood methodology. In practice, EDDA and RDA have comparable performances while EDDA proposes more readable classifiers. Moreover, the interest of EDDA increases when the number of classes is large. Other regularized discriminant methods have been presented [MKH 97], including Bayesian methods and methods for qualitative predictors.

6.4.1.3. *Mixture discriminant analysis*

Contrary to situations where regularization is desirable, there exist situations, with a reasonable learning sample size, where the Gaussian assumption for the class-conditional densities is too restrictive. As proposed by Hastie and Tibshirani [HAS 96], it is therefore possible to model the class-conditional densities with a mixture of Gaussian distributions

$$p(\boldsymbol{x}|G_k) = \sum_{\ell=1}^{r_k} \pi_{k\ell} \Phi(\boldsymbol{x}|\boldsymbol{\mu}_{k\ell}, \boldsymbol{\Sigma}_{k\ell}).$$

The mixture proportions $\pi_{k\ell}$ (checking $\sum_{\ell=1}^{r_k} \pi_{k\ell} = 1$) and $\Phi(\mu, \Sigma)$ represent the density of a Gaussian distribution with mean μ and covariance matrix Σ. The parameters of the g mixture distributions can be estimated with the EM algorithm (Chapter 8).

The drawback of this type of model is its large number of parameters to be estimated. Mixture discriminant analysis (MDA) therefore needs a large sample size for each class to be efficient. For this reason, [HAS 96] restricted attention to a mixture model with a common covariance matrix for all mixture components of each class: $\Sigma_{k\ell} = \Sigma$, $(1 \leq \ell \leq r_k; 1 \leq k \leq g)$, the number of mixture components r_k being chosen *a priori* by the user. Indeed, it allows the number of parameters to be estimated to be dramatically diminished, but the price to be paid is the great loss of the flexibility of MDA.

We therefore recommend a slightly different model. Instead of assuming a common covariance matrix for each mixture component in each class, we suggest considering covariance matrices of the form $\Sigma_{k\ell} = \sigma_{k\ell}^2 \mathbf{I}_p$, where $\mathbf{I}_p$ is the identity matrix. Acting in such a way, we model the class-conditional densities $p(x|G_k)$ by an union of Gaussian balls with free volumes. It is a flexible model since Gaussian balls can easily fit the cloud of n_k points of class G_k, and it remains parsimonious since each covariance matrix is parametrized with a single parameter. Moreover, in this context, the number of components can be successfully estimated by a Bayesian entropy criterion (BEC), a criterion related to BIC and which takes the classification purpose into account [BOU 06] or by minimization of the cross-validated error rate.

6.4.2. Radial basis functions

Radial basis functions involve a neural network structure different from the structure of the single hidden layer back-propagation network.

The analytic form of a radial basis function is

$$y_k(x) = \sum_{j=0}^{m} w_{kj} F_j(x), \ k = 1, \ldots, g.$$

where $F_0(x) = 1$. A radial basis function ensuring that the network approximates any function in an optimal way is

$$F_j(x) = \exp\left(\frac{-||x - \mu_j||^2}{2\sigma_j^2}\right),$$

where the parameters μ_j (centers) and σ_j (distances to the centers) are to be estimated. The learning involves two steps. In a first step, the radial basis functions are estimated independently of the outputs, then the weights of the hidden layer are computed by linear programming to minimize a quadratic loss function analogous to equation (6.3).

An efficient use of radial basis functions needs a good choice of the centers and of the distances to the centers. These choices can be achieved with a *k-means* algorithm. The most sensible approach, however, is to use radial basis function. These consist of modeling the marginal distribution of the data with a mixture of Gaussian distributions

$$p(\boldsymbol{x}) = \sum_{j=1}^{m} P_j F_j(\boldsymbol{x}),$$

estimated with the EM algorithm, where the proportions P_j are fixed and estimated in the second step.

The main purpose of radial basis functions is to simplify the learning task by incorporating a first exploratory phase. The exploratory phase is expected to provide a parsimonious representation of the input data. Thus, radial basis functions provide a simpler tool than the single hidden layer back-propagation network if the radial functions F_j are well chosen. In practice, this kind of network can outperform the single hidden layer back-propagation network. Even more than for the latter method, radial basis functions require important know-how heuristics. See [BIS 95] for a detailed description and good analysis of radial basis functions.

6.4.3. *Boosting*

Boosting is a family of algorithms consisting of combining many versions of an unstable classifier as a decision tree or a neural network in order to (dramatically) decrease the error rate of the resulting classifier. The underlying idea of boosting is to repeat a discriminant analysis method on the same learning sample by increasing the weights of misclassified vectors during the algorithm. The algorithm is therefore making efforts to correctly assign the vectors difficult to assign to their own class. Basic boosting *discrete adaboost* of Freund and Schapire [FRE 97] is described for a two-class problem, for which $y_i = 1$ if $z_i = 1$ and $y_i = -1$ if $z_i = 2$:

1) Set initial weights $\omega_i = 1/n, i = 1, \dots, n$.

2) Repeat for $j = 1, \dots, J$

    a) Compute the decision function $\chi_j(\boldsymbol{x})$ *with weights* ω_i for learning sample A,

    b) Compute $e_j = \mathbb{E}_\omega[\mathbb{1}_{y \neq \chi_j(\boldsymbol{x})}]$ and $c_j = \log[(1 - e_j)/e_j]$,

    c) Set $\omega_i \leftarrow \omega_i \exp[c_j \mathbb{1}_{y_i \neq \chi_j(\boldsymbol{x}_i)}], \ i = 1, \dots, n$ and normalize weights such that $\sum_i \omega_i = 1$.

3) The final classifier is $\text{sign}[\sum_{j=1}^{J} c_j \chi_j(\boldsymbol{x})]$.

This procedure, which is not dependent upon any tuning parameter other than the number of iterations J, dramatically improves the classification performances for unstable methods giving poor or very poor performances with the learning sample A. As a matter of fact, it allows the performances of decision trees such as CART to be dramatically improved, but the resulting classifier is no easier to interpret. Friedman, Hastie and Tibshirani [FRI 00] (see also [HAS 01]) have shown that boosting could be interpreted as an additive model $F(\mathbf{x}) = \sum_{j=1}^{J} c_j \chi_j(\mathbf{x})$ and that it could be regarded as a sequential estimation algorithm minimizing the following exponential loss function for the equation of a logistic regression:

$$L(F) = \mathbb{E}[(\exp(-yF(\boldsymbol{x})))].$$

The optimal solution is the logistic transformation of $p(y = 1 \mid \boldsymbol{x})$ defined by

$$F(\boldsymbol{x}) = \frac{1}{2} \frac{p(y = 1 \mid \boldsymbol{x})}{p(y = -1 \mid \boldsymbol{x})},$$

hence

$$p(y = 1 \mid \boldsymbol{x}) = \frac{e^{F(\boldsymbol{x})}}{e^{-F(\boldsymbol{x})} + e^{F(\boldsymbol{x})}}.$$

Starting from this point of view, Hastie *et al.* [HAS 01] recommend, in Chapter 10 of their book, that it is possible to obtain different boosting methods more or less sensitive to outliers by changing the loss function.

To close this section on stabilization methods of unstable classifiers, it is worthwhile mentioning another combining technique named *bagging*, based on [BRE 96]. It simply consists of rebuilding the classifier on bootstrap samples and assigning a vector to one of the g classes by a majority rule. Contrary to boosting, bagging does not penalize the vectors according to the classification results. In practice, bagging improves unstable classifiers but in a less dramatic way than boosting. Moreover, this technique has a tendency to deteriorate (contrary to boosting) the performances of poor classifiers. Finally, note the existence of a promising new classification methodology named *random forests*, proposed by Breiman [BRE 08], which takes advantage of the idea of combining unstable simple methods (random decision trees) with bagging-like algorithms.

6.4.4. *Support vector machines*

Support vector machines (SVM) is the name of a family of classifiers, first introduced by Vapnik [VAP 96], which aim to find a linear separation between classes in space which can be different from the space where the learning data lie. SVM methods have been defined for two-class classification problems and essentially use two ingredients:

1) First, for a fixed investigation space, SVM methods aim to compute an optimal hyperplane to separate the two classes which maximizes the margin between the two classes in a sense that it is made more precise.

2) Second, SVM methods use *the kernel trick* which consists of replacing the standard dot product by a non-linear kernel function. This allows the algorithm to fit the maximum margin hyperplane in the transformed feature space. Alhough the classifier is defined by a hyperplane in the high-dimensional feature space, it may be non-linear in the original input space.

Considering a two-class problem, we denote (as previously) $y_i = 1$ if $z_i = 1$ and $y_i = -1$ if $z_i = 2$. To describe the first phase of SVM methods, namely the computation of the maximum margin hyperplane, we will stay in the original input space where the observations are characterized by p descriptors. If the classes are linearly separable in the input space, the maximum margin hyperplane is a solution of the optimization problem

$$\max_{\beta,\beta_0,||\beta||=1} C, \quad \text{with} \quad y_i(x_i'\beta + \beta_0) \geq C, \ i = 1,\ldots,n,$$

which is equivalent to the problem

$$\min_{\beta,\beta_0} ||\beta||, \quad \text{with} \quad y_i(x_i'\beta + \beta_0) \geq 1, \ i = 1,\ldots,n.$$

In the general case, where the two classes are not linearly separable, we introduce dummy variables $(s_1,\ldots,s_n)$. The role of each s_i is to measure the magnitude of the classification error of i. Since we are obviously aiming to minimize the misclassification error, this leads to the following optimization problem being considered:

$$\max_{\beta,\beta_0,||\beta||} C, \quad \text{with} \quad y_i(x_i'\beta + \beta_0) \geq C(1 - s_i), s_i \geq 0 \ i = 1,\ldots,n,$$

and $\sum_{i=1}^{n} s_i \leq$ constant, which is equivalent to the problem

$$\min_{\beta,\beta_0} \frac{1}{2}||\beta||^2 + \gamma \sum_{i=1}^{n} s_i \quad \text{with} \quad y_i(x_i'\beta x_i + \beta_0) \geq 1 - s_i, \ i = 1,\ldots,n$$

where parameter γ, chosen by the user, is related to the constant of the previous problem. It is a convex optimization problem that can be solved by standard optimization tools [BON 97], with a solution of the form $\beta^* = \sum_{i=1}^{n} a_i y_i x_i$. The non-zero coefficients a_i are related to observations such that $y_i(x_i'\beta x_i + \beta_0) = 1 - s_i$, which are called *support vectors*. These support vectors are therefore the only vectors needed to define the maximum margin hyperplane. They are the vectors which are difficult to classify correctly. From this point of view, SVM strategy can be related to the boosting strategy.

Using the kernel trick, SVM methods achieve the above-described computations on the transformed vectors $\mathbf{h}(\boldsymbol{x}_i) = (h_1(\boldsymbol{x}_i), \ldots, h_m(\boldsymbol{x}_i))$ and provide a classifier of the form $\mathrm{sign}(\mathbf{h}(\boldsymbol{x}'\boldsymbol{\beta}^* + \beta_0^*))$. Setting

$$\mathbf{h}(\boldsymbol{x})'\boldsymbol{\beta}^* + \beta_0^* = \sum_{i=1}^{n} a_i y_i K(\boldsymbol{x}, \boldsymbol{x}_i) + \beta_0,$$

where function K is the dot product $K(\boldsymbol{x}, \boldsymbol{y}) = <\mathbf{h}(\boldsymbol{x}, \mathbf{h}(\boldsymbol{y}))>$, it can be seen that it is sufficient to characterize kernel K to define the SVM classifier. In practice, common kernels are polynomial kernel $K(\boldsymbol{x}, \boldsymbol{y}) = (1+ <\boldsymbol{x}, \boldsymbol{y}>)^r$ with, in general, $r = 2$. This means that the transformation space where the maximum margin space is researched is the space of all the two-degree polynomials generated with the input vectors.

Other favorite kernels are Gaussian kernels. Many applications of SVM show that it is a successful family of methods which can achieve quasi-optimal misclassification rates in many high-dimensional classification problems. The extension of SVM methods for multi-class problems $g > 2$ can be carried out using several strategies. A standard strategy is to reduce the multi-class problem into multiple binary problems. A method suggested by Friedman [FRI 96] consists of considering all the SVMs between every pair of classes (one-versus-one) and to assign a vector to the class with most votes.

SVM methods are widely used in the machine learning community and are quite a promising family of classification methods; this is an exciting topic which receives a lot of contributions. The reader is referred to the seminal book of [VAP 96] and also to [HAS 01] for a synthetic introduction to SVM with an interpretation of SVM as a regularization method in a reproducing kernel Hilbert space (RKHS). Another good introduction to SVM is [CRI 00], and the reader will find a detailed and readable presentation of kernel methods for classification and related topics in [SCH 02].

6.5. Conclusion

Discriminant analysis gathers a rich set of problems and methods in which many scientific domains are concerned: for example, statistical decision, exploratory data analysis, pattern recognition and machine learning. As a matter of fact, discriminant analysis receives contributions from researchers of different cultures (statistics, computer science including artificial intelligence, functional approximation and physics). Discriminant analysis is currently a collection of various and complex methods and, as stated above, the research in discriminant analysis is quite active. This vitality is of course beneficial and the toolkit of discriminant analysis methods is large.

In reality, it is important to keep in mind that there is no miracle method. A properly achieved discriminant analysis generally needs simple pre-treatment and not always expensive and complex methods. From that point of view, we think it is important to first use simple and well-known methods such as those presented in section 6.3, and especially linear discriminant analysis. Such methods provide sensible performances to be compared with more complex or more specific methods. Quite often, these simple methods provide satisfactory misclassification error rates and produce stable classifiers.

6.6. Bibliography

[AKA 74] AKAIKE H., "A new look at the statistical model identification", *IEEE Trans. Automat. Control*, vol. 19, p. 716–723, 1974.

[BEN 96] BENSMAIL H., CELEUX G., "Regularized Gaussian discriminant analysis through eigenvalue decomposition", *Journal of the American Statistical Association*, vol. 91, p. 1743–1748, 1996.

[BIS 95] BISHOP C., *Neural Network for Pattern Recognition*, Clarendon Press, Oxford, 1995.

[BIS 06] BISHOP C., *Pattern Recognition and Machine Learning*, Springer, London, 2006.

[BON 97] BONNANS F., GILBERT J., LEMARÉCHAL C., SAGASTIZÁBAL C., *Optimisation Numérique: Aspects Théoriques et Pratiques*, Springer-Verlag, Paris, 1997.

[BOU 06] BOUCHARD G., CELEUX G., "Selection of generative models in classification", *EEE Trans. Pattern Anal. Mach. Intell.*, vol. 28, p. 544–554, 2006.

[BRE 84] BREIMAN L., FRIEDMAN J., OHLSEN R., STONE C., *Classification and Regression Trees*, Wadsworth, Belmont, 1984.

[BRE 96] BREIMAN L., "Bagging predictors", *Machine Learning*, vol. 26, p. 123–140, 1996.

[BRE 08] BREIMAN L., CUTLER A., Random Forest: Classification Manual, Report, Utah State University, 2008, http://www.math.usu.edu/adele/forests.

[CEL 91] CELEUX G., *Analyse Discriminante sur Variables Continues*, Inria, Le Chesnay, 1991.

[CEL 94] CELEUX G., NAKACHE J.-P., *Analyse Discriminante sur Variables Qualitatives*, Polytechnica, Paris, 1994.

[CRI 00] CRISTIANINI N., SHAWE-TAYLOR J., *An Introduction to Support Vector Machines*, Cambridge University Press, Cambridge, 2000.

[DEV 96] DEVROYE L., GYÖRFI L., LUGOSI G., *A Probabilistic Theory of Pattern Recognition*, Springer-Verlag, New York, 1996.

[EFR 97] EFRON B., TIBSHIRANI R., "Improvement on cross-validation: the 632+ bootstrap", *Journal of the American Statistical Association*, vol. 92, p. 548–560, 1997.

[FIX 51] FIX E., HODGES J., Discriminatory Analysis-nonparametric discrimination: Consistency properties, Report num. 21-49004,4, US Air Force, Scholl of Aviation Medicine, Randolph Field, Texas, 1951.

[FRE 97] FREUND Y., SCHAPIRE R., "A decision-theoretic generalization of online learning and an application to bootsting", *Journal of Computer and System Sciences*, vol. 55, p. 119–139, 1997.

[FRI 89] FRIEDMAN J. H., "Regularized discriminant analysis", *Journal of the American Statistical Association*, vol. 84, p. 165–175, 1989.

[FRI 96] FRIEDMAN J. H., Another approach to polychotomous classification, Report, Stanford University, 1996.

[FRI 00] FRIEDMAN J., HASTIE T., TIBSHIRANI R., "Additive logistic regression: a statistical view of boosting (with discussion)", *Annals of Statistics*, vol. 28, p. 337–407, 2000.

[FUR 74] FURNIVAL O., WILSON R., "Regression by leaps and bounds", *Technometrics*, vol. 16, p. 499–511, 1974.

[GEI 66] GEISSER S., "Predictive discrimination", KRISHNAM P. R., Ed., *Multivariate Analysis*, New York, Academic Press, p. 149–163, 1966.

[HAN 96] HAND D., *Discrimination and Classification*, Wiley, New York, 1996.

[HAN 01] HAND D., MANNILA H., SMYTH P., *Principles of Data Mining*, The MIT Press, New York, 2001.

[HAS 96] HASTIE T., TIBSHIRANI R., "Discriminant analysis by Gaussian mixtures", *Journal of the Royal Statistics Society B.*, vol. 58, p. 155–176, 1996.

[HAS 01] HASTIE T., TIBSHIRANI R., FRIEDMAN J., *The Elements of Statistical Learning*, Springer-Verlag, New York, 2001.

[HOS 89] HOSMER D. W., LEMESHOW S., *Applied Logistic Regression*, Wiley, New York, 1989.

[JOH 96] JOHNSON R., MOUHAB A., "A Bayesian decision theory approach to classification problems", *Journal of Multivariate Analysis*, vol. 56, p. 232–244, 1996.

[KOO 97] KOOPERBERG C., BOSE S., C. S., "Polychotomous regression", *Journal of the American Statistical Association*, vol. 92, p. 117–127, 1997.

[MCC 89] MCCULLAGH P., NEDLER J., *Generalised Linear Models*, Chapman and Hall, London, 2nd edition, 1989.

[MCL 92] MCLACHLAN G., *Discriminant Analysis and Statistical Pattern Recognition*, Wiley, New York, 1992.

[MKH 97] MKHADRI A., CELEUX G., A. N., "Regularization in discriminant analysis: an overview", *Computational Statistics & Data Analysis*, vol. 23, p. 403–423, 1997.

[PRE 79] PRENTICE R., PYKE R., "Logistic disease incidence models and case-control studies", *Biometrika*, vol. 66, p. 403–411, 1979.

[RIP 95] RIPLEY B., *Pattern Recognition and Neural Network*, Cambridge University Press, Cambridge, 1995.

[RIS 83] RISSANEN J., "A universal prior for integers and estimation by minimum description length", *Annals of Statistics*, vol. 11, p. 416–431, 1983.

[SAP 90] SAPORTA G., *Probabilités, Analyse des Données et Statistique*, Technip, Paris, 1990.

[SCH 78] SCHWARZ G., "Estimating the dimension of a model", *Annals of Statistics*, vol. 6, p. 461–464, 1978.

[SCH 02] SCHÖLKOPF B., SMOLA J. A., *Learning with Kernel*, The MIT Press, New York, 2002.

[SIL 86] SILVERMAN B., *Density Estimation for Statistics and Data Analysis*, Chapman and Hall, London, 1986.

[STO 77] STONE M., "An asymptotic equivalence of choice of model by cross-validation and Akaike's criterion", *Journal of the Royal Statistics Society B.*, vol. 39, p. 44–47, 1977.

[VAP 96] VAPNIK V., *The Nature of Statistical Learning Theory*, Springer-Verlag, New York, 1996.

[VEN 98] VENDITTI V., Aspects du principe de maximum d'entropie en modélisation statistique, PhD thesis, Joseph Fourier University, Grenoble, 1998.

Chapter 7

Cluster Analysis

7.1. Introduction

Cluster analysis or *clustering*, which is an important tool in a variety of scientific areas including pattern recognition, information retrieval, micro-arrays and data mining, is a family of exploratory data analysis methods which can be used to discover structures in data. These methods aim to obtain a reduced representation of the initial data and, like principal components analysis, factor analysis or multidimensional scaling, are one form of data reduction. The aim of cluster analysis is the organization of the set into *homogenous classes* or *natural classes*, in a way which ensures that objects within a class are similar to one another.

For example in statistics, cluster analysis can identify several populations within a heterogenous initial population, thereby facilitating a subsequent statistical study. In natural science, the clustering of animal and plant species first proposed by Linnaeus (an 18th century Swedish naturalist) is a celebrated example of cluster analysis. In the study of social networks, clustering may be used to recognize communities within large groups of people and, at a more general level, the simple naming of objects can be seen as a form of clustering. Moreover, some problems bearing no apparent relation to classical data analysis methods can be formalized using clustering. We can cite, for instance, center location problems in operational research, segmentation and vector quantization in image processing and data compression in computer science.

There is some confusion about the use of the terms *classification* and *cluster analysis*. Classification, whose task is to assign objects to classes or groups on the

Chapter written by Mohamed NADIF and Gérard GOVAERT.

basis of measurements of the objects, is more general and can be divided into two parts:

– *unsupervised classification* (or cluster analysis) seeks to discover groups from the data, whereas

– *supervised classification* (or discrimination) seeks to create a classifier for the classification of future observations, starting from a set of labeled objects (a training or learning set).

This chapter is restricted to unsupervised classification methods.

The particular terminology depends on the field: *taxonomy* is the branch of biology concerned with the classification of organisms into groups; in medicine, *nosology* is the branch of medicine concerned with the classification of diseases; in machine learning, clustering is known as *unsupervised classification* or *unsupervised learning*; in marketing we tend to talk about *segmentation*; and in archaeology, anthropology or linguistics, the most frequently-used term is *typology*.

Attempting to define clustering formally, as a basis for an automated process, raises a number of questions. How can we define the objects (elements, cases, individuals or observations) to be classified? How can the concept of similarity between objects be defined? What is a cluster? How are clusters structured? How can different partitions be compared?

To carry out cluster analysis, two steps are generally undertaken:

– Objects which share certain characteristics are placed in the same class. Let us consider numbers of fingers and compare monkeys and humans: on this comparison criterion the two species will be considered to be similar. This kind of step leads to a *monothetic* classification which is the basis of the aristotelician approach [SUT 94]. All the objects in the same class then share a certain number of characteristics (eg: 'All men are mortal').

– The alternative is to use a *polythetic* approach where objects having close characteristics are placed in clusters. Generally, in this situation, the concept of proximity (distance, dissimilarity) will be used to define clusters. The polythetic approach is the subject of this chapter.

Section 7.2 will be devoted to the general principles of cluster analysis. Section 7.3 is concerned with hierarchical approaches, with particular emphasis on agglomerative hierarchical approaches. Section 7.4 looks at partitional methods and more specifically at k-means algorithms and their extensions. Some miscellaneous clustering methods and block clustering methods are studied in sections 7.5 and 7.6, and prospects for the future are briefly described in the final section.

7.2. General principles

7.2.1. *The data*

Many clustering methods require the data to be presented as a set of proximities. This notion of proximity, which is a quantitative measure of closeness, is a general term for similarity, dissimilarity and distance: two objects are close when their dissimilarity or distance is small or their similarity large.

More formally, a *dissimilarity* on the set Ω can be defined as a function d from $\Omega \times \Omega$ to the real numbers such that:

1) $d(x, y) > 0$ for all $x \neq y$ belonging to Ω
2) $d(x, x) = 0$ for all x belonging to Ω
3) $d(x, y) = d(y, x)$ for all x, y belonging to Ω.

A dissimilarity satisfying the triangle inequality

$$d(x, z) \leqslant d(x, y) + d(y, z) \quad \forall x, y, z \in \Omega$$

is a *distance*.

Sometimes these proximities are the form in which the data naturally occur. In most clustering problems, however, each of the objects under investigation will be described by a set of *variables* or *attributes*. The first step in clustering, possibly the most important, is to define these proximities. Different kinds of definitions depending on the type of variables (continuous, binary, categorical or ordinal) are to be found in the literature (see e.g. [GOW 86]). For instance, in the absence of information allowing the appropriate distance to be employed, the Euclidean distance between two vectors $x = (x^1, \ldots, x^p)'$ and $y = (y^1, \ldots, y^p)'$ in $\mathbb{R}^p$, defined by

$$d(x, y) = ||x - y|| = \sqrt{\sum_{j=1}^{p} (x^j - y^j)^2},$$

is the most frequently used distance for continuous data. Moreover, before computing these proximities, it is often necessary to consider scaling or transforming the variables, since variables with large variances tend to impact the resulting clusters more than those with small variances. Other transformations can be used according to the nature of data.

To illustrate the importance of the metric we have used a data matrix $x = (x_i^j)$ comprising a set of 900 objects described by 3 continuous variables reported partially in Figure 7.1. Here, two distances can be considered: the Euclidean distance on standardized data and, working on the row profiles, the χ^2 distance, which is the metric used in correspondence analysis (see Chapter 2). To illustrate the difference between the two choices we show in Figure 7.2 the scatter plot of the first two principal components obtained by principal component analysis (PCA), which implicitly used Euclidean distance. The scatter plot of the first two principal components obtained by correspondence analysis (CA), which implicitly used χ^2 distance on profiles, is also shown. A structure in three clusters is clearly visible in both cases, but the two classifications are actually completely different.

i	x_i^1	x_i^2	x_i^3	i	x_i^1	x_i^2	x_i^3	i	x_i^1	x_i^2	x_i^3
1	37	31	40	301	117	132	142	601	201	240	194
2	35	26	29	302	166	118	117	602	205	266	205
3	42	44	25	303	115	126	153	603	178	212	256
...				...				...			
298	44	32	27	598	109	109	91	898	190	223	203
299	31	38	29	599	136	150	95	899	277	206	190
300	28	47	49	600	100	132	152	900	212	198	259

Figure 7.1. *The data*

7.2.2. *Visualizing clusters*

The first method of identifying a cluster is to use graphical views of the data. For instance, when the data are objects described by one continuous variable, a histogram showing the existence of several modes indicates a heterogenous population. For two variables, a scatter plot of the data or a plot of the contours of kernel density estimators can reveal the existence of clusters. When there are more than two variables, different tools can be used, such as the scatterplot matrix, which is defined as a square grid of bivariate scatter plots. An alternative approach is to project the data onto two dimensions: see, for instance, principal component analysis in Chapter 1, correspondence analysis in Chapter 2, multidimensional scaling in Chapter 4 or projection pursuit in Chapter 3.

7.2.3. *Types of classification*

Classification procedures can take different forms such as partitioning, hierarchical clustering, overlapping clustering, high density clustering and fuzzy clustering, but this section focuses on the main types of classification: partition and hierarchy.

7.2.3.1. *Partition*

Given a finite set Ω, a set $P = \{P_1, \ldots, P_g\}$ of non-empty subsets of Ω is a *partition* if $\forall k \neq \ell$, $P_k \cap P_\ell = \emptyset$ and if $\cup_k P_k = \Omega$. With this kind of partition of P

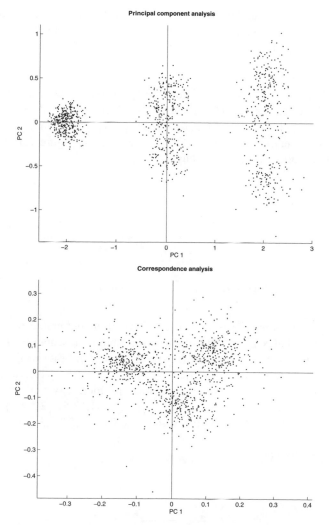

Figure 7.2. *Scatter plot of first two principal components for PCA and CA*

into g subsets or clusters $P_1, \ldots, P_g$, each element of Ω belongs to one and only one cluster and so P can be represented by the classification matrix:

$$
\mathbf{Z} = \begin{pmatrix} z_{11} & \cdots & z_{1g} \\ \vdots & \ddots & \vdots \\ z_{n1} & \cdots & z_{ng} \end{pmatrix}
$$

where $z_{ik} = 1$ if $i \in P_k$ and 0 otherwise. Note that the sum of the ith row values is equal to 1 (each element belongs to one and only one cluster) and the sum of the kth column values is equal to n_k, representing the cardinality of P_k.

7.2.3.2. *Indexed hierarchy*

Given a finite set Ω, a set H of non-empty subsets of Ω is a hierarchy on Ω if $\Omega \in H$, $\forall x \in \Omega$, $\{x\} \in H$ and $\forall h, h' \in H$ either $h \cap h' = \emptyset$, $h \subset h'$ or $h' \subset h$. The set $\{\{1\}, \{2\}, \{3\}, \{4\}, \{5\}, \{2,4\}, \{3,5\}, \{2,3,4,5\}, \{1,2,3,4,5\}\}$ is an example of a hierarchy defined on $\Omega = \{1, 2, 3, 4, 5\}$. Figure 7.3 illustrates two kinds of representations of this hierarchy.

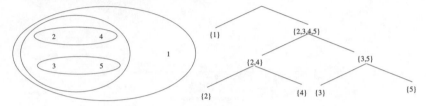

Figure 7.3. *Representations of a hierarchy*

These representations are seldom used. It is often preferable to associate the hierarchy with an index in order to obtain a readable representation. This *index* on a hierarchy H is a mapping, denoted i, from H to $\mathbb{R}^+$ and satisfying the following properties: $h \subset h'$ and $h \neq h' \Rightarrow i(h) < i(h')$ (i is a strictly increasing function) and $\forall x \in \Omega$, $i(\{x\}) = 0$. We shall denote the hierarchy with the index i as (H, i). Associating clusters $\{1\},\{2\},\{3\},\{4\},\{5\},\{2,4\},\{3,5\},\{2,3,4,5\}$, $\{1,2,3,4,5\}$ from the previous hierarchy with the values $0,0,0,0,0,1,2,2.5,3.5$ gives rise to an *indexed hierarchy* (H, i) which can be represented by a tree data structure often known as a *dendrogram* (see Figure 7.4).

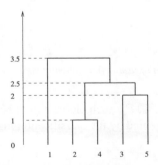

Figure 7.4. *An indexed hierarchy represented as a dendrogram*

Note that if $P = (P_1, P_2, \ldots, P_g)$ is a partition of Ω, the set H formed by the clusters P_k, the singletons of Ω and Ω itself defines a hierarchy. Conversely, it is also possible to associate a partition with each level of an indexed hierarchy, thus defining a sequence of nested partitions.

The number of hierarchies and partitions for a set Ω quickly becomes very large as the number of objects to be classified increases. For instance, the number of partitions of n objects into g clusters is given by the formula

$$S(n,g) = \frac{1}{g!} \sum_{k=0}^{g} (-1)^{k-1} C_g^k k^n,$$

and, for higher values of n and g, we have the approximation $S(n,g) \approx \frac{g^n}{g!}$. The exact numbers for the first few values are reported in Figure 7.5 and we have, for instance, approximately 10^{67} partitions with five clusters for a set of 100 objects.

(n,g)	1	2	3	4	5	6	7	8
1	1							
2	1	1						
3	1	3	1					
4	1	7	6	1				
5	1	15	25	10	1			
6	1	31	90	65	15	1		
7	1	63	301	350	140	21	1	
8	1	127	966	1701	1050	266	28	1

Figure 7.5. *Exact numbers of clusters*

7.2.3.3. *Indexed hierarchy and ultrametric*

Given a dissimilarity measure d on Ω, it is interesting to seek partitions which may be deduced from it in a 'natural' way. To this end, a neighbor relation can be defined for each real α, positive or null, by the relation:

$$x N_\alpha y \Leftrightarrow d(x,y) \leqslant \alpha \quad \forall x, y \in \Omega. \tag{7.1}$$

Starting from this neighbor relation, we seek to define conditions for the existence of a partition of Ω such that all the objects belonging to the same cluster are neighbors, and all elements belonging to distinct clusters are not neighbors. This property is satisfied if and only if N_α is an equivalence relation. The clusters are then the equivalence classes of N_α. As the function d is a dissimilarity measure, the relation N_α is reflexive and symmetric. It is then necessary and sufficient that transitivity should be satisfied, i.e.

$$x N_\alpha y \quad \text{and} \quad y N_\alpha z \quad \Rightarrow \quad x N_\alpha z$$

which leads to the relation

$$d(x,y) \leqslant \alpha \quad \text{and} \quad d(y,z) \leqslant \alpha \Rightarrow d(x,z) \leqslant \alpha. \tag{7.2}$$

Using the definition of *ultrametric*, which is a special kind of distance in which the triangle inequality is replaced with the ultrametric inequality

$$d(x,z) \leqslant \max\{d(x,y), d(y,z)\} \quad \forall x, y, z \in \Omega,$$

it is clear that equation (7.2) is satisfied if d is an ultrametric. Reciprocally, it can be shown that if N_α is an equivalence relation for all α, d must be an ultrametric: for each triplet x, y and z in Ω, taking $\alpha = \max(d(x,y), d(y,z))$, we have $d(x,y) \leqslant \alpha$ and $d(y,z) \leqslant \alpha$ and therefore $d(x,z) \leqslant \alpha$, which implies the ultrametric inequality.

We have just seen that the concept of ultrametric is closely related to that of partition. This type of property goes further in the case of hierarchies, since it can be shown that there is equivalence between an indexed hierarchy and an ultrametric. To show this property, we can define the following two functions:

1) Function φ associates an ultrametric with an indexed hierarchy: given an indexed hierarchy (H, i) on Ω, we can define a function δ from $\Omega \times \Omega$ to $\mathbb{R}^+$, by assigning for each couple x, y the smallest index of all clusters of H, including x and y. As the function i is increasing with the inclusion relation ($h_1 \subset h_2 \Rightarrow i(h_1) \leqslant i(h_2)$), $\delta(x,y)$ can also be considered as the index of the smallest cluster in relation to H and containing x and y. It can easily be shown that δ is an ultrametric.

2) Function Ψ associates an indexed hierarchy to an ultrametric: we consider the relations N_α (7.1) obtained by replacing the dissimilarity measure d by the ultrametric δ. We know that the N_αs are equivalence relations for each $\alpha \geqslant 0$. Let D_δ be the set of values taken by δ on Ω. We define the set H as the set of all equivalence classes of N_αs when α covers D_δ. Taking the *diameter* $i(h) = \max_{x,y \in h} \delta(x,y)$ as function i on H, we can show that (H, i) forms an indexed hierarchy on Ω.

Finally, it can be shown that these two functions are reciprocal, implying that the notions of indexed hierarchy and ultrametric are equivalent.

The example in Figure 7.6 illustrates this property. It is easy to check that the function φ associates the ultrametric δ_1 with the indexed hierarchy (H_1, i_1). With the function Ψ applied to the ultrametric δ_1, we obtain the following by symmetry: D_δ is the set $\{0, 1, 2, 3\}$; the equivalence classes of the four relations R_α are R_0 : $\{1\}, \{2\}, \{3\}, \{4\}$, R_1 : $\{1\}, \{2\}, \{3, 4\}$, R_2 : $\{1\}, \{2, 3, 4\}$ and R_3 : $\{1, 2, 3, 4\}$ which give the hierarchy $\{\{1\}, \{2\}, \{3\}, \{4\}, \{3, 4\}, \{2, 3, 4\}, \{1, 2, 3, 4\}\}$; the indexes corresponding to subsets of the hierarchy are $(0, 0, 0, 0, 1, 2, 3)$, respectively, which leads to the indexed hierarchy (H_1, i_1).

7.2.4. Objectives of clustering

The aim of clustering is to organize the objects of Ω into homogenous clusters. To define the notion of homogenity, we often use similarity or dissimilarity measures on Ω. For example, if d is a dissimilarity measure, we can characterize this homogenity by requiring that clusters in the partition satisfy the property:

$$\forall x, y \in \text{ same cluster cluster and } \forall z, t \in \text{ distinct clusters} \Rightarrow d(x,y) < d(z,t).$$

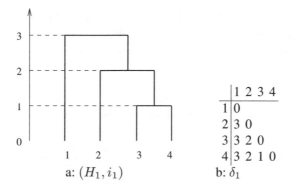

Figure 7.6. *Equivalence between indexed hierarchy and ultrametric*

The above property is designed to ensure greater similarity between two objects in the same cluster than between two objects in different clusters. In practice, this objective is impractical. For instance, two clusters can clearly be distinguished in Figure 7.7, but the distance between points 1 and 3 belonging to the same cluster is greater than the distance between the points 1 and 2 which have been classified separately.

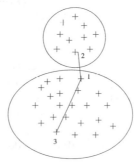

Figure 7.7. *Ambivalent clustering objective*

To overcome these difficulties, this excessively stringent condition will often be replaced by a numerical function usually known as a *criterion*, measuring the homogenity of the clusters. The problem can then appear simpler. For instance, in the case of a partition, the problem will be to select from among the finite set of partitions that which optimizes the numerical criterion. Unfortunately, the number of partitions is too large (see section 7.2.3.2) for them to be enumerated in a realistic time frame because of combinatorial complexity. Generally, heuristics are used which, rather than giving the best solution, give a 'good' one close to the optimal solution and lead to local optimization.

In the case of hierarchical clustering, we aim to obtain clusters which will be more homogenous the closer they are to the root of the tree structure. The criterion is less easy to define, but we will see that the definition can be based on the concept of optimal ultrametric.

A classical example of a criterion used when the objects $x_1, \ldots, x_n$ are described by p continuous variables is the *within-group sum of squares*, also *named within-group inertia*:

$$I_W(P) = \frac{1}{n} \sum_{k=1}^{g} \sum_{i \in P_k} d^2(x_i, \overline{x}_k) = \frac{1}{n} \sum_{k=1}^{g} \sum_{i \in P_k} ||x_i - \overline{x}_k||^2 \qquad (7.3)$$

where P is a partition of Ω into g clusters $\{P_1, \ldots, P_g\}$ of sizes $n_1, \ldots, n_k$, $\overline{x}_k$ denotes the mean vector of the kth cluster and d is the Euclidean distance. Using the within-group covariance matrix $\mathbf{S}_W = \frac{1}{n} \sum_k n_k \mathbf{S}_k$ where $\mathbf{S}_k$ is the covariance matrix of the kth cluster $\mathbf{S}_k = \frac{1}{n_k} \sum_{i \in P_k} (x_i - \overline{x}_k)(x_i - \overline{x}_k)'$, this criterion can be written

$$I_W(P) = \text{tr}(\mathbf{S}_W).$$

The closer the within-group sum-of-squares criterion is to 0, the more homogenous the partition will be. In particular, this criterion will be equal to 0 for a partition where each object is a cluster.

Some clustering algorithms have been empirically defined with no reference to an optimization problem. It is easy to put such algorithms forward, but less easy to prove that the obtained clusters are interesting and correspond to our needs. In fact, the algorithmic and numerical approaches often converge. A number of algorithms, first proposed without any reference to a criterion, appear to optimize a numerical criterion. This is the case for the classical k-means algorithm described in section 7.4.1.

7.3. Hierarchical clustering

In this section it will generally be assumed that all the relevant relationships within the set Ω to be classified are summarized by a dissimilarity d. The aim of hierarchical methods is to construct a sequence of partitions of a set Ω varying from partitions of singletons to the whole set. A hierarchical clustering algorithm transforms the similarity d into a sequence of nested partitions, or an indexed hierarchy, such that the closest objects are grouped in the clusters with the smallest indexes. There are two principal approaches:

– Divisive approach: this approach, also called the top-down approach, starts with just one cluster containing all the objects. In each successive iteration clusters are split into two or more further clusters, usually until every object is alone within a cluster.

Note that other stop conditions can be used, and the division into clusters is governed by whether or not a particular property is satisfied. For example, in taxonomy, animals may be separated into vertebrates and invertebrates.

– Agglomerative approach: in contrast to the divisive approach, this approach starts out from a set of n clusters, with each object forming a singleton cluster. In each successive iteration, the closest clusters are merged until just one cluster remains: the set Ω. From this point onwards we focus on this approach, which is the most frequently used.

7.3.1. *Agglomerative hierarchical clustering (AHC)*

7.3.1.1. *Construction of the hierarchy*

Using a dissimilarity D between groups, the different steps of agglomerative hierarchical algorithms are the following:

1) Initialization: start from the partition where each object is a cluster and compute the dissimilarity D between these clusters.

2) Repeat until the number of clusters is equal to 1:
  - merge the two clusters which are the *closest* with respect to D,
  - compute the dissimilarity between the new cluster thus obtained and the old (as yet unmerged) clusters.

Two remarks should be made: first, it is easy to show that the set of clusters defined during the successive iterations form a hierarchy; secondly, at each stage, the methods merge the groups of objects which are the closest, and different definitions of the dissimilarity D between groups will lead to different agglomerative methods. This dissimilarity D, computed starting from the initial dissimilarity d, will from now on be referred to as *agglomerative criterion*. Several agglomerative criteria will be studied.

7.3.1.2. *Construction of the index*

Once a hierarchy is defined, it is necessary to define an index. For the singleton clusters, this index is necessarily equal to 0. For the other clusters, this index is generally defined by associating with each new agglomerated cluster the dissimilarity D, which evaluates the proximity between the merged clusters to form this new cluster. Note that in order to have the property of being an index, the proposed clusters must *increase strictly* with the level of hierarchy. Several difficulties may arise:

Inversion problem: For some agglomerative criteria D, the index thus defined is not necessarily a strictly increasing function; this leads to the *inversion* problem and, in this situation, the AHC algorithm does not give an index hierarchy. For example, if the data are the three vertices of an equilateral triangle with sides equal to 1, and if the distance between the centroids of the different clusters is taken as the agglomerative criterion D, the inversion illustrated in Figure 7.8 is obtained. With the agglomerative criteria studied in this chapter, it can be shown that there is no inversion.

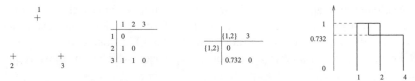

Figure 7.8. *Inversion problem*

Not strictly increasing: When indexes are equal for several nested levels, it is sufficient to 'filter' the hierarchy, i.e. to keep only one cluster containing all the nested classes having the same index. In the example of Figure 7.9, the cluster $A \cup B$ having the same index as the cluster $A \cup B \cup C$ can be removed. This problem can occur with the family of D examined below, and the associated algorithms will therefore require this filtering operation to be undertaken.

Figure 7.9. *Filtering the hierarchy*

7.3.2. *Agglomerative criteria*

There exist several agglomerative criteria, but the following are the most commonly used:

– single linkage or nearest-neighbor criterion [SIB 73]

$$D_{\min}(A, B) = \min\{d(i, i'), i \in A \text{ and } i' \in B\};$$

– complete linkage or furthest-neighbor criterion [SOR 48]

$$D_{\max}(A, B) = \max\{d(i, i'), i \in A \text{ and } i' \in B\};$$

– average linkage criterion [SOK 58]

$$D_{\text{average}}(A, B) = \frac{\sum_{i \in A} \sum_{i' \in B} d(i, i')}{n_A . n_B}$$

where n_E represents the cardinality of the cluster E.

7.3.3. *Example*

We now consider four aligned points (Figure 7.10), separated by the distances 2, 4 and 5. We take the standard Euclidean distance as the dissimilarity measure between these points, and perform the AHC algorithm according to the three agglomerative criteria. The results are reported in Figures 7.11–7.13. Note that in the latter case we obtain two different solutions, depending on whether we choose to merge the clusters {1,4} and {3} or the clusters {2} and {4}.

Figure 7.10. *Example*

	1	2	3	4
1	0			
2	11	0		
3	6	5	0	
4	2	9	4	0

	{1,4}	2	3
{1,4}	0		
2	9	0	
3	4	5	0

	{1,4,3}	2
{1,4,3}	0	
2	5	0

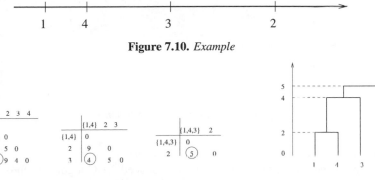

Figure 7.11. $D_{\min}$ *criterion*

	1	2	3	4
1	0			
2	11	0		
3	6	5	0	
4	2	9	4	0

	{1,4}	2	3
{1,4}	0		
2	11	0	
3	6	5	0

	{1,4}	{2,3}
{1,4}	0	
{2,3}	11	0

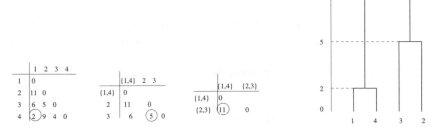

Figure 7.12. $D_{\max}$ *criterion*

7.3.4. *Ward's method or minimum variance approach*

When the objects $\boldsymbol{x}_1, \ldots, \boldsymbol{x}_n$ are described by p continuous variables, it is possible to use the following agglomerative criterion

$$D_{\text{ward}}(A, B) = \frac{n_A . n_B}{n_A + n_B} d^2(\boldsymbol{\mu}_A, \boldsymbol{\mu}_B)$$

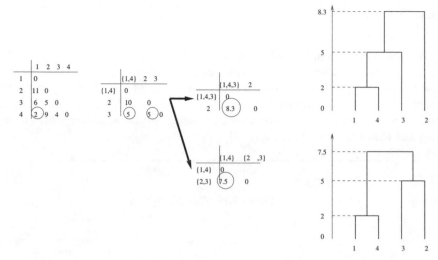

Figure 7.13. D_{average} *criterion*

where d is the Euclidean distance and μ_E and n_E represent the centroid and the cardinality of the set E, respectively. The associated AHC algorithm thus obtained is known as Ward's method [WAR 63].

7.3.5. *Optimality properties*

We have seen that the concept of indexed hierarchy is equivalent to the concept of ultrametric. The AHC algorithm therefore transforms an initial dissimilarity measure d into a new dissimilarity measure δ which is ultrametric. Consequently, the aim of hierarchical clustering may be stated in the following terms: find the 'closest' ultrametric δ to the dissimilarity measure d. This notion of 'closest' can be defined by a distance on the set of dissimilarity measures on Ω. We can use for instance $\Delta(d, \delta) = \sum_{i,i' \in \Omega} (d(i, i') - \delta(i, i'))^2$ or $\Delta(d, \delta) = \sum_{i,i' \in \Omega} |d(i, i') - \delta(i, i')|$. Unfortunately, this problem has no general solution. We shall now turn our attention to the optimality properties of the various algorithms previously described.

If the data dissimilarities d exhibit a strong clustering tendency, that is if d is ultrametric, it can easily be shown that the three algorithms – single, complete and average linkage – give the same result, which is simply the hierarchy associated with the initial ultrametric.

7.3.5.1. *Hierarchy of single linkage*

Let U be the set of all ultrametrics smaller than the initial dissimilarity measure d ($\delta \in U \Leftrightarrow \forall i, i' \in \Omega, \delta(i, i') \leqslant d(i, i')$) and δ_m the higher hull of U, i.e. the application from $\Omega \times \Omega$ to $\mathbb{R}$ satisfying $\forall i, i' \in \Omega, \delta_m(i, i') = \sup\{\delta(i, i'), \delta \in U\}$. It can be shown that δ_m remains an ultrametric, namely the ultrametric obtained by the AHC algorithm with the single linkage criterion. Among all the ultrametrics less than d it is, moreover, the closest to d with respect to every distance Δ. This ultrametric is commonly called the *subdominant* ultrametric. This result also shows that the single linkage agglomerative criteria gives a unique indexed hierarchy.

This kind of hierarchical clustering is analogous to the problem of determining the minimum spanning tree, a well-known problem in graph theory. We consider the complete graph on Ω, each edge (a, b) of this graph being valued by the dissimilarity $d(a, b)$. We can show that the search for this tree is equivalent to the search for the subdominant ultrametric. Then, to find the tree corresponding to the single linkage, it is possible to use the algorithms which were developed to find a minimum spanning tree for a connected weighted graph, for instance Prim's or Kruskal's algorithms [KRU 56, PRI 57]. The link between the minimum spanning tree and the subdominant ultrametric is illustrated by the small example in Figure 7.14. The corresponding dendrogram and subdominant ultrametric are described in Figure 7.15. By retaining from the initial complete graph only 3 edges used in the algorithm, the edge (2,4) of length 2, the edge (1,2) of length 3 and the edge (2,3) of length 7, we obtain the minimum spanning tree (Figure 7.16).

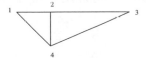

Figure 7.14. *Data*

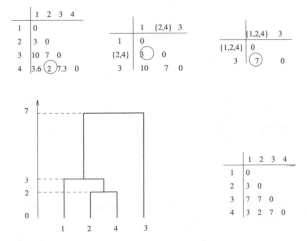

Figure 7.15. *Construction of the subdominant ultrametric*

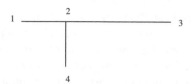

Figure 7.16. *Minimum spanning tree*

This link with the minimum spanning tree highlights one of the weaknesses of this criterion: when a cluster is elongated, two distant points can be agglomerated together early in the hierarchy if there exists a chain of points connecting them.

7.3.5.2. *Hierarchy of complete linkage*

The results obtained by the single linkage cannot be extended to this situation. For instance, if we try to define the closest ultrametric higher than d we need to consider the lower hull of the ultrametrics greater than d. Unfortunately, this hull is not necessarily an ultrametric, and Figure 7.17 provides a counterexample. We can verify that δ_1 (middle table) and δ_2 (right table) are two ultrametrics higher than d (left table) defined with the 3 points a, b and c, and that the lower hull of these ultrametrics is just d. Consequently, this hull of all the ultrametrics higher than d is necessarily d, which is not an ultrametric.

	a	b	c
a	0		
b	1	0	
c	2	1	0

	a	b	c
a	0		
b	1	0	
c	2	2	0

	a	b	c
a	0		
b	2	0	
c	2	1	0

Figure 7.17. *Distance d and ultrametrics δ_1 and δ_2*

The properties of this agglomerative criterion are therefore not as interesting as those of the criterion discussed previously, particularly since it does not necessarily yield a unique result. For example, because of the presence of two or more equalities in the intermediate dissimilarity matrices, different results are obtained by changing the order of the agglomerations of clusters. This can also be observed in the previous example of the average criterion (see Figure 7.13).

7.3.5.3. *Hierarchy of average linkage*

The average criterion does not check any property of optimality, but in practice it has been shown that it is close to the ultrametric that minimizes

$$\sum_{i,i' \in \Omega} (d(i, i') - \delta(i, i'))^2.$$

Average linkage can be viewed as a compromise between single and complete linkages.

7.3.5.4. *Ward's method*

Let $P = (P_1, \ldots, P_g)$ be a partition and P' be the partition obtained from P by merging the clusters P_k and P_ℓ. Knowing that the centroid of $P_k \cup P_\ell$ is equal to

$$\frac{n_k \overline{x}_k + n_\ell \overline{x}_\ell}{n_k + n_\ell},$$

we have the equality

$$I_W(P') - I_W(P) = \frac{n_k n_\ell}{n_k + n_\ell} d^2(\overline{x}_k, \overline{x}_\ell)$$

where I_W is the within-group sum of squares (equation (7.3)).

Since merging two classes necessarily increases the within-group sum-of-squares criterion, it is possible to propose an AHC algorithm which merges the two classes that increase the within-group sum of squares by the lowest possible amount at each stage, i.e. that minimize the expression:

$$D(A, B) = \frac{n_k n_\ell}{n_k + n_\ell} d^2(\overline{x}_k, \overline{x}_\ell)$$

which is identical to the agglomerative criterion in Ward's method. This AHC algorithm therefore has a local optimum property: it minimizes the within-group sum-of-squares criterion at each step of the algorithm, but there is no global optimality.

7.3.6. *Using hierarchical clustering*

The first difficulty when using hierarchical clustering is choosing the dissimilarity on Ω, as described in section 7.2.1. Once this difficulty has been overcome, there remains the choice of the agglomerative criterion D. We have seen that several possible criteria exist, and that the dendrograms so obtained may not be identical. This choice is crucial and has been extensively studied in the context of different approaches for hierarchical and non-hierarchical methods. We remark that the sizes, shapes and the presence of outliers influence the obtained results. We now provide some practical guidelines for selecting the appropriate criterion:

– First, when there is clearly a clustering structure without outliers and the clusters are well separated, the different criteria can give the same dendrogram.

– As we have seen, from a theoretical point of view the single linkage satisfies a certain number of desirable mathematical properties. In practice, however, it outperforms other studied criteria only when the clusters are elongated or irregular and there is no chain of points between the clusters. It is very prone to chaining effects. To have an idea of the shapes of clusters, methods of visualization such as principal component analysis or multidimensional scaling are useful.

– Complete linkage means that outliers can be integrated into the process of training clusters, and so clusters with a single outlying object are avoided.

– Even though it has no theoretical basis, the average criterion tends to produce clusters that accurately reflect the structure present in the data, but this can require a great deal of computation.

– Generally, when the variables are continuous, the recommended criterion is the within-group sum-of-squares criterion used by Ward's method, even if this entails extensive computation. The results are then used jointly with those of PCA. However, this criterion tends to give spherical clusters of nearly equal size.

Tools are often needed to facilitate the interpretation, as well as tools to decrease the number of levels of the dendrogram, such as the classical *semi-partial R-square*

$$\frac{I_W(P_k \cup P_\ell) - I_W(P_k) - I_W(P_\ell)}{I},$$

where $I = \frac{1}{n} \sum_{i \in \Omega} d^2(x_i, \bar{x})$ which expresses the loss of homogenity when the clusters P_k and P_ℓ are agglomerated. This loss decreases with the number of clusters, and so a scree plot with one or several *elbows* may be used to propose a cut of the dendrogram. Note that methods of visualization such as PCA can assist in assessing the number of clusters.

Finally, note that the problems arising from the time complexity of the AHC algorithms, which depend on the linkage chosen, are in practice solved first using the relations

$$D_{\min} : D(A, B \cup C) = \min\{D(A, B), D(A, C)\},$$

$$D_{\max} : D(A, B \cup C) = \max\{D(A, B), D(A, C)\},$$

$$D_{\text{average}} : D(A, B \cup C) = \frac{n_B.D(A, B) + n_C.D(A, C)}{n_B + n_C},$$

$$D_{\text{ward}} : D(A, B \cup C) = \frac{(n_A + n_B)D(A, B) + (n_A + n_C)D(A, C) - n_A D(B, C)}{n_A + n_B + n_C}$$

known as the *recurrence formulae of Lance and Williams* [LAN 67]. Algorithms such as that of De Rham [DER 80], based on the construction of nearest neighbor chains and carrying out aggregations whenever reciprocally nearest neighbors are encountered, can then be used.

Because of their tree structure, hierarchical methods have proved very successful. Unfortunately, given the complexity of the AHC algorithm, is not suitable for large volumes of data. In addition, in the merge process, once a cluster is formed it does not undo what was previously done; no modifications of clusters or permutations of objects are therefore possible. Finally, the AHC algorithm with the fourth criterion studied generally gives convex clusters that are fragile in the presence of outliers. Other approaches can be used to correct these weaknesses, such as CURE (clustering using representatives) [GUH 98], CHAMELEON based on a k-nearest neighbor graph [KAR 99] and BIRCH (balanced iterative reducing and clustering using hierarchies) [ZHA 96]. CHAMELEON and BIRCH are particularly interesting for large datasets. In this context, we shall see that non-hierarchical methods, used either alone or in combination with hierarchical methods, are preferable.

7.4. Partitional clustering: the k-means algorithm

The objective of partitional (or non-hierarchical clustering) is to define a partition of a set of objects into clusters, such that the objects in a cluster are more 'similar' to each other than to objects in other clusters. This section is devoted to the k-means algorithm, which is the classical method in partitional clustering when the data Ω are a set of objects $x_1, \ldots, x_n$ described by p continuous variables.

7.4.1. *The algorithm*

k-means involves the following steps when partitioning the data into g clusters:

1) Randomly select g objects from Ω as cluster centers.

2) While there is no convergence:

   a) Assign each object in Ω to the nearest cluster center. If two or more cluster centers are equally close, the object is assigned to the cluster with the smallest subscript.

   b) Use the centroids of the different clusters as the new cluster centers.

We illustrate the different steps of k-means by applying it with $g = 2$ to a simple set Ω of 10 objects located in $\mathbb{R}^2$. The different steps of the k-means algorithm are summarized in Figure 7.18, where the centers are depicted by os. The process terminates and this algorithm will modify the results no further. Note that the obtained partition corresponds to the observable structure in two clusters.

Using the two-parameter criterion $C(P, \lambda)$,

$$C(P, \lambda) = \frac{1}{n} \sum_{k=1}^{q} \sum_{i \in P_k} ||x_i - \lambda_k||^2, \tag{7.4}$$

where $P = (P_1, \ldots, P_g)$ is a partition of Ω and $L = (\lambda_1, \ldots, \lambda_g)$ with $\lambda_k \in \mathbb{R}^p$ represents the center of the cluster P_k. The alternated optimization of $C(P, L)$ for

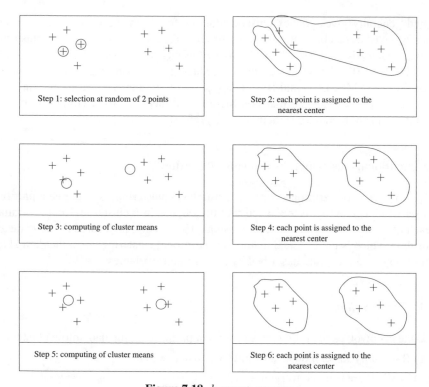

Figure 7.18. *k-means process*

fixed L and then for fixed P leads precisely to the two steps of the k-means algorithm. This algorithm yields a sequence $L^{(0)}, P^{(1)}, L^{(1)}, P^{(2)}, \ldots$ of centers and partitions with decreasing values of the criterion C and, as the number of partitions of Ω in g clusters is finite, the sequence is stationary. Moreover, as $\boldsymbol{\lambda}_k^{(c)}$ is the centroids $\overline{\boldsymbol{x}}_k^{(c)}$ of the cluster $P_k^{(c)}$, we have the relation

$$C(P^{(c)}, L^{(c)}) = I_W(P^{(c)})$$

where I_W is the within-group sum-of-squares criterion (equation (7.3)). Finally, the k-means algorithm defines a sequence of partitions decreasing the within-group sum-of-squares criterion.

7.4.2. *k-means: a family of methods*

The term k-means actually covers a whole family of methods, and the algorithm described in the previous section is only one example. Bock [BOC 07] has carried out an interesting survey of some historical issues related to the k-means algorithm.

All these methods assume that the data come from a probability distribution. In some situations this distribution will be known, but usually the probability P is unknown and we have a sample of fixed size. More than the algorithm itself, what unifies this family is the form of the problem which, if P is known, involves minimizing the continuous criterion

$$I_W(P) = \sum_k \int_{P_k} ||\boldsymbol{x} - \mathbb{E}_{P_k}(\boldsymbol{X})||^2 f(\boldsymbol{x}) d\boldsymbol{x}$$

where $\boldsymbol{X}$ is a random vector in $\mathbb{R}^p$ with pdf f and P is a partition of $\mathbb{R}^p$ and, if we have only one sample, to minimize discrete criterion (7.3),

$$I_W(P) = \frac{1}{n} \sum_k \sum_{i \in P_k} ||\boldsymbol{x}_i - \overline{\boldsymbol{x}}_k||^2$$

where $\boldsymbol{x}_1, \ldots, \boldsymbol{x}_n$ is a sample in $\mathbb{R}^p$ and P is a partition of this sample.

Regarding minimizing the continuous criterion, we should cite Dalenius [DAL 50, DAL 51], Lloyd's algorithm [LLO 57] in the context of scalar quantization in the 1D case and Steinhaus [STE 56] for data in $\mathbb{R}^p$ in the multidimensional case. Regarding the discrete criterion, different strategies have been used: first, we have batch algorithms that process all the objects of the sample at each iteration and incremental algorithms that process only one object at each iteration. Secondly, we have on-line or off-line training. In on-line training, each object is discarded after it has been processed (on-line training is always incremental) and in off-line training, all the data are stored and can be accessed repeatedly (batch algorithms are always off-line). The term *sequential* is ambiguous, referring sometimes to incremental algorithms and sometimes to on-line learning. On-line k-means variants are particularly suitable when all the data to be classified are not available at the outset.

The k-means algorithm previously described (section 7.4.1) is an example of a batch algorithm. This batch version, which is the most often used, was proposed by Forgy [FOR 65], Jancey [JAN 66] and Linde *et al.* [LIN 80] in vector quantization. The algorithm of MacQueen [MAC 67], who was the first to use the name 'k-means', is an example of on-line k-means. The objects of the assignation step are randomly selected, and the update of cluster means is performed immediately after every assignation of one object. More precisely, at the cth iteration, the object $\boldsymbol{x}$ is randomly selected. We then determine the nearest center $\boldsymbol{\mu}_k^{(c)}$ which becomes, after assignation of $\boldsymbol{x}$, equal to

$$\boldsymbol{\mu}_k^{(c+1)} = \frac{n_k^{(c)} \cdot \boldsymbol{\mu}_k^{(c)} + \boldsymbol{x}}{n_k^{(c)} + 1},$$

where $n_k^{(c)}$ represents the cardinality of the cluster $P_k^{(c)}$. Note that the expression of the new cluster mean can take the form

$$\boldsymbol{\mu}_k^{(c+1)} = \boldsymbol{\mu}_k^{(c)} + \frac{1}{n_k^{(c)} + 1}(\boldsymbol{x} - \boldsymbol{\mu}_k^{(c)}),$$

or, more generally,

$$\boldsymbol{\mu}_k^{(c+1)} = \boldsymbol{\mu}_k^{(c)} + \varepsilon(c)(\boldsymbol{x} - \boldsymbol{\mu}_k^{(c)}),$$

where $\varepsilon(c)$ is a decreasing learning coefficient. The usual hypothesis on the adaptation parameter to obtain almost certain results is then (conditions of Robbins–Monro) $\sum_c \varepsilon(c) = +\infty$ and $\sum_c \varepsilon(c)^2 < +\infty$.

Variants have been proposed, including the *neural gas* algorithm [MAR 91], where all the centers are modified at each step according to the following procedure:

$$\boldsymbol{\mu}_k^{(c+1)} = \boldsymbol{\mu}_k^{(c)} + \varepsilon(c)h(r_k)(\boldsymbol{x} - \boldsymbol{\mu}_k^{(c)}),$$

where h is a decreasing function and r_k is the rank of $\boldsymbol{\mu}_k$ ordered according to the decreasing values of $||\boldsymbol{x} - \boldsymbol{\mu}_k||^2$.

7.4.3. *Using the k-means algorithm*

We now summarize briefly the principal features of the k-means algorithm and its use.

Before performing a cluster analysis on coordinate data, it is necessary to consider scaling or transforming the variables since variables with large variances tend to have a greater effect on the resulting clusters than variables with small variances. Other transformations can be used according to the nature of data. This scaling step is very important and may be present in the metric.

To illustrate the importance of the metric in clustering, we can use the Figure 7.1 data, consisting of a set of 900 objects described by 3 continuous variables. If we have applied k-means using the Euclidean distance, we obtain the partition of 3 clusters $a1, a2$ and $a3$, and with the metric χ^2 we obtain another partition of 3 clusters $b1, b2$ and $b3$. To evaluate the influence of the two distances, we project the clusters onto the first factorial planes by PCA and CA (Figures 7.19 and 7.20). We can easily evaluate the influence of the metrics and therefore the risk entailed by applying one metric rather than another.

The k-means algorithm can also use an L_m clustering criterion instead of the least-squares L_2 criterion. Note that values of m less than 2 reduce the effect of outliers on the cluster centers in relation to the least-squares criterion.

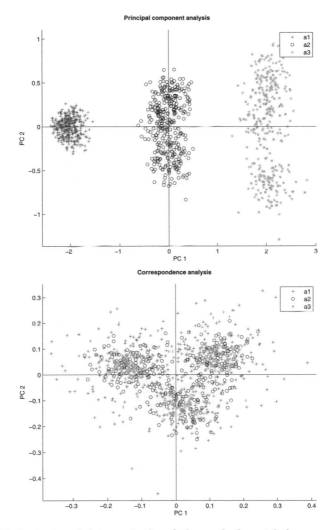

Figure 7.19. *Projection of clusters a1, a2 and a3 onto the factorial planes spawned by the first and second axes by PCA and CA*

If our aim is to find the partition P minimizing the criterion I_W, the k-means algorithm does not necessarily provide the best result. It simply provides a sequence of partitions, the values of whose criterion will decrease, giving a local optimum. Since convergence is reached very quickly in practice (often in fewer than 10 iterations even with a large dataset), the user can run k-means several times with different random initializations, in order to obtain an interesting solution. Several strategies are then possible. We can either retain the best partition, i.e. that which optimizes the criterion, or we can use a set of results to deduce stable clusters (the 'strong forms' method).

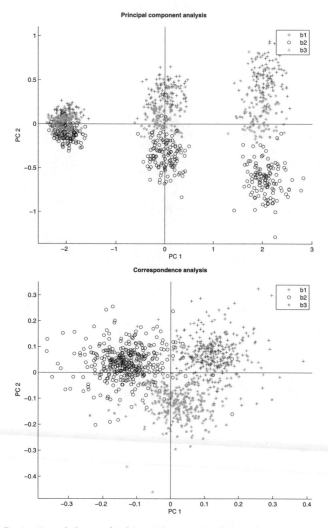

Figure 7.20. *Projection of clusters* b1, b2 *and* b3 *onto the factorial planes spawned by the first and second axes by PCA and CA*

We can also select a 'good' initialization with supplementary information or with an automatic procedure (points separated by large distances, regions with high densities, etc.). There is, however, a tradeoff between the time required to establish the initial configuration and the time taken by the algorithm itself. It is also possible to use a certain number of stochastic methods such as simulated annealing which, without guaranteeing a global optimum, have properties of asymptotic convergence.

In general, the criterion is not independent of the number of classes. For example, the partition into n classes, where each object forms a singleton cluster, has a null within-group sum-of-squares criterion and therefore the optimal partition is without interest. It is therefore necessary to fix *a priori* the number of classes. If this number is not known, several solutions are possible to solve this very difficult problem. For example, the best partition is sought for several numbers of classes. We observe how the criterion decreases with the number of classes, and then select the number of classes using a scree plot and choosing an elbow. It is also possible to add additional constraints relating, for example, to the number of objects by cluster or to the volume of a cluster (see e.g. the Isodata algorithm [BAL 67]).

Finally, note that there are other approaches using statistical methods, such as hypothesis tests or selection model criteria. This latter approach, apparently the most interesting, consists of penalizing the criterion by a function depending on the number of classes, making the criterion 'independent' of this number of classes. For instance, minimizing the within-group sum of squares can be viewed as maximizing a likelihood (see Chapter 8), and selection model criteria such as the Akaike information criterion (AIC) or the Bayesian information criterion (BIC) can be used to determine the number of clusters.

It can be interesting to use k-means and Ward's method simultaneously. The two methods are similar in that they both attempt to minimize the within-group sum-of-squares criterion. This leads us to propose strategies using the two approaches, such as, for example, the hybrid method proposed by Wong [WON 82].

7.5. Miscellaneous clustering methods

7.5.1. *Dynamic cluster method*

The *dynamic cluster method* or *méthode des nuées dynamiques* (MND) proposed by Diday [DID 71, DID 76] is a generalization of the k-means algorithm based on the quite powerful idea that the cluster centers are not necessarily centroids of clusters in $\mathbb{R}^p$. The proposition is to replace them by centers that may take a variety of forms, depending on the problem to be solved. Let $\mathbb{L}$ be the set of centers and $D : \Omega \times L \to \mathbb{R}^+$, a measure of dissimilarity between objects in Ω and the centers of $\mathbb{L}$. The aim is to seek a partition of Ω of g clusters minimizing the criterion:

$$C(P, L) = \sum_{k=1}^{g} \sum_{x \in P_k} D(x, \lambda_k)$$

where $P = (P_1, \ldots, P_g)$ and $L = (\lambda_1, \ldots, \lambda_g)$ with $\lambda_k \in \mathbb{L}$. Note that if $\Omega \subset \mathbb{R}^p$, $\mathbb{L} = \mathbb{R}^p$ and $D(\mathbf{x}, \boldsymbol{\lambda}) = d^2(\boldsymbol{x}, \boldsymbol{\lambda})$, then $C(P, L)$ is precisely the criterion C used in k-means algorithm (7.4). As with k-means, to tackle the minimization of $C(P, L)$ we can employ an alternating optimization method using the steps:

1) Compute $P^{(c+1)}$ minimizing $C(., L^{(c)})$.

2) Compute $L^{(c+1)}$ minimizing $C(P^{(c+1)}, .)$.

These steps yield the sequence

$$L^{(0)}, P^{(1)}, L^{(1)}, P^{(2)}, L^{(2)}, \ldots, P^{(c)}, L^{(c)}, \ldots,$$

where $L^{(0)}$ is an arbitrary initialization. The conditions of existence of this type of algorithm relate only to the second step. The first step is equivalent to the assignment step in k-means, and is quite straightforward. In contrast, the second step depends on the particular situation. If the minimizing in step (2) can be defined and leads to a single value, as in the case of k-means, then the sequence $(P^{(c)}, L^{(c)})$ defined is stationary and the sequence of associated criterion $W(P^{(c)}, L^{(c)})$ is decreasing. We now take a brief look at some examples of applications:

– The k-medoids algorithm [KAU 87] is a typical dynamic cluster method minimizing the criterion C used in k-means, but in contrast to k-means the cluster centers are objects in Ω. Different versions have been proposed: PAM (partition around medoids) [KAU 90] and CLARANS (Clustering Large Applications based upon RANdomized Search) [NG 94] which is more efficient for large volumes of data. When the data are given as a dissimilarity matrix d, this method is particularly well adapted. In this context, it is generally preferable to use square dissimilarity.

– The k-modes algorithm [HUA 97, NAD 93] for categorical data is another example. In this approach, the centers are vectors of categories and the dissimilarity between an object and a center, which are two vectors of categories, is expressed as $D(\boldsymbol{x}_i, \boldsymbol{\lambda}_k) = \sum_{j=1}^{p} \delta(x_i^j, \lambda_k^j)$, where $\delta(x_i^j, \lambda_k^j)$ is equal to 1 if $x_i^j = \lambda_k^j$ and 0 otherwise. This dissimilarity measure reflects the number of different categories between the vector $\boldsymbol{x}_i$ and the center $\boldsymbol{\lambda}_k$. When the data are binary, the distance $D(\boldsymbol{x}, \boldsymbol{\lambda}_k) = \sum_{j=1}^{p} |x_i^j - \lambda_k^j|$ is the Manhattan distance and the vector centers belong to $\{0, 1\}^p$.

– A final example, *adaptive distance* [DID 74, DID 77], concerns the choice of the distance which, as we have already seen, is crucial in clustering. When processing dynamic clusters, the metric can be adapted instead of being fixed. As for the k-means situation, this approach can be used if the set Ω is in $\mathbb{R}^p$. To train this metric, it may be considered that $\mathbb{L} = \mathbb{R}^p \times \Delta$ where Δ is the set of distances defined on $\mathbb{R}^p$ and $D(\boldsymbol{x}, (\boldsymbol{\lambda}, d)) = d(\boldsymbol{x}, \boldsymbol{\lambda})$. The method therefore performs clustering and distance metric learning simultaneously, and the process allows the shapes of clusters, for example, to be taken into account.

7.5.2. *Fuzzy clustering*

This section describes another extension of the k-means algorithm that can handle overlapped clusters. The definition of partition is based on the *classical ensemblist concept*. The notion of a *fuzzy partition* is derived directly from the notion of a *fuzzy set.* Fuzzy clustering, developed towards the beginning of the 1970s [RUS 69], generalizes the classical approach in clustering by extending the notion of membership to a cluster. Let us assume that each element x_i can belong to more than one cluster with different levels. In clustering, if $\mathbf{Z}$ is the classification matrix (section 7.2.3.1) corresponding to a partition, it is a question of relaxing the constraint on the $z_{ik} \in \{0, 1\}$ and observing the membership degree coefficients $c_{ik} \in [0, 1]$. A fuzzy partition can then be represented by a fuzzy classification matrix $\mathbf{C} = \{c_{ik}\}$ satisfying the following conditions: $\forall i, k, c_{ik} \in [0, 1]$, $\forall k, \sum_i c_{ik} > 0$ and $\forall i, \sum_k c_{ik} = 1$. Note that the second condition excludes empty clusters and that the third expresses the concept of total membership.

Bezdeck [BEZ 81] proposed the *fuzzy k-means* algorithm, which can be viewed as a fuzzy version of k-means. The objective of this method is to minimize the criterion

$$W(\mathbf{C}) = \sum_{i=1}^{n} \sum_{k=1}^{g} c_{ik}^{\gamma} d^2(x_i, \mathbf{g}_k),$$

where $\mathbf{g}_k \in \mathbb{R}^p$ and d is the Euclidean distance. The value $\gamma > 1$ determines the degree of fuzziness of the final solution, i.e. the degree of overlap between groups. This value must be strictly greater than 1, otherwise the function W is minimal for values of $c_{ik} = 0$ or 1 and we obtain the classical within-group sum of squares. Values used are generally between 1 and 2. This criterion can be minimized by the fuzzy c-means algorithm, which is very similar to the k-means algorithm. This algorithm alternates between the two following steps:

1) compute the centers: $\mathbf{g}_k = \sum_i c_{ik}^{\gamma} x_i / \sum_i c_{ik}^{\gamma}$;

2) compute the fuzzy partition: $c_{ik} = \dfrac{D_i}{\|x_i - \mathbf{g}_k\|^{\frac{2}{\gamma-1}}}$ where $D_i = \sum_{\ell} \dfrac{1}{\|x_i - \mathbf{g}_\ell\|^{\frac{2}{\gamma-1}}}$.

It can be shown that this algorithm minimizes the fuzzy criterion. It has the same weaknesses as k-means: the solution is a local minimum, and the results depend on the initial choice of the algorithm.

The parameter estimation of a mixture model (see Chapter 8) can also be viewed as a fuzzy clustering, and the associated expectation–maximization algorithm is a more statistically formalized method which includes the notion of partial membership in classes. It has better convergence properties and is in general preferred to fuzzy k-means.

7.5.3. Constrained clustering

It is often the case that the user possesses some background knowledge (about the domain or the dataset) that could be useful in clustering the data. Traditional clustering algorithms have no way of making use of such information when it exists. The aim of constrained algorithms is to integrate background information into clustering algorithms. The information in question might be pairwise constraints such as 'must-link' or 'cannot-link' constraints, contiguity constraints or order constraints, which can sometimes be used to improve the clustering algorithm. We have seen previously that iterative clustering algorithms such as k-means give local optima, but there exist situations for which we have effective algorithms allowing a global optimum to be identified. When the criterion is additive over the set of clusters, if there is an order structure on the finite Ω that has to be respected by the partition, it is possible to use dynamic programming [FIS 58] to obtain an algorithm giving the optimal solution. These conditions can be implicit e.g. when the data are in $\mathbb{R}$ and the criterion is the within-group sum of squares, or explicit e.g. with constraints imposed by the situation such as contiguity constraints in geographical data.

7.5.4. *Self-organizing map*

The *self-organizing map* (SOM) or *Kohonen map* [KOH 82] was first inspired by the adaptive formation of topology-conserving neural projection in the brain. Its aim is to generate a mapping of a set of high-dimensional input signals onto a 1D or 2D array of formal neurons. Each neuron becomes representative of some input signals such that the topological relationship between input signals in the input space is reflected as faithfully as possible in the arrangement of the corresponding neurons in the array (also called output space). When using this method for clustering, it is possible either to match each neuron with a unique cluster or to match many neurons to one cluster. In the latter case, the Kohonen algorithm produces a reduced representation of the original dataset and clustering algorithms may operate on this new representation.

In the SOM literature, we refer to the clusters by the nodes or neurons, each of which has a weight in $\mathbb{R}^p$. The weights refer to the cluster means. The principal advantage of SOM is that it preserves the topology clustering. Generally, the neurons are arranged as a 1D or 2D rectangular grid preserving relations between the objects, also referred to as units. SOM is therefore a useful tool for visualizing clusters and evaluating their proximity in a reduced space.

The Kohonen map can be viewed as an extension of the on-line k-means algorithm described in section 7.4.3. The expression of the cluster means or the weight of a neuron k becomes, in the SOM context,

$$\boldsymbol{\mu}_k^{(c+1)} = \boldsymbol{\mu}_k^{(c)} + \varepsilon(c) \times h(k, \ell)(\boldsymbol{x} - \boldsymbol{\mu}_k^{(c)}), \tag{7.5}$$

where $h(k, \ell)$ is the neighborhood function between the neuron k whose weight $\boldsymbol{\mu}_k^{(c)}$ is the most similar to $\boldsymbol{x}$ (the best matching unit or the winner) and the other neurons

ℓ with weights μ_ℓ close enough to $\mu_k^{(c)}$. This function can take different forms. It evaluates the proximity between the winner k and the neuron ℓ located in a reduced space, generally in $\mathbb{R}^2$, with position r_k and r_ℓ. A common choice is a Gaussian function

$$h(k, \ell) = \exp\left(-\frac{\alpha\|r_k - r_\ell\|^2}{2\sigma_h(c)}\right)$$

where $\sigma_h(c)$, which controls the width of the neighborhood of h, decreases during the training process, as does the learning rate $\varepsilon(c)$. Like k-means, SOM requires the number of clusters (nodes of the grid) to be fixed and initial values to be selected. Different strategies can be used, as for k-means, but initialization using PCA would appear to be an attractive and interesting approach. In addition, a batch version can be employed.

To illustrate this approach, we have reported its results on the well-known Iris flower data [FIS 36]. The dataset in question consists of the lengths and the widths of sepals and petals of 50 samples from each of three species of Iris flowers (*Iris setosa*, *Iris virginica* and *Iris versicolor*). SOM was applied with a 10×10 grid, that is 100 clusters. Since SOM does not directly show the inter-neuron distances on the map, different techniques are available for visualizing the results of SOM. For example, a traditional mapping method gives the 10×10 grid shown in Figure 7.21, highlighting the three species.

Due to its simplicity, SOM has been very successful over several decades, even though no criterion has been found whose optimization implies expression (7.5). Convergence has not in fact been proved in the general case, but only for special cases and in particular for 1D data.

7.5.5. Clustering variables

When the data consist of a set of objects described by a set of variables, the clustering carried out on the objects can often, without difficulty, be extended to a set of variables. Compared to factor analysis, clustering variables identifies the key variables which explain the principal dimensionality in the data, rather than just abstract factors, greatly simplifying interpretation.

Nevertheless, we should add a word of caution: when the variables are continuous, the principal aim is often to provide clusters where each cluster includes correlated variables. The correlation matrix can then be converted to a dissimilarity matrix by replacing each correlation $\rho_{jj'}$ between two variables j and j' by $1 - \rho_{jj'}$, which is a Euclidean distance. However, unless all correlations are positive, this distance is not appropriate. Otherwise, two variables that are highly negatively correlated may be considered to be very similar. Consequently, $1 - |\rho_{jj'}|$ would appear to be more suitable. On the other hand, since the correlation has an interpretation as the cosine

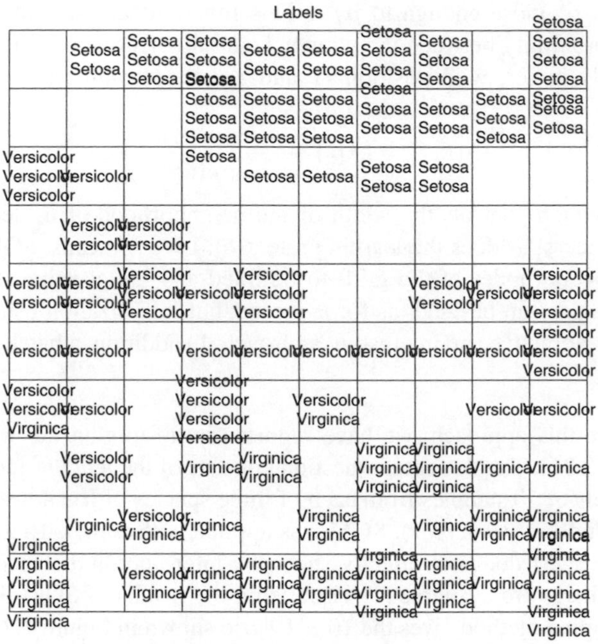

Labels

					Setosa			Setosa
Setosa Setosa	Setosa Setosa Setosa	Setosa Setosa **Setosa**	Setosa Setosa Setosa	Setosa Setosa Setosa	Setosa Setosa Setosa	Setosa Setosa Setosa		Setosa Setosa Setosa
			Setosa Setosa Setosa	Setosa Setosa Setosa	Setosa Setosa Setosa	Setosa Setosa	Setosa Setosa Setosa	Setosa Setosa Setosa
Versicolor Versicolor Versicolor	Versicolor		Setosa		Setosa Setosa	Setosa Setosa		
	Versicolor Versicolor	Versicolor Versicolor						
Versicolor Versicolor Versicolor Versicolor	Versicolor Versicolor Versicolor	Versicolor	Versicolor Versicolor Versicolor		Versicolor Versicolor	Versicolor Versicolor Versicolor Versicolor		
Versicolor Versicolor		Versicolor Versicolor Versicolor Versicolor	Versicolor Versicolor Versicolor	Versicolor Versicolor	Versicolor			
Versicolor Versicolor Virginica	Versicolor Versicolor Versicolor Versicolor	Versicolor Virginica		Versicolor Versicolor				
Versicolor Versicolor	Virginica	Virginica Virginica	Virginica Virginica Virginica Virginica	Virginica Virginica Virginica Virginica				
Virginica	Versicolor Virginica	Versicolor Virginica	Virginica Virginica	Virginica	Virginica Virginica Virginica Irginica			
Virginica Virginica Virginica Virginica	Versicolor Virginica Virginica Virginica	Versicolor Virginica Virginica Virginica Virginica	Virginica Virginica Virginica Virginica Virginica	Virginica Virginica Virginica Virginica	Virginica Virginica Virginica Virginica			Virginica

Figure 7.21. *Iris on grid* 10×10

of the angle between two vectors corresponding to two variables, the dissimilarity expressed as $\arccos(|\rho_{jj'}|)$ can be used. Once the dissimilarity matrix has been defined, hierarchical or partitioning algorithms can be applied.

Furthermore, there exists another commonly-used approach based on factor analysis. The following steps represent a divisive hierarchical algorithm:

1) start by applying a factor analysis;

2) apply an ortho-oblique transformation according to the quartimax criterion [HAR 76];

3) if the second eigenvalue is higher than 1, then two clusters are formed: the first cluster is formed by the variables more closely correlated with the first component and the second cluster by the other variables;

4) apply the previous steps recursively on each cluster until the second eigenvalue is less than 1.

7.5.6. *Clustering high-dimensional datasets*

High-dimensional data present a particular challenge to clustering algorithms. This is because of the so-called curse of dimensionality which leads to the sparsity of the data: in high-dimensional space, all pairs of points tend to be almost equidistant from one another. As a result, it is often unrealistic to define distance-based clusters in a meaningful way. Usually, clusters cannot be found in the original feature space because several features may be irrelevant for clustering due to correlation or redundancy. However, clusters are usually embedded in lower dimensional subspaces, and different sets of features may be relevant for diffcrent sets of objects. Thus, objects can often be clustered differently in varying subspaces of the original feature space. Different approaches have been proposed.

Subspace clustering seeks to find clusters in different subspaces within a dataset. An example of this kind of approach is the CLIQUE (CLustering In QUEst) algorithm [AGR 98]. It is a density-based method that can automatically find subspaces of the highest dimensionality such that high-density clusters exist in those subspaces. *Subspace ranking* aims to identify all subspaces of a (high-dimensional) feature space that contain interesting clustering structures. The subspaces should be ranked according to this level of interest. *Projected clustering* is a method whereby the subsets of dimensions selected are specific to the clusters themselves. We can also cite the HDDC approach (high-dimensional data clustering) of Bouveyron [BOU 07], where a family of Gaussian mixture models designed for high-dimensional data are used to develop a clustering method based on the expectation–maximization algorithm. This method combines the ideas of subspace clustering and parsimonious modeling.

7.6. Block clustering

Although many clustering procedures such as hierarchical clustering and k-means aim to construct an optimal partition of objects or (sometimes) variables, there are other methods known as block clustering methods which consider the two sets simultaneously and organize the data into homogenous blocks. In recent years block clustering, also referred to as co-clustering or bi-clustering, has become an important challenge in the context of data mining. In the text mining field, Dhillon [DHI 01] has proposed a spectral block clustering method by exploiting the duality between rows (documents) and columns (words). In the analysis of microarray data, where data are often presented as matrices of expression levels of genes under different conditions, block clustering of genes and conditions has been used to overcome the problem of choosing the similarity of the two sets found in conventional clustering methods [CHE 00].

The aim of block clustering is to try to summarize this matrix by homogenous blocks. A wide variety of procedures have been proposed for finding patterns in data matrices. These procedures differ in the pattern they seek, the types of data they apply to and the assumptions on which they rest. In particular, we should mention the work of [ARA 90, BOC 79, DUF 91, GAR 86, GOV 83, GOV 84, GOV 95, HAR 75,

MAR 87, VAN 04], who have proposed some algorithms dedicated to different kinds of matrices.

Here, we restrict ourselves to block clustering methods defined by a partition of rows and a partition of columns of the data matrix. If $\mathbf{X}$ denotes an $n \times p$ data matrix defined by $\mathbf{X} = \{(x_i^j); i \in I, j \in J\}$, where I is a set of n rows (objects, observations, cases) and J is a set of d columns (variables, attributes), the basic principle of these methods is to make permutations of objects and variables in order to construct a correspondence structure on $I \times J$. In Figure 7.22, we report a binary dataset described by the set of $n = 10$ objects $I = \{A, B, C, D, E, F, G, H, I, J\}$ and a set of $d = 7$ binary variables $J = \{1, 2, 3, 4, 5, 6, 7\}$ (1). Array (2) consists of data reorganized by a partition of I into $g = 3$ clusters $a = \{A, C, H\}$, $b = \{B, F, J\}$ and $c = \{D, G, I, E\}$. Array (3) consists of data reorganized by the same partition of I and a partition of J into $m = 3$ clusters $I = \{1, 4\}$, $II = \{3, 5, 7\}$ and $III = \{2, 6\}$. Array (3) clearly reveals an interesting pattern.

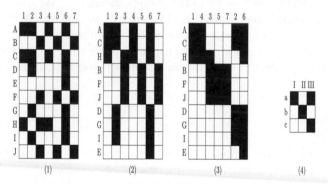

Figure 7.22. *(1) Binary dataset, (2) data reorganized by a partition on I, (3) by partitions on I and J simultaneously and (4) summary binary data*

In this example, the initial (10×7) binary data matrix has been reduced to a $(g \times m) = (3 \times 3)$ summary binary data matrix (Figure 7.22 (4)). This general principle is retained in the block clustering methods described below, which reduce the initial data matrix $\mathbf{X}$ to a simpler data matrix $\mathbf{A}$ having the same structure according to the scheme shown in Figure 7.23.

This problem can be studied under the simultaneous partition approach where the two sets I and J are partioned into g and m clusters. Govaert [GOV 83, GOV 95] has proposed algorithms to perform block clustering on several kinds of data. These algorithms consist of optimizing a criterion $C(P, Q, \mathbf{A})$, where $P = (P_1, \ldots, P_g)$ is a partition of I into g clusters, $Q = (Q_1, \ldots, Q_m)$ is a partition of J into m clusters and $\mathbf{A} = (a_k^\ell)$ is a $g \times m$ matrix which can be viewed as a summary of data matrix $\mathbf{X}$. A more precise definition of this summary and criterion W will depend on the

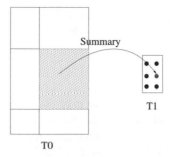

Figure 7.23. *Objective of block algorithms*

nature of the data. The search for the optimal partitions P and Q was made using an iterative algorithm. We shall focus on binary data, contingency tables and continuous data below.

7.6.1. *Binary data*

When the data are binary ($x_i^j = 1$ or 0), the values a_k^ℓ of the summary matrix **A** are also chosen in order to be binary. In this situation the problem consists of minimizing the following criterion:

$$W(P, Q, \mathbf{A}) = \sum_{k=1}^{g} \sum_{\ell=1}^{m} \sum_{i \in P_k} \sum_{j \in Q_\ell} |x_i^j - a_k^\ell| \qquad \text{where} \quad a_k^\ell \in \{0, 1\}.$$

The minimization of this criterion can be carried out by the alternated minimization of two conditional criteria:

$$W(P, \mathbf{A}|Q) = \sum_{k=1}^{g} \sum_{i \in P_k} \sum_{\ell=1}^{m} |u_i^\ell - \#Q_\ell \, a_k^\ell|$$

and

$$W(Q, \mathbf{A}|P) = \sum_{\ell=1}^{m} \sum_{j \in Q_\ell} \sum_{k=1}^{g} |v_k^j - \#P_k \, a_k^\ell|$$

where $u_i^\ell = \sum_{j \in Q_\ell} x_i^j$, $v_k^j = \sum_{i \in P_k} x_i^j$ and $\#B$ is the cardinal of the set B. The different steps of the *Crobin* algorithm performing this optimization are then:

1) Start from an initial position $(P^{(0)}, Q^{(0)}, \mathbf{A}^{(0)})$.

2) Iterate the following two steps until convergence:
   a) Compute $(P^{(c+1)}, \mathbf{A}')$ starting from $(P^{(c)}, \mathbf{A}^{(c)})$ by minimizing $W(P, \mathbf{A}|Q^{(c)})$.
   b) Compute $(Q^{(c+1)}, \mathbf{A}^{(c+1)})$ starting from $(Q^{(c)}, \mathbf{A}')$ by minimizing $W(Q, \mathbf{A}|P^{(c+1)})$.

Steps 2(a) and 2(b) can be performed by applying the dynamic cluster algorithm (see section 7.5.1) to $n \times m$ matrix (u_i^ℓ) with the L_1 distance and centers of the form $(\#Q_1 a_k^1, \ldots, \#Q_m a_k^m)$ and to the $g \times p$ matrix (v_k^j) with the L_1 distance and centers of the form $(\#P_1 a_1^\ell, \ldots, \#P_g a_g^\ell)$, respectively.

To illustrate the *Crobin* algorithm, we propose running it on a simple example. Let $\mathbf{X}$ be a 20×10 matrix of binary data (Table 7.1a). In Table 7.1b, we report the best result obtained at convergence by rearranging rows and columns of the initial data matrix x according to the best partitions P and Q. The centers a_k^ℓ and degree of homogenity corresponding to the proportion of binary values in the block (k, ℓ), equal to the binary center a_k^ℓ, are presented in Table 7.1c and d, respectively.

	abcdefghij
y_1	1010001101
y_2	0101110011
y_3	1000001100
y_4	1010001100
y_5	0111001100
y_6	0101110101
y_7	0111110111
y_8	1100111011
y_9	0100110000
y_{10}	1010101101
y_{11}	1010001100
y_{12}	1010000100
y_{13}	1010001101
y_{14}	0010011100
y_{15}	0010010100
y_{16}	1111001100
y_{17}	0101110011
y_{18}	1010011101
y_{19}	1010001000
y_{20}	1100101100

(a) The data

	a c g h	b d e f i j
y_2	0 0 0 0	1 1 1 1 1 1
y_6	0 0 0 1	1 1 1 1 0 1
y_7	0 1 0 1	1 1 1 1 1 1
y_8	1 0 1 0	1 0 1 1 1 1
y_9	0 0 0 0	1 0 1 1 0 0
y_{17}	0 0 0 0	1 1 1 1 1 1
y_1	1 1 1 1	0 0 0 0 0 1
y_3	1 0 1 1	0 0 0 0 0 0
y_4	1 1 1 1	0 0 0 0 0 0
y_5	0 1 1 1	1 1 0 0 0 0
y_{10}	1 1 1 1	0 0 1 0 0 1
y_{11}	1 1 1 1	0 0 0 0 0 0
y_{12}	1 1 0 1	0 0 0 0 0 0
y_{13}	1 1 1 1	0 0 0 0 0 1
y_{14}	0 1 1 1	0 0 0 1 0 0
y_{15}	0 1 0 1	0 0 0 1 0 0
y_{16}	1 1 1 1	1 1 0 0 0 0
y_{18}	1 1 1 1	0 0 0 1 0 1
y_{19}	1 1 1 0	0 0 0 0 0 0
y_{20}	1 0 1 1	1 0 1 0 0 0

(b) Reorganized data

0 1
1 0

(c) Centers a

0.79 0.86
0.86 0.83

(d) Degree of homogenity

Table 7.1. *Example of binary data*

7.6.2. Contingency table

Block clustering methods have practical importance in a wide variety of applications such as text, weblog and market basket data analysis, where the data are arranged in a 2D $n \times p$ contingency or co-occurrence table $\mathbf{X} = \{(x_{ij}); i \in I, j \in J\}$, where I is a categorical variable with n categories and J a categorical variable with p categories. We shall denote the row and column totals of x by $x_{i.} = \sum_{j=1}^{p} x_{ij}$ and $x_{.j} = \sum_{i=1}^{n} x_{ij}$, and the overall total by $N = \sum_{ij} x_{ij}$. We shall also use the frequency table $\{(f_{ij} = x_{ij}/N); i \in I, j \in J\}$, the marginal frequencies

$f_{i.} = \sum_j f_{ij}$ and $f_{.j} = \sum_i f_{ij}$, the row profiles $f_J^i = (f_{i1}/f_{i.}, \ldots, f_{ip}/f_{i.})$ and the average row profile $f_J = (f_{.1}, \ldots, f_{.p})$.

The information contributed by such a table can be evaluated by several measures of association, the most commonly employed being chi-square χ^2. This criterion, used for example in correspondence analysis, is defined as:

$$\chi^2(I, J) \;=\; \sum_{i,j} \frac{(x_{ij} - \frac{x_{i.}x_{.j}}{N})^2}{\frac{x_{i.}x_{.j}}{N}} = N \sum_{i,j} \frac{(f_{ij} - f_{i.}f_{.j})^2}{f_{i.}f_{.j}}.$$

It is a measure that generally provides statistical evidence of a significant association, or dependency, between rows and columns of the table. This quantity represents the deviation between the theoretical frequencies $f_{i.}f_{.j}$ that we would have if I and J were independent, and the observed frequencies f_{ij}. If I and J are independent, the χ^2 will be zero, and if there is a strong relationship between I and J, χ^2 will be high. Consequently, a significant chi-square indicates a departure from row or column homogenity and can be used as a measure of heterogenity. The chi-square can then be used to evaluate the quality of a partition P of I and Q of J. We associate with these partitions P and Q the chi-square of the $g \times m$ contingency table obtained from the initial table. The sum of rows and columns of each cluster is denoted $\chi^2(P, Q)$ and equal to

$$N \sum_{k,\ell} \frac{(f_{k\ell} - f_{k.}f_{.\ell})^2}{f_{k.}f_{.\ell}}.$$

It follows that we have $\chi^2(I, J) \geqslant \chi^2(P, Q)$, which shows that the proposed regrouping necessarily leads to a loss of information.

In this situation, the objective of block clustering is to find the partitions P and Q which minimize this loss, i.e. which maximize $\chi^2(P, Q)$. Govaert [GOV 83, GOV 95] has proposed the *Croki2* algorithm, which is an alternated maximization of $\chi^2(P, J)$ and $\chi^2(I, Q)$. These maximizations can be performed using the *Mndki2* algorithm [GOV 07], which is a dynamic cluster method with the χ^2 metric. The *Mndki2* algorithm is based on the same geometrical representation of a contingency table as that used in correspondence analysis. In this representation, each row i corresponds to a vector of $\mathbb{R}^p$ defined by the profile f_{iJ} weighted by the marginal frequency $f_{i.}$. The distances between profiles are not defined by the usual Euclidean metric but instead by the weighted Euclidean metric, known as the chi-squared metric, defined by the diagonal matrix $(\frac{1}{f_{.1}}, \ldots, \frac{1}{f_{.p}})$.

7.6.3. *Continuous data*

For continuous data, the summary a_k^ℓ associated with each block k, ℓ is a real number and the most commonly used criterion to measure the deviation between the data matrix $\mathbf{X} = (x_i^j)$ and the structure described by P, Q and $\mathbf{A} = (a_k^\ell)$ is:

$$W(P, Q, \mathbf{A}) = \sum_{k=1}^{g} \sum_{\ell=1}^{m} \sum_{i \in P_k} \sum_{j \in Q_\ell} (x_i^j - a_k^\ell)^2,$$

which, by considering the classification matrices $\mathbf{Z}$ and $\mathbf{W}$ associated with the partitions P and Q, can also be written $||\mathbf{X} - \mathbf{Z}\mathbf{A}\mathbf{W}^T||^2$.

In this situation, the objective of block clustering is to find the partitions P and Q minimizing this criterion. Govaert [GOV 83, GOV 95] has proposed the *Croeuc* algorithm which is an alternated minimization of the criterion using reduced intermediate matrices $\mathbf{U} = (u_i^\ell)$, where $u_i^\ell = \sum_{j \in Q_\ell} x_i^j / \#Q_\ell$ and $\mathbf{V} = (v_k^j)$ where $v_k^j = \sum_{i \in P_k} x_i^j / \#P_k$. These two alternated minimizations can be performed using the k-means algorithm. In the first case, k-means is applied to the $n \times m$ matrix $\mathbf{U}$ with the Euclidean distance and the mean values of block clusters. The second step is carried out by the application k-means on the $g \times p$ matrix $\mathbf{V}$ with the Euclidean distance and the mean values of block clusters.

7.6.4. Some remarks

These methods are fast and can process large datasets. Far less computation is required than for processing the two sets separately. Consider the cases of the well-known k-means algorithm and the block clustering which applies an adapted version of k-means alternately to $(n \times m)$ and $(p \times g)$ reduced intermediate matrices. It is clear from Table 7.2, when we compare the number of distance computations in two spaces (I and J) for the two situations (separate as opposed to simultaneous clustering under the same conditions of convergence), that block clustering requires less computation. Consequently, these methods are of interest in data mining.

n	p	g	m	Separately	Simultaneously
100	100	5	5	5×10^5	1.25×10^5
1000	1000	10	5	7.5×10^6	1.375×10^6
1000	1000	10	10	100×10^6	5×10^6

Table 7.2. *Number of distance computations, processing I and J separately, then simultaneously (n and p are the cardinals of I and J, g and m are the numbers of row and column clusters)*

We shall see in Chapter 8 that some of the most popular clustering methods can be viewed as approximate estimations of probability models. For instance, the variance criterion optimized by the k-means algorithm corresponds to the hypothesis of a population arising from a Gaussian mixture. Then, from the expressions minimized by the three algorithms *Crobin*, *Croki2* and *Croeuc*, the question which naturally arises is which probabilistic model these criterion might correspond to. The answer to this question will, on the one hand, provide further insight into this criterion and, on the other hand, suggest new criteria that might be used. In this way, adapted block latent models have been proposed [GOV 03, GOV 05, GOV 06, GOV 08].

7.7. Conclusion

Other methods not studied here can provide solutions for particular kinds of data. *Kernel clustering* methods have attracted much attention in recent years [FIL 08]. These involve transforming a low-dimensional input space into a high-dimensional kernel-deduced feature space in which patterns are more likely to be linearly separable [SCH 02]. They have certain advantages when handling non-linear separable datasets. We can also cite *spectral clustering* techniques which make use of the spectrum of the similarity matrix of the data to perform dimensionality reduction for clustering in fewer dimensions and, more generally, graph clustering techniques.

Cluster analysis is an important tool in a variety of scientific areas. In this chapter, we have presented a state-of-the-art of well-established (as well more recent) methods. We have reviewed different approaches, essentially classical and generally based on numerical criteria. Unfortunately, defining these criteria and using them successfully, as we have seen, is not always easy. To overcome these difficulties there exist other approaches, and the mixture model approach is undoubtedly a very useful contribution to clustering. It offers considerable flexibility, gives a meaning to certain criteria and sometimes leads to replacing criteria with new ones with fewer drawbacks. In addition, it provides solutions to the problem of the number of clusters. The following chapter is devoted to this approach.

7.8. Bibliography

[AGR 98] AGRAWAL R., GEHRKE J., GUNOPULOS D., RAGHAVAN P., "Automatic subspace clustering of high dimensional data for data mining applications", *Proceedings of the ACM SIGMOD Conference*, Seattle, WA., p. 94–105, 1998.

[ARA 90] ARABIE P., HUBERT L. J., "The Bond energy algorithm revisited", *IEEE Transactions on Systems, Man, and Cybernetics*, vol. 20, p. 268–274, 1990.

[BAL 67] BALL G. H., HALL D. J., "A clustering technique for summarizing multivariate data", *Behavorial Science*, vol. 12, num. 2, p. 153–155, March 1967.

[BEZ 81] BEZDEK J., *Pattern Recognition with Fuzzy Objective Function Algorithms*, Plenum Press, New York, 1981.

[BOC 79] BOCK H.-H., "Simultaneous Clustering of Objects and Variables", DIDAY E., Ed., *Analyse des Donnés et Informatique*, INRIA, p. 187–203, 1979.

[BOC 07] BOCK H., "Clustering methods: a history of k-means algorithms", BRITO P., CUCUMEL G., BERTRAND P., CARVALHO F., Eds., *Selected Contributions in Data*

Analysis and Classification, Studies in Classification, Data Analysis, and Knowledge Organization, p. 161–172, Springer Verlag, Heidelberg, 2007.

[BOU 07] BOUVEYRON C., GIRARD S., SCHMID C., "High-dimensional data clustering", *Computational Statistics & Data Analysis*, vol. 52, num. 1, p. 502–519, 2007.

[CHE 00] CHENG Y., CHURCH G., "Biclustering of expression data", *ISMB2000, 8th International Conference on Intelligent Systems for Molecular Biology*, San Diego, California, p. 93–103, August 19–23 2000.

[DAL 50] DALENIUS T., "The problem of optimum stratification", *Skandinavisk Aktuarietidskrift*, vol. 33, p. 203–213, 1950.

[DAL 51] DALENIUS T., GURENEY M., "The problem of optimum stratification. II", *Skandinavisk Aktuarietidskrift*, vol. 34, p. 133–148, 1951.

[DER 80] DE RHAM C., "La classification hiérarchique selon la méthode des voisins réciproques", *Cahier de l'Analyse des Données*, vol. 5, p. 135–144, 1980.

[DHI 01] DHILLON I., "Co-clustering documents and words using bipartite spectral graph partitioning", *Seventh ACM SIGKDD Conference*, San Francisco, California, USA, p. 269–274, 2001.

[DID 71] DIDAY E., "Une nouvelle méthode de classification automatique et reconnaissance des formes: la méthode des nuées dynamiques", *Revue de statistique appliquée*, vol. 19, num. 2, p. 19–34, 1971.

[DID 74] DIDAY E., GOVAERT G., "Classification avec distance adaptative", *Comptes Rendus Acad. Sci.*, vol. 278, p. 993–995, 1974.

[DID 76] DIDAY E., SIMON J.-C., "Clustering analysis", FU K. S., Ed., *Digital Pattern Recognition*, p. 47–94, Spinger-Verlag, Berlin, 1976.

[DID 77] DIDAY E., GOVAERT G., "Classification avec distances adaptatives", *RAIRO Information/Computer Science*, vol. 11, num. 4, p. 329–349, 1977.

[DUF 91] DUFFY D., A.J. Q., "A permutation-based algorithm for block clustering", *Journal of Classification*, vol. 8, p. 65–91, 1991.

[FIL 08] FILIPPONE M., CAMASTRA F., MASULLI F., ROVETTA S., "A survey of kernel and spectral methods for clustering", *Pattern Recognition*, vol. 41, num. 1, p. 176–190, 2008.

[FIS 36] FISHER R.-A., "The use of multiple measurements in taxonomic problems", *Annals of Eugenics*, vol. 7, p. 179–188, 1936.

[FIS 58] FISHER W.-D., "On regrouping for maximum homgeneity", *Journal of the American Statistical Association*, vol. 53, p. 789–598, 1958.

[FOR 65] FORGY E., "Cluster analysis of multivariate data: efficiency versus interpretability of classification", *Biometrics*, vol. 21, num. 3, p. 768–769, 1965.

[GAR 86] GARCIA H., PROTH J. M., "A new cross-decomposition algorithm: the GPM comparison with the bond energy method", *Control and Cybernetics*, vol. 15, p. 155–165, 1986.

[GOV 83] GOVAERT G., Classification croisée, PhD thesis, University of Paris 6, France, 1983.

[GOV 84] GOVAERT G., "Classification de tableaux binaires", DIDAY E., Ed., *Data Analysis and Informatics 3*, Amsterdam, North-Holland, p. 223–236, 1984.

[GOV 95] GOVAERT G., "Simultaneous clustering of rows and columns", *Control and Cybernetics*, vol. 24, num. 4, p. 437–458, 1995.

[GOV 03] GOVAERT G., NADIF M., "Clustering with block mixture models", *Pattern Recognition*, vol. 36, p. 463–473, 2003.

[GOV 05] GOVAERT G., NADIF M., "An EM algorithm for the block mixture model", *IEEE Transactions on Pattern Analysis and Machine Intelligence*, vol. 27, p. 643–647, April 2005.

[GOV 06] GOVAERT G., NADIF M., "Fuzzy clustering to estimate the parameters of block mixture models", *Soft Computing*, vol. 10, p. 415–422, March 2006.

[GOV 07] GOVAERT G., NADIF M., "Clustering of contingency table and mixture model", *European Journal of Operational Research*, vol. 183, p. 1055–1066, 2007.

[GOV 08] GOVAERT G., NADIF M., "Block clustering with Bernoulli mixture models: comparison of different approaches", *Computational Statistics and Data Analysis*, vol. 52, p. 3233–3245, 2008.

[GOW 86] GOWER J. C., "Metric and Euclidean properties", *Journal of Classification*, vol. 3, p. 5–48, 1986.

[GUH 98] GUHA S., RASTOGI R., SHIM K., "CURE: an efficient clustering algorithm for large databases", *ACM SIGMOD International Conference on Management of Data*, p. 73–84, June 1998.

[HAR 75] HARTIGAN J. A., *Clustering Algorithms*, Wiley, New York, 1975.

[HAR 76] HARMAN H.-H., *Modern Factor Analysis*, University Of Chicago Press, Chicago, 1976.

[HUA 97] HUANG Z., "Clustering large data sets with mixed numeric and categorical values", *In The First Pacific-Asia Conference on Knowledge Discovery and Data Mining*, p. 21–34, 1997.

[JAN 66] JANCEY R. C., "Multidimensional group analysis", *Australian Journal of Botany*, vol. 14, num. 1, p. 127–130, 1966.

[KAR 99] KARYPIS G., HAN E.-H., KUMAR V., "CHAMELEON: A hierarchical clustering algorithm using dynamic modeling", *Computer*, vol. 32, p. 68–75, 1999.

[KAU 87] KAUFMAN L., ROUSSEEUW P., "Clustering by means of medoids", DOGE Y., Ed., *Statistical Data Anlaysis Based on the L1-norm and Related Methods*, Amsterdam, North-Holland, 1987.

[KAU 90] KAUFMAN L., ROUSSEEUW P., *Finding Groups in Data*, Wiley, New York, 1990.

[KOH 82] KOHONEN T., "Self organized formation of topological correct feature maps", *Biological Cybernetics*, vol. 43, p. 59–69, 1982.

[KRU 56] KRUSKAL J.-B., "On the shortest spanning subtree of a graph and the traveling salesman problem", *Proceedings of American Mathematical Society*, vol. 7, p. 48–50, 1956.

[LAN 67] LANCE G. N., WILLIAMS W. T., "A general theory of classificatory sorting strategies, 1: hierarchical systems", *Computer Journal*, vol. 9, p. 373–380, 1967.

[LIN 80] LINDE Y., BUZO A., GRAY R., "An algorithm for vector quantizer design", *IEEE Transactions on Communications*, vol. 28, num. 1, p. 84–95, 1980.

[LLO 57] LLOYD S., "Least squares quantization in PCM", *Telephone Labs Memorandum, Murray Hill, NJ. Reprinted (1982) in: IEEE Trans. Information Theory*, vol. 2, p. 129–137, 1957.

[MAC 67] MACQUEEN J., "Some methods of classification and analysis of multivariate observations", *Proceedings of 5th Berkeley Symposium on Mathematical Statistics and Probability*, p. 281–297, 1967.

[MAR 87] MARCHOTORCHINO F., "Block seriation problems: a unified approach", *Applied Stochastic Models and Data Analysis*, vol. 3, p. 73–91, 1987.

[MAR 91] MARTINETZ T., SCHULTEN K., "A 'neural-gas' network learns topologies", KOHONEN T., MÄKISARA K., SIMULA O., KANGAS J., Eds., *Artificial Neural Networks*, Amsterdam, North-Holland, p. 397–402, 1991.

[NAD 93] NADIF M. ET MARCHETTI F., "Classification de données qualitatives et modèles", *Revue de Statistique Appliquée*, vol. 41, num. 1, p. 55–69, 1993.

[NG 94] NG R., HAN J., "Efficient and effective clustering methods for spatial data mining", BOCCA J. B., JARKE M., ZANIOLO C., Eds., *VLDB'94, Proceedings of 20th International Conference on Very Large Data Bases*, September 12–15, 1994, Santiago de Chile, Chile, Morgan Kaufmann, p. 144–155, 1994.

[PRI 57] PRIM R. C., "Shortest connection network and some generalizations", *Bell System Technical Journal*, vol. 36, p. 1389–1401, 1957.

[RUS 69] RUSPINI E. H., "A new approach to clustering", *Information and Control*, vol. 15, p. 22–32, 1969.

[SCH 02] SCHÖLKOPF B. SMOLA A.-J., *Learning with Kernels: Support Vector Machines, Regularization, Optimization, and Beyond*, The MIT Press, London, 2002.

[SIB 73] SIBSON R., "SLINK: An optimally efficient algorithm for the single-link cluster method", *The Computer Journal*, vol. 16, num. 1, p. 30–45, 1973.

[SOK 58] SOKAL R.-R., MICHENER C.-D., "A statistical method for evaluating systematic relationships", *Univ. Hansas Sci. Bull*, vol. 38, p. 1409–1438, 1958.

[SOR 48] SORENSEN T., "A method of establishing groups of equal amplitude in plant sociology based on similarity of species content and its applications to analyse vegetation on Danish commons", *Bioligiske Skrifter*, vol. 5, p. 1–34, 1948.

[STE 56] STEINHAUSS H., "Sur la division des corps matériels en parties", *Bulletin de Académie Polonaise des Sciences*, vol. III, num. IV, 12, p. 801–804, 1956.

[SUT 94] SUTCLIFFE J., "On the logical necessity and priority of a monothetic conception of class, and on the consequent inadequacy of polythetic accounts of category and categorisation", DIDAY E., Ed., *New Approaches in Classification and Data Analysis*, Berlin, Springer-Verlag, p. 53–63, 1994.

[VAN 04] VAN MECHELEN I., BOCK H.-H., BOECK P., "Two-mode clustering methods: A structured overview", *Statistical Methods in Medical Research*, vol. 13, p. 363–394, 2004.

[WAR 63] WARD J., "Hierarchical grouping to optimize an objective function", *Journal of the American Statistical Association"*, vol. 58, p. 236–244, 1963.

[WON 82] WONG M., "A hybrid clustering method for identifying high density clusters", *Journal of the American Statistical Association*, vol. 77, p. 841–847, 1982.

[ZHA 96] ZHANG T., RAMAKRISHNAN R., LIVNY M., "BIRCH: an efficient data clustering method for very large databases", *SIGMOD '96: Proceedings of the 1996 ACM SIGMOD International Conference on Management of Data*, ACM, p. 103–114, 1996.

Chapter 8

Clustering and the Mixture Model

8.1. Probabilistic approaches in cluster analysis

8.1.1. *Introduction*

Clustering methods, introduced in Chapter 7, are used for organizing a set of objects or individuals into homogenous classes. They are for the most part heuristic techniques derived from empirical methods, usually optimizing measurement criteria. The two most commonly used clustering algorithms, namely the k-means algorithm for obtaining partitions and Ward's hierarchical clustering method for obtaining hierarchies both use the within-group sum of squares criterion $\mathrm{tr}(\mathbf{S}_W)$ derived from the within-group covariance matrix $\mathbf{S}_W$.

Implementing these solutions involves choosing not only a metric reflecting the dissimilarity between the objects in the set to be segmented, but also a criterion derived from this metric capable of measuring the degree of cohesion and separation between classes. A rapid perusal of the lists of metrics and criteria proposed in the clustering literature will be enough to convince most readers that these are not easy choices.

In recent years, what used to be an algorithmic, heuristic and geometric focus has become a more statistical approach using probabilistic clustering models [BOC 89] to formalize the intuitive notion of a natural class. This approach allows precise analysis and can provide a statistical interpretation of certain metrical criteria whose different variants are not always clear (such as the within-group sum of squares criterion $\mathrm{tr}(\mathbf{S}_W)$), as well as yielding new variants corresponding to precise hypotheses. It also

---

Chapter written by Gérard GOVAERT.

represents a formal framework for tackling difficult problems such as determining the number of classes or validating the obtained clustering structure. We should bear in mind that, in many cases, the set to be segmented is merely a sample drawn from a much larger population, and that the conclusions drawn from clustering the sample are extrapolated to the entire population. Here, clustering becomes meaningless in the absence of a probabilistic model justifying this extrapolation.

All probabilistic approaches to clustering first assume that the data represent a random sample $x_1, \ldots, x_n$ from among a population, and then use an analysis of the probability distribution of this population to define a clustering. A number of different probabilistic clustering methods have been proposed, which we can separate into two broad categories: parametric approaches and non-parametric approaches.

8.1.2. *Parametric approaches*

One way of proceeding is to formalize the concept of 'natural' classes by making hypotheses about the probability distribution that yield a segmentation. Several types of parametric clustering models may be distinguished, the most important being finite probability mixture models, functional fixed effect models and point processes.

8.1.2.1. *Finite mixture models*

Finite mixture models, which assume that every class is characterized by a probability distribution, are highly flexible models which can take account of a variety of situations including heterogenous populations and outlier elements. Due to the EM algorithm, which is particularly well suited to this kind of context, a number of mixture models have been developed in the field of statistics. The use of mixture models in clustering has been studied by authors including Scott and Symons [SCO 71], Marriott [MAR 75], Symons [SYM 81], McLachlan [MCL 82] and McLachlan and Basford [MCL 88]. The mixture model approach is attractive for several reasons. It corresponds to our intuitive idea of a population composed of several classes, it is strongly linked to reference methods such as the k-means algorithm, and it is able to handle a wide variety of special situations in a more or less natural way. It is this approach which forms the subject of this chapter.

8.1.2.2. *Functional fixed effect models*

Functional fixed effect models, characterized by the equation

$$\text{Data} = \text{Structure} + \text{Error}$$

where the structure is unknown but fixed and the error random, may be applied in clustering using an adapted structure. If the data are vectors $x_1, \ldots, x_n$ of $\mathbb{R}^p$, the most simple example is the model $x_i = y_i + \varepsilon_i$ where y_i are required to belong to a set of g centers $\{a_1, \ldots, a_g\}$ and the errors ε_i are required to follow a centered Gaussian distribution of equal variance. We may also apply this type of model to similarity data and assume, for example, that the dissimilarity d between two objects in the set to be segmented may be expressed as $d(\mathbf{a}, \mathbf{b}) = \delta(\mathbf{a}, \mathbf{b}) + \varepsilon(\mathbf{a}, \mathbf{b})$, where δ is an ultrametric distance. Degens [DEG 83] has shown that the agglomerative hierarchical clustering UPGMA (unweighted pair-groups method using arithmetic averages) is a local maximum of likelihood for this model when the error is Gaussian.

8.1.2.3. *Point processes*

In spatial statistics [CRE 93, RIP 81], data such as the distribution of trees in a forest or stars in space are considered as point data resulting from point processes. Some of these processes correspond to an aggregate organization and can be treated as probabilistic models associated with a clustering. The most frequently used is the Neyman–Scott model [NEY 58], which can be seen as a data-generation process in three stages:

1) g points $\mathbf{a}_1, \ldots, \mathbf{a}_g$ are drawn at random with a uniform distribution from a convex region;

2) the sizes $n_1, \ldots, n_g$ of classes are drawn at random, for example with a Poisson distribution; and

3) for each class k, n_k points are drawn at random using a spherical distribution centered at $\mathbf{a}_k$, for example a Gaussian distribution with mean $\mathbf{a}_k$.

8.1.3. *Non-parametric methods*

This second approach includes those probabilistic clustering methods which make no assumptions concerning the probability distribution. These methods may be extremely varied in nature, but all will be based on the shape of this distribution. Where data are continuous, the distribution is characterized by its density function, so methods will use this density function in defining classes (e.g. high-density classes or modal classes). Hartigan [HAR 75] therefore defines a high-density class as a connected subset of points whose density exceeds a certain threshold. By varying this threshold he obtains a hierarchical tree of classes. The existence of several density maxima may be interpreted as the presence of heterogenous data and consequently of classes. By seeking these maxima and allocating points within the reference space to each one of them we are able to define modal classes.

Applying these methods clearly entails estimating the unknown distribution from the data. The most usual methods are based on a non-parametric estimation of density such as nearest-neighbor estimation, kernel estimation or simply by using a histogram. This strategy has given rise to a large number of algorithms (e.g. [HAR 75, WON 83]) and correspondences have been drawn with classical algorithms such the single-link hierarchical clustering algorithm. These methods will not be given any further coverage in this chapter.

We might also include among these non-parametric methods Lerman's hierarchical clustering algorithms [LER 81], based on the notion of link likelihood.

8.1.4. *Validation*

Another important use of probabilistic tools in clustering is in the validation of results. Given that every clustering algorithm yields a result, we need to know whether this result corresponds to a real structural feature, or whether it is simply due to chance. To this end, a number of statistical validation tools have been developed. When results have been obtained using the sort of parametric models described above, a straightforward approach is to design tests based on the model used: for example, verifying the normality of a class in the case of Gaussian mixture models. Other statistical validation tools are independent of the clustering algorithms. These include using Ling's random graphs [LIN 73] to test the significance of a class. Van Cutsem and Ycart [VAN 98] looked at clustering structures resulting from different algorithms under conditions of random non-classifiability, and proposed a collection of explicit tests of non-classifiability. We refer interested readers to Bock's article [BOC 96] with its detailed bibliographical synthesis of these validation problems.

8.1.5. *Notation*

Throughout this chapter, we shall assume that data take the form of a matrix $\mathbf{X}$ of dimension (n, p) defined by values x_i^j, where i denotes a set I of n individuals and j denotes a set of p variables which may be either continuous or qualitative.

Since each individual is characterized by a vector $\boldsymbol{x}_i = (x_i^1, \ldots, x_i^p)$, the aim is to obtain a partition of I of g classes. The number of classes g is assumed to be known. We shall use the notation $\mathbf{Z} = (\boldsymbol{z}_1, \ldots, \boldsymbol{z}_n)$ with $\boldsymbol{z}_i = (z_{i1}, \ldots, z_{ig})$ where $z_{ik} = 1$ if $\boldsymbol{x}_i$ arose from cluster k, and $z_{ik} = 0$ otherwise. Thus, the clustering will be represented by a matrix $\mathbf{Z}$ of n vectors in $\mathbb{R}^g$ satisfying $z_{ik} \in \{0, 1\}$ and $\sum_{k=1}^{g} z_{ik} = 1$: the kth class corresponds to the set of objects i such that $z_{ik} = 1$ and $z_{i\ell} = 0 \ \forall \ell \neq k$. For example, if the set I comprises 5 elements, for the partition $\mathbf{z}$ made up of the two classes $\{1, 3, 4\}$ and $\{2, 5\}$ we have:

$$
\mathbf{Z} = \begin{pmatrix} z_{11} & z_{12} \\ z_{21} & z_{22} \\ z_{31} & z_{32} \\ z_{41} & z_{52} \\ z_{51} & z_{52} \end{pmatrix} = \begin{pmatrix} 1 & 0 \\ 0 & 1 \\ 1 & 0 \\ 1 & 0 \\ 0 & 1 \end{pmatrix}.
$$

Moreover, probability distributions will always be denoted as a density f, the discrete case corresponding to a discrete measure.

Sections 8.2 and 8.3 are devoted to probability mixture models and the EM algorithm, the standard tool for estimating the parameters of such models, respectively. Section 8.4 describes how clustering may be carried out using a mixture model. Sections 8.5–8.7 look at several classical situations including Gaussian mixture models for continuous variables, and the latent class model for binary variables and qualitative variables in general. In section 8.8, we study the implementation of these different methods. In the final section we shall briefly describe some prospects for the future which this approach to clustering has opened up.

8.2. The mixture model

8.2.1. *Introduction*

Since their first use by Newcomb in 1886 for the detection of outlier points, and then by Pearson in 1894 to identify two separate populations of crabs, finite mixtures of distributions have been employed to model a wide variety of random phenomena. These models assume that measurements are taken from a set of individuals, each of which belongs to one of a number of different classes, while any individual's particular class is unknown. We might, for instance, know the sizes of fish in a sample, but not their sex, which is difficult to ascertain. Mixture models can therefore address the heterogenity of a population, and are especially well suited to the problem of clustering. This is an area where much research has been done. McLachlan and Peel's book [MCL 00] is a highly detailed reference for this domain which has seen considerable developments over the last few years. We shall first briefly recall the model and the problems of estimating its parameters.

8.2.2. *The model*

In a finite probability mixture model, the data $\mathbf{X} = (\boldsymbol{x}_1, \ldots, \boldsymbol{x}_n)$ are taken to constitute a sample of n independent instances of a random variable $\boldsymbol{X}$ in $\mathbb{R}^p$. Density can be expressed as

$$f(\boldsymbol{x}_i) = \sum_{k=1}^{g} \pi_k f_k(\boldsymbol{x}_i), \quad \forall i \in I$$

where g is the number of components, f_k are the density of each component and π_k are the mixture proportions ($\pi_k \in]0, 1[\; \forall k$ and $\sum_k \pi_k = 1$). The principle of a mixture model is to assume, given the proportions $\pi_1, \ldots, \pi_g$ and the distributions f_k of each class, that the data are generated according to the following mechanisms:

– z: each individual is allocated to a class according to a multinomial distribution with parameters $\pi_1, \ldots, \pi_g$;

– x: each $\boldsymbol{x}_i$ is assumed to arise from a random vector with probability density function f_k.

In addition, it is usually assumed that the components' density f_k belongs to a parametric family of densities $f(., \alpha)$. The density of the mixture can therefore be written as

$$f(\boldsymbol{x}_i, \boldsymbol{\theta}) = \sum_{k=1}^{g} \pi_k f(\boldsymbol{x}_i; \boldsymbol{\alpha}_k), \quad \forall i \in I$$

where $\boldsymbol{\theta} = (\pi_1, \ldots, \pi_g, \boldsymbol{\alpha}_1, \ldots, \boldsymbol{\alpha}_g)$ is the parameter of the model. For example, the density of a mixture model for two univariate Gaussian distributions of variance 1 in $\mathbb{R}$ is written as

$$f(x_i; \pi, \mu_1, \mu_2) = \pi\varphi(x_i; \mu_1, 1) + (1 - \pi)\varphi(x_i; \mu_2, 1)$$

where $\varphi(.; \mu, \sigma^2)$ is the density of the univariate Gaussian distribution of mean μ and variance σ^2.

Figure 8.1 uses the density obtained from a mixture of three Gaussian components in $\mathbb{R}^2$ to illustrate this concept of a probability mixture.

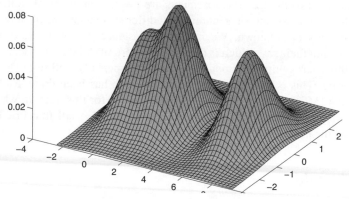

Figure 8.1. *Gaussian mixture in* $\mathbb{R}^2$

8.2.3. *Estimation of parameters*

Much effort has been devoted to the estimation of parameters for the mixture model following the work of Pearson, whose use of the method of moments to estimate the five parameters $(\mu_1, \mu_2, \sigma_1^2, \sigma_2^2, \pi)$ of a univariate Gaussian mixture model with two components required him to solve polynomial equations of degree 9. There have been a number of studies [MCL 88, TIT 85] and different estimation methods have been envisaged. Apart from the method of moments, we also find graphic methods, the maximum likelihood method and Bayesian approaches. In this chapter we shall restrict ourselves to examining the maximum likelihood method using the EM algorithm, which is currently the most widely used. Before examining this method in section 8.3, we first draw the reader's attention to certain difficulties which the estimation of parameters of a mixture model presents.

8.2.4. *Number of components*

In certain situations, such as in the case of the fish described in section 8.2.1 where the idea of component has a precise physical basis, the number of components may be completely determined. Most often, however, the number of components is not known and must itself be estimated.

It should be noted that if the number of components is considered as a parameter, the mixture model may be seen as a semi-parametric compromise between a classical parametric estimation problem when the number of components corresponds to a fixed constant, and a non-parametric estimation problem (in this case via the kernel method) when the number of components is equal to the size of the sample.

We assume from here onwards that the number g of components is known. Later we shall look at the proposed solutions for making this difficult choice.

8.2.5. *Identifiability*

If the problem is to be of any interest, the density of the mixture needs to be identifiable, meaning that any two mixtures whose densities are the same must have the same parameters. A number of studies have addressed this problem; several difficulties arise. The first difficulty is due to the numbering of the classes. For example, in the case of a mixture with two components, the parameters $(\pi_1, \pi_2), (\alpha_1, \alpha_2)$ and $(\pi_2, \pi_1), (\alpha_2, \alpha_1)$, although different, obviously yield the same pdf: a mixture is consequently never identifiable. The difficulties to which this situation gives rise will depend on the estimation algorithms. In the case of the EM algorithm which we shall use, it simply does not matter. This cannot be said of the Bayesian approach, however, where this situation is known as the *switching problem*. The second, considerably more awkward difficulty may arise from the very nature of component pdf. It may easily be established that a mixture of uniform or binomial distributions is not identifiable. Mixtures of Gaussian, exponential and Poisson distributions, however, are identifiable.

8.3. EM algorithm

8.3.1. *Introduction*

Maximizing the log-likelihood of a mixture model

$$L(\boldsymbol{\theta}) = \log \left(\prod_{i=1}^{n} \sum_{k=1}^{g} \pi_k f(\boldsymbol{x}_i, \boldsymbol{\alpha}_k) \right)$$

leads to likelihood equations which usually have no analytical solution. It may nevertheless be shown that if the parameter α_k is a vector of real numbers α_{kr}, the solution of these likelihood equations must satisfy

$$\pi_k = \frac{1}{n} \sum_{i=1}^{n} t_{ik} \ \forall k \quad \text{and} \quad \sum_{i=1}^{n} t_{ik} \frac{\partial \log f_k(\boldsymbol{x}_i, \boldsymbol{\alpha}_k)}{\partial \alpha_{kr}} = 0 \ \forall k, r \qquad (8.1)$$

where

$$t_{ik} = \frac{\pi_k f_k(\boldsymbol{x}_i, \boldsymbol{\alpha}_k)}{\sum_{\ell=1}^{g} \pi_\ell f_\ell(\boldsymbol{x}_i, \boldsymbol{\alpha}_\ell)}. \qquad (8.2)$$

These equations suggest the following iterative algorithm:

1) start from an initial solution θ;

2) calculate the values t_{ik} from this parameter using relation (8.2);

3) update the parameter θ on the basis of these values t_{ik} using equations (8.1);

4) continue from (2).

If this algorithm converges, then the fixed point obtained will satisfy the likelihood equations. The procedure corresponds, in fact, to the application of the Dempster *et al.* [DEM 77] EM algorithm to the mixture model. Before describing this algorithm, we shall define the concept of complete data on which it relies.

8.3.2. *Complete data and complete-data likelihood*

At the outset, we consider that the observed data $\mathbf{X}$ correspond to what is merely a partial knowledge of unknown data $\mathbf{Y}$ which are termed *complete data*, the two being linked by a function $\mathbf{X} = T(\mathbf{Y})$. The complete data might for instance be of the form $\mathbf{Y} = (\mathbf{X}, \mathbf{Z})$, in which case $\mathbf{Z}$ is known as *missing information*. This idea of complete data may either be meaningful for a model, which is the case for the mixture model, or it may be completely artificial. The likelihood $f(\mathbf{Y}; \boldsymbol{\theta})$ calculated from these complete data is termed *complete-data likelihood* or, in the case of the mixture model, *classification likelihood*. Starting from the relation $f(\mathbf{Y}; \boldsymbol{\theta}) = f(\mathbf{Y}|\mathbf{X}; \boldsymbol{\theta})f(\mathbf{X}; \boldsymbol{\theta})$ between the densities, we obtain the relation

$$L(\boldsymbol{\theta}) = L_c(\boldsymbol{\theta}) - \log f(\mathbf{Y}|\mathbf{X}; \boldsymbol{\theta}) \qquad (8.3)$$

between the initial log-likelihood $L(\boldsymbol{\theta})$ and the complete-data log-likelihood $L_c(\boldsymbol{\theta})$.

8.3.3. *Principle*

The EM algorithm rests on the hypothesis that maximizing the complete-data likelihood is simple. Since this likelihood cannot be calculated – $\mathbf{Y}$ is unknown –

an iterative procedure based on the conditional expectation of the log-likelihood for a value of the current parameter θ' is used as follows. First, calculating the conditional expectation for the two members of relation (8.3), we obtain the fundamental relation of the EM algorithm:

$$L(\theta) = Q(\theta, \theta') - H(\theta, \theta')$$

where $Q(\theta, \theta') = \mathbb{E}(L_c(\theta)|\mathbf{X}, \theta')$ and $H(\theta, \theta') = \mathbb{E}(\log f(\mathbf{Y}|\mathbf{X}; \theta)|\mathbf{X}, \theta')$.

Introducing the parameter θ' allows us to define an iterative algorithm to increase the likelihood. Using Jenssen's inequality it can be shown that, for fixed θ', the function $H(\theta, \theta')$ is maximum for $\theta = \theta'$. The value θ which maximizes $Q(\theta, \theta')$ therefore satisfies the relation

$$L(\theta) \geqslant L(\theta'). \tag{8.4}$$

The EM algorithm involves constructing, from an initial solution $\theta^{(0)}$, the sequence $\theta^{(p)}$ satisfying $\theta^{(q+1)} = \arg\max Q(\theta, \theta^{(q)})$. Relation (8.4) shows that this sequence causes the criterion $L(\theta)$ to grow.

8.3.4. *Application to mixture models*

For the mixture model, the complete data are obtained by adding the original component $\mathbf{z}_i$ to each individual member of the sample:

$$\mathbf{Y} = (\mathbf{y}_1, \ldots, \mathbf{y}_n) = ((\mathbf{x}_1, \mathbf{z}_1), \ldots, (\mathbf{x}_n, \mathbf{z}_n)).$$

Coding $\mathbf{z}_i = (z_{i1}, \ldots, z_{ig})$ where, we recall, z_{ik} equals 1 if i belongs to component k and 0 otherwise, we obtain the relations:

$$f(\mathbf{Y}; \theta) = \prod_{i=1}^{n} f(\mathbf{y}_i; \theta) = \prod_{i=1}^{n} \sum_{k=1}^{g} z_{ik} \pi_k f(\mathbf{x}_i; \alpha_k),$$

$$L_c(\theta) = \log(f(\mathbf{Y}; \theta)) = \sum_{i=1}^{n} \sum_{k=1}^{g} z_{ik} \log\left(\pi_k f(\mathbf{x}_i; \alpha_k)\right)$$

and

$$Q(\theta|\theta') = \sum_{i=1}^{n} \sum_{k=1}^{g} \mathbb{E}(z_{ik}|\mathbf{X}, \theta') \log\left(\pi_k f(\mathbf{x}_i; \alpha_k)\right).$$

Denoting the probabilities $\mathbb{E}(z_{ik}|\mathbf{X}, \theta') = P(z_{ik} = 1|\mathbf{X}, \theta')$ that x_i belongs to component k as t_{ik}, the EM algorithm takes the form:

1) initialize: arbitrarily select an initial solution $\theta^{(0)}$;

2) repeat the following two steps until convergence:

- step E (expectation): calculate the probabilities of x_i belonging to the classes, conditionally on the current parameter:

$$t_{ik}^{(q)} = \pi_k^{(q)} f(\boldsymbol{x}_i, \alpha_k^{(q)})/(\sum_{\ell=1}^{g} \pi_\ell^{(q)} f(\boldsymbol{x}_i, \alpha_\ell^{(q)}));$$

- step M (maximization): maximize the log-likelihood conditionally on $t_{ik}^{(q)}$; the proportions are then obtained simply by the relation $\pi_k^{(q+1)} = 1/n \sum_{i=1}^{n} t_{ik}^{(q)}$, while the parameters $\alpha_k^{(q+1)}$ are obtained by solving the likelihood equations that depend on the mixture model employed.

8.3.5. *Properties*

Under certain conditions of regularity, it has been established that the EM algorithm always converges to a local likelihood maximum. It shows good practical behavior, but may nevertheless be quite slow in some situations. This is the case, for instance, when classes are very mixed. This algorithm was proposed by Dempster *et al.* in a seminal paper [DEM 77], and is often simple to implement. It has gained widespread popularity and given rise to a large number of studies which are thoroughly covered in McLachlan and Krishnan's book [MCL 97].

8.3.6. *EM: an alternating optimization algorithm*

Hathaway [HAT 86] has shown that the EM algorithm applied to a mixture model may be interpreted as an alternating algorithm for optimizing a fuzzy clustering criterion. We make use of this fact below when we examine the links between estimating the parameters of a mixture model and fuzzy clustering. It is a finding not limited to mixture models, and has in fact been extended in a more general fashion to the EM algorithm by Neal and Hinton [NEA 98] as follows. Denoting the known data as $\mathbf{X}$, the missing data as $\mathbf{Z}$ and, where $\hat{P}$ is a distribution over the domain of the missing data, we define the criterion

$$W(\hat{P}, \boldsymbol{\theta}) = \mathbb{E}_{\hat{P}}(L_c(\boldsymbol{\theta}, \mathbf{Z})) + H(\hat{P}) \qquad (8.5)$$

where $H(\hat{P}) = -\mathbb{E}_{\hat{P}}(\log \hat{P})$ is the entropy of the distribution $\hat{P}$. Moreover, if we denote the distribution $P(\mathbf{Z}|\mathbf{X}, \boldsymbol{\theta})$ as $P_{\boldsymbol{\theta}}$ it can be shown, using relation (8.3), that the criterion W can also be expressed

$$W(\hat{P}, \boldsymbol{\theta}) = L(\boldsymbol{\theta}) - \text{KL}(\hat{P}, P_{\boldsymbol{\theta}}) \qquad (8.6)$$

where KL is the Kullback–Liebler divergence between two distributions.

The alternating algorithm for optimizing the criterion W then becomes simple to implement:

1) Minimizing for fixed $\boldsymbol{\theta}$: relation (8.6) implies that $\hat{P}$ must minimize $\mathrm{KL}(\hat{P}, P_{\boldsymbol{\theta}})$ and consequently $\hat{P} = P_{\boldsymbol{\theta}}$;

2) Minimizing for fixed $\hat{P}$: relation (8.5) shows that $\boldsymbol{\theta}$ must maximize the expectation $\mathbb{E}_{\hat{P}}(L_c(\boldsymbol{\theta}, \mathbf{Z}))$.

We are therefore dealing with what are precisely the two steps of the EM algorithm. In addition, after each first step we have $W(\hat{P}, \boldsymbol{\theta}) = W(P_{\boldsymbol{\theta}}, \boldsymbol{\theta}) = L(\boldsymbol{\theta})$, demonstrating that the EM algorithm increases the likelihood.

8.4. Clustering and the mixture model

8.4.1. *The two approaches*

Mixture models may be used in two different ways to obtain a partition of the initial data.

– The first, known as the *mixture approach*, estimates the parameters of the model and then determines the partition by allocating each individual to the class that maximizes the *a posteriori* probability t_{ik} computed using these estimated parameters; this allocation is known as the MAP (maximum *a posteriori* probability) method.

– The second, the *classification approach*, was first presented by Scott and Symons [SCO 71] and developed further by Schroeder [SCH 76]. This approach involves creating a partition of the sample such that each class k is made to correspond to a sub-sample respecting the distribution $f(., \alpha_k)$. This requires simultaneous estimation of the model parameters and the desired partition.

In this section we describe the criterion which this latter approach optimizes, as well as the optimization algorithm usually employed in this situation. We then briefly compare the two approaches and examine links between these types of methods and the more classical metrical approaches to clustering. We conclude the section by looking at how the mixture model may be interpreted in terms of fuzzy clustering.

8.4.2. *Classification likelihood*

Introducing the $\mathbf{Z}$ partition in the likelihood criterion is not an obvious step, and various ideas have been proposed. Scott and Symons [SCO 71] defined the criterion

$$L_{CR}(\boldsymbol{\theta}, \mathbf{Z}) = \sum_{k=1}^{g} \sum_{i/z_{ik}=1} \log f(\boldsymbol{x}_i, \alpha_k)$$

in which the proportions do not appear. Symons [SYM 81], realizing that this criterion tends to yield classes of similar proportions, modified it in order to use the complete-data (or classification) log-likelihood described above:

$$L_C(\boldsymbol{\theta}, \mathbf{Z}) = \sum_{k=1}^{g} \sum_{i/z_{ik}=1} \log \pi_k f(\boldsymbol{x}_i, \alpha_k) = \sum_{k=1}^{g} \sum_{i=1}^{n} z_{ik} \log \pi_k f(\boldsymbol{x}_i, \alpha_k).$$

This is linked to the previous criterion by the relation

$$L_C(\boldsymbol{\theta}, \mathbf{Z}) = L_{CR}(\boldsymbol{\theta}, \mathbf{Z}) + \sum_{k=1}^{g} n_k \log \pi_k,$$

where n_k is the cardinal of the class k. The quantity $\sum_k n_k \log \pi_k$ is a penalty term which disappears if all the proportions are set as identical. The criterion $L_{CR}(\boldsymbol{\theta}, \mathbf{Z})$ can therefore be seen as a variant of classification likelihood, restricted to a mixture model where all classes have the same proportion.

8.4.3. The CEM algorithm

When seeking to maximize classification likelihood, it is possible to use a clustering version of the EM algorithm, obtained by adding a clustering step. This yields the very general clustering algorithm known as CEM (classification EM) [CEL 92] defined as follows:

– step 0: arbitrarily select an initial solution $\theta^{(0)}$;

– step E: compute $t_{ik}^{(q)}$ as in the EM algorithm;

– step C: obtain the $\mathbf{Z}^{(q+1)}$ partition by allocating each $\boldsymbol{x}_i$ to the class that maximizes $t_{ik}^{(q)}$ (MAP); this is equivalent to modifying the $t_{ik}^{(q)}$ by replacing them with the nearest 1 or 0 values;

– step M: maximize the likelihood depending on the $z_{ki}^{(q+1)}$. The estimations of the maximum likelihood among the π_k and the α_k are obtained using the classes of the partition $\mathbf{Z}^{(q+1)}$ as sub-samples. The proportions are given by the formula $\pi_k^{(q+1)} = \frac{1}{n} n_k^{(q+1)}$. The $\alpha_k^{(q+1)}$ are computed according to the particular mixture model selected.

Here we have an alternating *dynamic cluster method* [DID 79] optimization algorithm, where the E and the C steps correspond to the allocation step and the M step corresponds to the representation step.

It can be shown that this algorithm is stationary and that it increases the complete-data likelihood at each iteration, given some very general assumptions.

8.4.4. *Comparison of the two approaches*

The clustering approach, which determines the parameters at each iteration using truncated mixture model samples, yields a biased and inconsistent estimation since the number of parameters to be estimated increases with the size of the sample. Different authors have studied this problem and shown that is is usually preferable to use the mixture approach.

However, when the classes are well separated and membership relatively small, the clustering approach can sometimes give better results [CEL 93, GOV 96]. Moreover, the CEM algorithm is considerably faster than the EM algorithm, and it may be necessary to use it when computation time is limited e.g. in real-time operations or for very large volumes of data.

Finally, the clustering approach has the advantage of being able to present a large number of clustering algorithms as special cases of the CEM algorithm, which allows it to incorporate them into a probabilistic clustering approach. We shall see in section 8.5.3, for example, that the k-means algorithm can be seen as a simple special case of the CEM algorithm. In particular, we shall show that the optimized criteria, the within-group sum of squares criterion for continuous data and the information criterion for qualitative data correspond to the classification likelihood of a particular mixture model. These correspondences, studied in [GOV 89, GOV 90b], can be formalized by the following theorem.

THEOREM 8.1.– *If the clustering criterion can be expressed as*

$$W(\mathbf{Z}, \boldsymbol{\lambda}, D) = \sum_{i=1}^{n} \sum_{k=1}^{g} z_{ik} D(\boldsymbol{x}_i, \boldsymbol{\lambda}_k)$$

(where $\mathbf{Z}$ is a partition of the set to be segmented, $\boldsymbol{\lambda} = (\boldsymbol{\lambda}_1, \ldots, \boldsymbol{\lambda}_g)$, $\boldsymbol{\lambda}_k$ is a representative of the class k and D is a measure of the dissimilarity between an object $\boldsymbol{x}$ and the representative of a class) and if there exists a real r such that the quantity $\int r^{-D(\boldsymbol{x}, \boldsymbol{\lambda})} d\boldsymbol{x}$ is independent of $\boldsymbol{\lambda}$, then it is equivalent to the classification likelihood criterion of a mixture model with densities of the form $f(\boldsymbol{x}, \boldsymbol{\lambda}) = \frac{1}{s} r^{-D(\boldsymbol{x}, \boldsymbol{\lambda})}$, where s is a positive constant.

This theorem may be used equally well for continuous as for discrete (binary or qualitative) data – either a Lebesgue measure or a discrete measure will be used accordingly. This theorem is important in that a great many clustering criteria can be put into this very general form. For example, this is the case for the intraclass inertia criterion whose class representative is its centroid and where the distance D is the square of the Euclidean distance. It can also help us to fix the fields of application of these criteria and to suggest others.

8.4.5. *Fuzzy clustering*

In fuzzy clustering it is no longer the case that an object either belongs or does not belong to a particular class. Instead there is a degree of belonging. Formally, a fuzzy clustering is characterized by a matrix $\mathbf{C}$ with terms c_{ik} satisfying $c_{ik} \in [0, 1]$ and $\sum_k c_{ik} = 1$. Bezdek's fuzzy k-means [BEZ 81], one of the most commonly encountered, involves minimizing the criterion

$$W(\mathbf{C}) = \sum_{i=1}^{n} \sum_{k=1}^{g} c_{ik}^{\gamma} d^2(\boldsymbol{x}_i, \mathbf{g}_k)$$

where $\gamma > 1$ is a coefficient for adjusting the degree of fuzziness, $\mathbf{g}_k$ is the center of the class and d is the Euclidean distance. It is required that γ is different to 1, otherwise the function W is minimal for values of $c_{ik} = 0$ or 1 and thus we have the usual within-group sum of squares criterion. The values usually recommended are between 1 and 2. Minimizing this criterion is achieved using an algorithm that alternates the two following steps:

1) compute the centers: $\mathbf{g}_k = \sum_{i=1}^{n} c_{ik}^{\gamma} \boldsymbol{x}_i / \sum_{i=1}^{n} c_{ik}$;

2) compute the fuzzy partition: $c_{ik} = \dfrac{D_i}{\|\boldsymbol{x}_i - \mathbf{g}_k\|^{\frac{2}{\gamma-1}}}$ where $D_i = \sum_{\ell} \dfrac{1}{\|\boldsymbol{x}_i - \mathbf{g}_\ell\|^{\frac{2}{\gamma-1}}}$.

Validating this kind of approach and, in particular, choosing the coefficient γ, can be tricky.

Estimating the parameters of a mixture model is an alternative, more natural, way of addressing this problem. The estimation of the *a posteriori* probabilities t_{ik} of objects belonging to each class directly provides a fuzzy clustering and the *EM* algorithm, applied to the mixture model, may be seen as a fuzzy clustering algorithm.

As mentioned above, Hathaway [HAT 86] went even further and showed that seeking a fuzzy partition and the parameter θ using an optimization alternated with a fuzzy clustering criterion leads precisely to the two steps of the EM algorithm, which can therefore be considered as a fuzzy clustering algorithm. We may obtain the same result simply by applying the results from section 8.3.6 to the mixture model. Given that here the probability distribution $\hat{P}$ is defined by the vector (c_{ik}) and that we simply have $E_{\hat{P}}(L_c(\boldsymbol{\theta}, \mathbf{Z})) = L_c(\boldsymbol{\theta}, \mathbf{C})$, we show that the EM algorithm alternately maximizes the criterion

$$W(\mathbf{C}, \boldsymbol{\theta}) = L_c(\boldsymbol{\theta}, \mathbf{c}) + H(\mathbf{C})$$

where L_c is the complete-data log-likelihood function where the partition $\mathbf{Z}$ has been replaced by the fuzzy partition $\mathbf{C}$:

$$L_c(\boldsymbol{\theta}, \mathbf{C}) = \sum_{i=1}^{n} \sum_{k=1}^{g} c_{ik} \log\left(\pi_k f_k(\boldsymbol{x}_i; \boldsymbol{\alpha})\right)$$

and H is the entropy function

$$H(\mathbf{C}) = -\sum_{i=1}^{n}\sum_{k=1}^{g} c_{ik} \log c_{ik}.$$

It is easy to show that if the entropy term of the criterion W is removed, then 'hard' partitions are obtained at each step. The resulting algorithm is simply the CEM algorithm. The difference between the EM and CEM algorithms is the presence of the entropy term. If the components are highly separated, the fuzzy partition $\mathbf{Z}(\boldsymbol{\theta})$ is close to a partition and we have

$$H(\mathbf{Z}(\boldsymbol{\theta})) \approx 0$$

and

$$L(\boldsymbol{\theta}) = W(\mathbf{Z}(\boldsymbol{\theta}), \boldsymbol{\theta}) = L_C(\boldsymbol{\theta}, \mathbf{Z}(\boldsymbol{\theta})) + H(\mathbf{Z}(\boldsymbol{\theta})) \approx L_C(\boldsymbol{\theta}, \mathbf{Z}(\boldsymbol{\theta})).$$

8.5. Gaussian mixture model

We shall now examine what happens to this approach when each class is modeled by a Gaussian distribution.

8.5.1. *The model*

The density of the mixture can be written as $f(\boldsymbol{x}_i; \boldsymbol{\theta}) = \sum_{k=1}^{g} \pi_k \varphi(\boldsymbol{x}_i; \boldsymbol{\mu}_k, \boldsymbol{\Sigma}_k)$, where φ is the density of the Gaussian multivariate distribution:

$$\varphi(\boldsymbol{x}_i; \boldsymbol{\mu}_k, \boldsymbol{\Sigma}_k) = \frac{1}{(2\pi)^{\frac{p}{2}} |\boldsymbol{\Sigma}_k|^{\frac{1}{2}}} \exp\{-\frac{1}{2}(\boldsymbol{x}_i - \boldsymbol{\mu}_k)' \boldsymbol{\Sigma}_k^{-1}(\boldsymbol{x}_i - \boldsymbol{\mu}_k)\}$$

and $\boldsymbol{\theta}$ is the vector $(\pi_1, \ldots, \pi_g, \boldsymbol{\mu}_1, \ldots, \boldsymbol{\mu}_q, \boldsymbol{\Sigma}_1, \ldots, \boldsymbol{\Sigma}_g)$ formed by the proportions π_k and the parameters $\boldsymbol{\mu}_k$ and $\boldsymbol{\Sigma}_k$ which are the mean vector and the covariance matrix of class k, respectively.

When the sample size is small, or when the dimension of the space is large, the number of parameters must be reduced in order to obtain more parsimonious models. To this end, the spectral decomposition of the matrices [BAN 93, CEL 95] may be used. This allows the covariance matrices to be parameterized uniquely as $\boldsymbol{\Sigma}_k = \lambda_k \mathbf{D}_k \mathbf{A}_k \mathbf{D}'_k$, where the diagonal matrix $\mathbf{A}_k$ with determinant 1 and decreasing values defines the *shape* of the class, the orthogonal matrix $\mathbf{D}_k$ defines the *direction* of the class and the positive real number λ_k represents the *volume* of the class. The mixture model is therefore parameterized by the centers $\boldsymbol{\mu}_1, \ldots, \boldsymbol{\mu}_g$, the proportions $\pi_1, \ldots, \pi_g$, the volumes $\lambda_1, \ldots, \lambda_g$, the shapes $\mathbf{A}_1, \ldots, \mathbf{A}_g$ and the directions $\mathbf{D}_1, \ldots, \mathbf{D}_g$ of each class.

For example, when the data are in a plane, $\mathbf{D}$ is a rotation matrix defined by an angle α and $\mathbf{A}$ is a diagonal matrix with diagonal terms a and $1/a$. Figure 8.2 represents the equidensity ellipse of this distribution depending on the values α, λ and a.

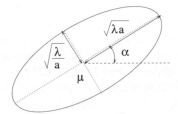

Figure 8.2. *Parameterization of a Gaussian class in the plane*

Using this parameterization, it becomes possible to propose solutions that can be seen as a middle way between, on the one hand, restrictive hypotheses (covariance matrices proportional to the identity matrix or covariance matrices identical for all classes) and, on the other hand, very general constraint-free hypotheses [BAN 93, CEL 95].

This parameterization also highlights two distinct notions that are often gathered under the rather vague heading of *size*. These are, firstly, the proportion of individuals present within a class and secondly, the volume that a class occupies in space. It is quite possible for a class to have a small volume and a high proportion or, alternatively, a large volume but a low proportion.

We shall now look at what happens to the CEM algorithm and the classification likelihood criterion in the case of the Gaussian mixture model. It should be noted that a similar approach could be applied to the EM algorithm.

8.5.2. *CEM algorithm*

8.5.2.1. *Clustering step*

Each $\boldsymbol{x}_i$ is allocated to the class that maximizes the probability of membership $t_{ik} = \pi_k \varphi(\boldsymbol{x}_i; \boldsymbol{\mu}_k, \boldsymbol{\Sigma}_k)/(\sum_{\ell=1}^{g} \pi_\ell \varphi(\boldsymbol{x}_i; \boldsymbol{\mu}_\ell, \boldsymbol{\Sigma}_\ell))$, i.e. $\pi_k \varphi(\boldsymbol{x}_i; \boldsymbol{\mu}_k, \boldsymbol{\Sigma}_k)$ or, equivalently, the class that minimizes $- \log(\pi_k \varphi(\boldsymbol{x}_i; \boldsymbol{\mu}_k, \boldsymbol{\Sigma}_k))$. We write this as:

$$d^2_{\boldsymbol{\Sigma}_k^{-1}}(\boldsymbol{x}_i, \boldsymbol{\mu}_k) + \log |\boldsymbol{\Sigma}_k| - 2 \log \pi_k \qquad (8.7)$$

where $d^2_{\boldsymbol{\Sigma}_k^{-1}}(\boldsymbol{x}_i, \boldsymbol{\mu}_k)$ is the quadratic distance $(\boldsymbol{x}_i - \boldsymbol{\mu}_k)' \boldsymbol{\Sigma}_k^{-1}(\boldsymbol{x}_i - \boldsymbol{\mu}_k)$.

8.5.2.2. *Step M*

Here, for a given partition z, we have to determine the parameter θ that maximizes $L_C(\theta, \mathbf{Z})$, which is equal to (ignoring one constant)

$$-\frac{1}{2}\sum_{k=1}^{g}\left(\sum_{i=1}^{n} z_{ik}(\boldsymbol{x}_i - \boldsymbol{\mu}_k)'\boldsymbol{\Sigma}_k^{-1}(\boldsymbol{x}_i - \boldsymbol{\mu}_k) + n_k \log|\boldsymbol{\Sigma}_k| - 2n_k \log \pi_k\right).$$

The parameter $\boldsymbol{\mu}_k$ is therefore necessarily the centroid $\bar{\boldsymbol{x}}_k = \frac{1}{n_k}\sum_{i=1}^{n} z_{ik}\boldsymbol{x}_i$ and the proportions, if they are not constrained, satisfy $\pi_k = n_k/n$. The parameters $\boldsymbol{\Sigma}_k$ must then minimize the function

$$F(\boldsymbol{\Sigma}_1, \ldots, \boldsymbol{\Sigma}_g) = \sum_{k=1}^{g} n_k \left(\text{tr}(\mathbf{S}_k\boldsymbol{\Sigma}_k^{-1}) + \log|\boldsymbol{\Sigma}_k|\right) \tag{8.8}$$

where $\mathbf{S}_k = \frac{1}{n_k}\sum_{i=1}^{n} z_{ik}(\boldsymbol{x}_i - \bar{\boldsymbol{x}}_k)(\boldsymbol{x}_i - \bar{\boldsymbol{x}}_k)'$ is the covariance matrix of the class k. We now examine three particular situations.

8.5.3. *Spherical form, identical proportions and volumes*

We now look at the most straightforward situation where all classes have a Gaussian spherical distribution with the same volume and the same proportion. The covariance matrices are written $\boldsymbol{\Sigma}_k = \lambda \mathbf{D}_k \mathbf{I}_p \mathbf{D}_k' = \lambda \mathbf{I}_p \quad \forall k$, and equation (8.7) shows that individuals can be allocated to the different classes simply by using the usual Euclidean distance $d^2(\boldsymbol{x}_i, \boldsymbol{\mu}_k)$. Function F therefore becomes

$$F(\lambda) = \frac{1}{\lambda}\sum_{k=1}^{g} n_k \text{tr}(\mathbf{S}_k) + np \log \lambda = \frac{1}{\lambda}n \text{tr}(\mathbf{S}_W) + np \log \lambda$$

where $\mathbf{S}_W = \frac{1}{n}\sum_{k=1}^{g} n_k \mathbf{S}_k$ is the within-group covariance matrix, giving us $\lambda = \frac{\text{tr}(\mathbf{S}_W)}{p}$. The classification likelihood is written

$$L_C(P, \theta) = -\frac{np}{2}\log \text{tr}(\mathbf{S}_W) + cste.$$

Maximizing the classification likelihood is therefore equivalent to minimizing the within-group sum of squares criterion $\text{tr}(\mathbf{S}_W)$. Moreover, the CEM algorithm is simply the k-means algorithm. This means that to use the within-group sum of squares criterion is to assume that classes are spherical and have the same proportion and the same volume.

8.5.4. *Spherical form, identical proportions but differing volumes*

We now take the model described above and modify it slightly to include classes with different volumes. The covariance matrices are now written $\mathbf{\Sigma}_k = \lambda_k \mathbf{I}_p$, and equation (8.7) shows that individuals are allocated to classes according to the distance

$$\frac{1}{\lambda_k} d^2(\boldsymbol{x}_i, \boldsymbol{\mu}_k) + p \log \lambda_k.$$

The distance from a point to the center of a class has been modified by an amount that depends on the volume of the class. This modification has important repercussions; for example the regions of separation, which in the previous case were hyperplanes, become hyperspheres. It may be shown that the minimized criterion can be written

$$\sum_k \log \operatorname{tr}(\mathbf{S}_k).$$

With this model we can very easily recognize situations such as that shown in Figure 8.3. Here the two classes have been simulated with two spherical Gaussian distributions having the same proportions but widely differing volumes. The result obtained using the classical intraclass inertia criterion corresponds to a separation of the population by a straight line and therefore bears no relation at all to the simulated partition. With the variable-volume model the obtained partition, shown by the circle, is very close to the initial clustering.

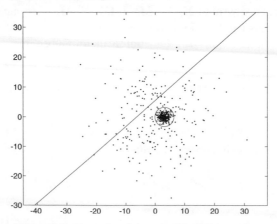

Figure 8.3. *Example of classes with different volumes*

It will be noted that without the help of the mixture model it would have been difficult, on the basis of a simple metrical interpretation, to derive the distance and the criterion used in this approach.

8.5.5. *Identical covariance matrices and proportions*

Our final example is where classes all have the same form and the same proportion. The covariance matrix of each class can therefore be written $\Sigma_k = \Sigma$. It can be shown that individuals are now allocated to classes on the basis of the distance $d^2_{\Sigma^{-1}}(\boldsymbol{x}_i, \boldsymbol{\mu}_k)$ and that the criterion to be minimized may be written as $|\mathbf{S}_W|$. This justifies the use of this criterion, sometimes proposed in a metrical context [FRI 67] without reference to the Gaussian model.

8.6. Binary variables

We now turn, still within a broad discussion of clustering methods based on probability distribution mixture models, to the clustering of sets of individuals measured using binary variables.

8.6.1. *Data*

Binary variables are widely used in statistics. Examples include presence-absence data in ecology, black and white pixels in image processing and the data obtained when recoding a table of qualitative variables. Data take the form of a sample $\mathbf{X} = (\boldsymbol{x}_1, \ldots, \boldsymbol{x}_n)$ where $\boldsymbol{x}_i$ is a vector $(x_i^1, \ldots, x_i^p)$ of values x_i^j belonging to the set $\{0, 1\}$. These data may be represented either as a table of individuals–variables as in the case of continuous variables that we encountered in the previous section, as a frequency table or as a contingency table. For example, the data might correspond to a set of ten squares of woodland in which the presence (1) or absence (0) of two types of butterfly P1 and P2 has been observed. Figure 8.4 illustrates the three alternative ways of presenting these data.

Square	P1	P2
1	1	1
2	1	0
3	1	1
4	0	1
5	0	1
6	1	1
7	1	0
8	0	0

State	Frequency
00	1
01	2
10	2
11	3

P1 \ P2	1	0
1	$n_{11} = 3$	$n_{10} = 2$
0	$n_{01} = 2$	$n_{00} = 1$

Figure 8.4. *Example of binary data*

In clustering, tables of binary data have been associated with a large number of distances, most of which are defined using the values n_{11}, n_{10}, n_{01} and n_{00} of the contingency table crossing the two variables. For example, the distances between two binary vectors i and i' measured using Jaccard's index and using the 'agreement coefficient' can be written

$$d(\boldsymbol{x}_i, \boldsymbol{x}_{i'}) = \frac{n_{11}}{n_{00} + n_{10} + n_{01}}$$

and

$$d(\boldsymbol{x}_i, \boldsymbol{x}_{i'}) = n_{11} + n_{00},$$

respectively.

8.6.2. *Binary mixture model*

Just as the Gaussian model is often chosen to model each component of the mixture when variables are continuous, the log-linear model [AGR 90, BOC 86] is a natural choice when variables are binary.

The complete or saturated log-linear model, where each class has a multinomial distribution with 2^p values, is not really applicable in the case of a mixture. Instead we use a log-linear model with sufficient constraints. The simplest example is the independence model which assumes that, conditionally on membership of a class, the binary variables are independent. From this we obtain the latent class model [GOO 74, LAZ 68], which we shall now examine. In the binary case, each x_i^j has a Bernoulli distribution whose probability distribution takes the form

$$f(x_i^j; \alpha_k^j) = (\alpha_k^j)^{x_i^j}(1 - \alpha_k^j)^{1-x_i^j}$$

where $\alpha_k^j = P(x_i^j = 1|k)$.

The distribution of the class k is therefore

$$f(\boldsymbol{x}_i; \boldsymbol{\alpha}_k) = \prod_{j=1}^{p}(\alpha_k^j)^{x_i^j}(1 - \alpha_k^j)^{1-x_i^j}$$

where $\boldsymbol{\alpha}_k = (\alpha_k^1, \ldots, \alpha_k^p)$.

The mixture model chosen considers that the data $\boldsymbol{x}_1, \ldots, \boldsymbol{x}_n$ constitute a sample of independent instances from a random $\{0, 1\}^p$ probability vector

$$f(\boldsymbol{x}_i; \boldsymbol{\theta}) = \sum_{k=1}^{g} \pi_k f(\boldsymbol{x}_i; \boldsymbol{\alpha}_k) = \sum_{k=1}^{g} \pi_k \prod_{j=1}^{p}(\alpha_k^j)^{x_i^j}(1 - \alpha_k^j)^{1-x_i^j}$$

where the parameter $\boldsymbol{\theta}$ is constituted by the proportions $\pi_1, \ldots, \pi_g$ and by the parameter vector $\boldsymbol{\alpha} = (\boldsymbol{\alpha}_1, \ldots, \boldsymbol{\alpha}_g)$ of each component.

The problem which arises is how to estimate these parameters and possibly also the origin class. As in the case of the Gaussian model, the estimation may be obtained using the EM or the CEM algorithms described above. The only differences concern the computation of the parameter α_k^j which becomes

$$\alpha_k^j = \frac{\sum_{i=1}^n t_{ik} x_i^j}{\sum_{i=1}^n t_{ik}}$$

for the first and

$$\alpha_k^j = \frac{\sum_{i=1}^n z_{ik} x_i^j}{n_k} = \% \text{ of } 1$$

for the second.

8.6.3. *Parsimonious model*

The number of parameters of this latent class model is equal to $(g-1)+g\times p$, where g, it will be recalled, is the number of classes and p is the number of binary variables. In the case of the complete log-linear model, however, the number of parameters is equal to 2^p. For example, when $g = 5$ and $p = 10$ the number of parameters for the two models is 54 and 1024, respectively. Given that one of the identifiability conditions of the model is that the number of states is greater than the number of parameters, there is a clear interest in being able to propose even more parsimonious models. To this end, the model may be restated as:

$$f(\boldsymbol{x}; \boldsymbol{\theta}) = \sum_{k=1}^g \pi_k \prod_{j=1}^p (\varepsilon_k^j)^{|x_i^j - a_k^j|} (1 - \varepsilon_k^j)^{1 - |x_i^j - a_k^j|}$$

where

$$\begin{cases} a_k^j = 0, \ \varepsilon_k^j = \alpha_k^j & \text{if } \alpha_k^j < 0.5 \\ a_k^j = 1, \ \varepsilon_k^j = 1 - \alpha_k^j & \text{if } \alpha_k^j > 0.5. \end{cases}$$

The parameter $\boldsymbol{\alpha}_k$ is therefore replaced by the two following parameters:

– a binary vector $\mathbf{a}_k$ representing the center of the class and which is the most frequent binary value for each variable;

– a vector $\boldsymbol{\varepsilon}_k$ belonging to the set $]0, 1/2[^p$ which defines the dispersion of the component and which represents the probability of any particular variable having a value different from that of the center.

We are therefore led to the parameters used by Aitchinson and Aitken [AIT 76] for discrimination with non-parametric estimation via the kernel method. Starting from their formulation, we arrive at parsimonious situations by stipulating certain constraints. The $[\varepsilon]$ model is defined by stipulating that the dispersion should not depend either on the component or on the variable, the $[\varepsilon_k]$ model by stipulating that it should depend only on the component and the $[\varepsilon^j]$ model by stipulating that it should depend only on the variable.

In the example of the simplest case of the $[\varepsilon]$ model, given identical proportions $(\pi_k = 1/g)$, the clustering approach results in the complete-data log-likelihood being maximized:

$$L_C(\mathbf{Z}, \boldsymbol{\theta}) = \log \frac{\varepsilon}{1-\varepsilon} \sum_{k=1}^{g} \sum_{i/z_{ik}=1} d(\boldsymbol{x}_i, \mathbf{a}_k) + np \log(1 - \varepsilon),$$

which is to say that the criterion

$$W(\mathbf{Z}, \boldsymbol{\theta}) = \sum_{k=1}^{g} \sum_{i/z_{ik}=1} d(\boldsymbol{x}_i, \mathbf{a}_k)$$

where $d(\boldsymbol{x}_i, \mathbf{a}_k) = \sum_{j=1}^{p} |x_i^j - a_k^j|$ is minimized.

The step E of the CEM algorithm then consists simply of allocating each individual to the class k that minimizes $d(\boldsymbol{x}_i, \mathbf{a}_k)$. At step M, the parameters a_k^j for each variable j correspond to the majority binary values in each class k. A class center therefore corresponds to a binary vector, and the criterion is easy to interpret: it is simply the number of differences between individuals and their representative in the partition $\mathbf{z}$. To use this binary clustering criterion proposed by different authors [GOV 90a, GOW 74] is therefore to assume that the data come from a particular latent class model.

S1	S2	S3	S4	frequencies
1	1	1	1	42
1	1	1	0	23
1	1	0	1	6
1	1	0	0	25
1	0	1	1	6
1	0	1	0	24
1	0	0	1	7
1	0	0	0	38
0	1	1	1	1
0	1	1	0	4
0	1	0	1	1
0	1	0	0	6
0	0	1	1	2
0	0	1	0	9
0	0	0	1	2
0	0	0	0	20

Table 8.1. *Stouffer–Toby data*

Class	p_k	a_k^1	a_k^2	a_k^3	a_k^4
1	0.279	0.993	0.940	0.927	0.769
2	0.721	0.714	0.330	0.354	0.132

Table 8.2. *Results obtained by EM for the latent class model*

8.6.4. *Example of application*

To illustrate this approach we have taken the Stouffer–Toby dataset [STO 51], analyzed within the framework of the latent class model by Goodman [GOO 74]. The table of data (Table 8.1) contains the reactions, classed as one of two possible attitudes, shown by 216 subjects placed in four different conflict situations. We compare the latent class model (here requiring 9 parameters) with the log-linear model with interaction of order 2, having a very similar number of parameters (11). It is interesting to note that, for this data, the deviance drops from 7.11 in the case of the linear model to a value of 2.72 in the case of the latent class model. The parameters obtained for the latent class model are shown in Table 8.2.

8.7. Qualitative variables

We now extend the results of the previous section to qualitative variables.

8.7.1. *Data*

Qualitative variables, sometimes known as questionnaire or categorical data, are a generalization of binary data to cases where there are more than two possible values. Here, each variable j may take an arbitrary finite set of values, usually referred to as *modalities*. m_j denotes the number of modalities of each of these variables j. As for binary data, qualitative variables may be represented in different ways: as a table of individuals–variables of dimension (n, p), as a frequency vector for the different possible states, as a contingency table with p dimensions linking the modalities or as a *complete disjunctive table* where modalities are represented by their indicators. In the latter form of representation, which we shall use here, the data are composed of a sample $(x_1, \ldots, x_n)$, where $x_i = (x_i^{jh}; j = 1, \ldots, p; h = 1, \ldots, m_j)$, with

$$\begin{cases} x_i^{jh} = 1 & \text{if } i \text{ takes the modality } h \text{ for the variable } j \\ x_i^{jh} = 0 & \text{otherwise.} \end{cases}$$

8.7.2. *The model*

As for binary data, we shall examine the latent class model and therefore assume that the jth qualitative variables are independent, conditionally on their membership of a class. If α_k^{jh} is the probability that the jth variable takes the modality h when an individual belongs to the class k, then the probability of the mixture can be written

$$f(\boldsymbol{x}_i; \boldsymbol{\theta}) = \sum_{k=1}^{g} \pi_k f(\boldsymbol{x}_i; \boldsymbol{\alpha}_k) = \sum_{k=1}^{g} \pi_k \prod_{j=1}^{p} \prod_{h=1}^{m_j} (\alpha_k^{jh})^{x_i^{jh}}$$

where the parameter $\boldsymbol{\theta}$ is defined by the proportions $\pi_1, \ldots, \pi_g$ and by the parameters $\boldsymbol{\alpha}_k = (\alpha_k^{jh}; j = 1, \ldots, p; h = 1, \ldots, m_j)$ of the density of each component.

As before, estimating the parameter $\boldsymbol{\theta}$ and possibly the native class of each of the $\boldsymbol{x}_i$ may be achieved by maximizing the likelihood $L(\boldsymbol{\theta})$ using the EM algorithm, or by maximizing the complete-data likelihood $L_C(\mathbf{Z}, \boldsymbol{\theta})$ using the CEM algorithm. In the case of the EM algorithm, the computation of the parameters $\boldsymbol{\alpha}_k$ at step M is defined by the relation $\alpha_k^{jh} = \sum_{i=1}^{n} t_{ik} x_i^{jh} / \sum_{i=1}^{n} t_{ik}$, where t_{ik} are the probabilities obtained in the usual fashion at step E. In the case of the CEM algorithm, the computation of the parameters $\boldsymbol{\alpha}_k$ becomes $\alpha_k^{jh} = \sum_{i=1}^{n} z_{ik} x_i^{jh} / n_k$, where $\mathbf{z}$ is the partition obtained by the MAP from the probabilities t_{ik}.

We now look at what happens to the complete-data likelihood criterion when the clustering approach is used with the assumption that the proportions π_k are constant. If we denote $s_k^{jh} = \sum_{i=1}^{n} z_{ik} x_i^{jh}$, $s_{\cdot}^{jh} = \sum_{i=1}^{n} x_i^{jh}$, $s_k^{\cdot\cdot} = \sum_{j=1}^{p} \sum_{h=1}^{m_j} s_k^{jh}$ and $s^{\cdot\cdot} = \sum_{i=1}^{n} \sum_{j=1}^{p} \sum_{h=1}^{m_j} x_i^{jh} = np$, then we can easily show that the relation

$$L_C(\mathbf{Z}, \boldsymbol{\theta}) = \sum_{k=1}^{g} \sum_{j=1}^{p} \sum_{h=1}^{m_j} s_k^{jh} \log \alpha_k^{jh}$$

is obtained.

Given that, at convergence $\alpha_k^{jh} = s_k^{jh} / n_k$, it can be shown [CEL 91] that the CEM algorithm maximizes the information criterion [BEN 73]:

$$H(\mathbf{Z}) = \sum_{k=1}^{g} \sum_{j=1}^{p} \sum_{h=1}^{m_j} \frac{s_k^{jh}}{s^{\cdot\cdot}} \log \frac{s^{\cdot\cdot} s_k^{jk}}{s_k^{\cdot\cdot} s_{\cdot}^{jh}}$$

which represents the information from the initial table retained by the partition $\mathbf{Z}$ and yielding results very close to the χ^2 criterion:

$$\chi^2(\mathbf{Z}) = \sum_{k=1}^{g} \sum_{j=1}^{p} \sum_{h=1}^{m_j} \frac{(s^{\cdot\cdot} s_k^{jh} - s_k^{\cdot\cdot} s_{\cdot}^{jh})^2}{s^{\cdot\cdot} s_k^{\cdot\cdot} s_{\cdot}^{jh}}.$$

It therefore follows that to seek a partition into g classes maximizing the information criterion or the χ^2 criterion (approximately equivalent) is to assume that the data derive from a latent class model.

A parallel may be drawn here with the analysis of multiple correspondences (see Chapter 2). It is not difficult to see that with the geometrical representation used in this factorial analysis, the χ^2 criterion is quite simply the familiar criterion of intraclass inertia.

8.7.3. *Parsimonious model*

The number of parameters $(g-1)+g\times\sum_{j=1}^{p}(m_j-1)$ required by the latent class model which we have just described is usually considerably smaller than the number of parameters $\prod_{j=1}^{p}m_j$ required by the complete log-linear model. For example, for a number of classes g equal to 5 and a number of qualitative variables p equal to 10, where the number of modalities m_j is 4 for all the variables, the number of parameters for the two models is 154 and 10^6, respectively. This number will be quite excessive in many cases, and more parsimonious models are called for.

To this end, we begin by remarking that if for each variable j the modality of highest probability is denoted as h^*, then the model may be re-parameterized as follows:

$$f(\boldsymbol{x}_i; \boldsymbol{\theta}) = \sum_{k=1}^{y} \pi_k \prod_{j-1}^{p} \left(\left(1 - \varepsilon_k^{jh^*}\right)^{1-d(x_i^j,a_k^j)} \prod_{h\neq h^*} (\varepsilon_k^{jh})^{|x_i^{jh}-a_k^{jh}|} \right)$$

where $\mathbf{a}_k^j = (a_k^{j1},\ldots,a_k^{jm_j})$ with $a_k^{jh} = 1$ if $h = h^*$ and 0 otherwise, $\varepsilon_k^j = (\varepsilon_k^{j1},\ldots,\varepsilon_k^{jm_j})$ where $\varepsilon_k^{jh} = 1-\alpha_k^{jh}$ if $h = h^*$ and α_k^{jh} otherwise, and $d(x_i^j,a_k^j) = 0$ if x_i^j and a_k^j take the same modality and 1 otherwise. As for binary data, the vector $\mathbf{a}_k = (\mathbf{a}_k^1,\ldots,\mathbf{a}_k^p)$ may be interpreted as the center of the class k and the vectors ε_k^j as dispersions. For example, if the parameter α_k^j is equal to the vector $(0.7,0.2,0.1)$, the new parameters become $\mathbf{a}_k^j = (1,0,0)$ and $\varepsilon_k^j = (0.3,0.2,0.1)$.

With the model restated in this way, it is possible to introduce simple constraints such as requiring non-majority modalities to have the same dispersion:

$$\begin{cases} \varepsilon_k^{jh^*} = \varepsilon_k^j \\ \varepsilon_k^{jh} = (1 - \varepsilon_k^j)/(m_j - 1) \text{ for } h \neq h^*. \end{cases}$$

This gives us a model used in discrimination, where the number of parameters has been reduced from $(g - 1) + g \times \sum_{j=1}^{p}(m_j - 1)$ to $(g - 1) + \sum_{j=1}^{p}(m_j - 1)$.

Even more parsimonious models may been obtained if additional constraints are placed on dispersions, for example by requiring that $\varepsilon_k^{jh^*}$ should not depend on the variable (model $[\varepsilon_k]$), on the class (model $[\varepsilon^j]$), or on neither the variable nor the class (model $[\varepsilon]$). If we restrict ourselves to this last model $[\varepsilon]$, and if we require proportions to be equal, the complete-data log-likelihood can be expressed quite simply:

$$L_C(\mathbf{Z}, \boldsymbol{\theta}) = \log\left(\frac{\varepsilon}{1-\varepsilon}\right) \sum_{k=1}^{g} \sum_{i=1}^{n} z_{ik} d(\boldsymbol{x}_i, \mathbf{a}_k) + np \log(1 - \varepsilon)$$

where $d(\boldsymbol{x}_i, \mathbf{a}_k)$ is a distance reflecting the number of different modalities between the vector $\boldsymbol{x}_i$ and the center $\mathbf{a}_k$. At the clustering step of the CEM algorithm, the individuals i are therefore allocated to the class k that minimizes $d(\boldsymbol{x}_i, \mathbf{a}_k)$, and at step M the coordinates a_k^j of the centers $\mathbf{a}_k$ are obtained by taking the majority modalities.

In practice, we suggest taking very simple models by class (for example latent class models with a single parameter) and then increasing the number of components if necessary.

When the qualitative variables are ordinals, it is possible either to convert the data into binary data or to use an approach similar to the one we have just described, taking into account the order that exists between the modalities.

8.8. Implementation

There are a number of software developments implementing the methods described in this chapter, including the MIXMOD [1] program. In this section, we give a rapid overview of the problems that software implementations need to address.

8.8.1. *Choice of model and of the number of classes*

Clustering methods are often justified heuristically, and choosing the 'right' method or the 'right' number of classes can be a difficult problem that is often badly stated. The use of clustering methods based on mixture models allows us to place the problem within the more general framework of the selection of probabilistic models.

In the Bayesian context, choosing the most probable model calls for frequently used selection criteria such as Schwarz's [SCH 78] BIC criterion comprising two terms. The first is likelihood which tends to favor the more complex model, and the second is a penalizing term, an increasing function of the number of the model's parameters. Worthy of mention is the ICL criterion [BIE 00] which, taking the objective of the clustering into account, generally provides good solutions.

---

1. http://www-math.univ-fcomte.fr/mixmod

8.8.2. *Strategies for use*

Maximizing the likelihood criterion via the EM algorithm, or maximizing the clustering likelihood via the CEM algorithm, always involves obtaining a series of solutions that see the criterion increase to a local maximum, and which are therefore dependent on the initial position selected by the algorithm. The strategy usually adopted for obtaining a 'good' solution is to run the algorithm several times from different starting points and to retain the best solution from among them. For example, see [BIE 03] where some subtle and effective strategies are examined, including an initial phase in which the algorithm is run a large number of times without waiting for complete convergence.

8.8.3. *Extension to particular situations*

We have seen that the mixture model in clustering can cope with a variety of situations (spherical or non-spherical classes, equal or unequal proportions, etc.) and can deal with both continuous and binary data.

In this section we briefly list some clustering problems that the mixture model approach addresses quite naturally, illustrating its adaptability to particular situations.

Noisy data Atypical or outlier data (measurement errors, etc.) generally perturb clustering methods quite considerably. Getting mixture models to take account of noise can be a simple matter, for example by adding a uniformly distributed class or by using distributions less sensitive to atypical elements such as Laplace distributions.

Incomplete labeling in discrimination In discrimination (see Chapter 6) we often have, in addition to the learning sample whose class is known, a (sometimes large) set of observations whose class is not known. Making use of these unlabeled observations, which can significantly improve the results of the discrimination, can by accomplished easily by introducing into the EM and CEM algorithms observations whose membership of a class is not brought into question during the iterations of the algorithm.

Spatial data The mixture model rests on the hypothesis that the vector $Z = (z_1, \ldots, z_n)$ grouping the classes of the different observations is an independent sample. However, there are more complex situations such as the segmentation of pixels in image processing where this hypothesis must be rejected. In these cases, the mixture model may be extended to the clustering of geographically localized multivariate observations such as hidden Markov fields, in order to include this type of data. These approaches are described in Chapter 9.

Block clustering The clustering methods described so far were all designed to classify individuals, or occasionally variables. There are other methods, however, often known as block or simultaneous clustering methods (see Chapter 7, section 7.6), which process the two sets simultaneously and organize the data into homogenous blocks. Here it is also possible to extend the use of mixture models by using a latent block model generalizing the mixture model [GOV 03, GOV 05, GOV 06, GOV 08].

8.9. Conclusion

In this chapter we have attempted to show the advantages of using mixture models in clustering. This approach provides a general framework capable of taking into account specificities in the data and in the problem. Moreover, a probabilistic model means being able to harness the entire set of statistical results in proposing solutions to difficult problems such as the choice of the model or the number of classes.

Obviously, one of the difficulties of this approach is deciding whether the selected mixture model is realistic for the data in question. However, as Everitt [EVE 93] has rightly observed, it is not a difficulty specific to this approach. We cannot avoid choosing a method's underlying hypotheses simply by 'concealing' them.

8.10. Bibliography

[AGR 90] AGRESTI A., *Categorical Data Analysis*, Wiley, New York, 1990.

[AIT 76] AITCHISON J., AITKEN C. G. G., "Multivariate binary discrimination by the kernel method", *Biometrika*, vol. 63, p. 413–420, 1976.

[BAN 93] BANFIELD J. D., RAFTERY A. E., "Model-based Gaussian and non-Gaussian clustering", *Biometrics*, vol. 49, p. 803–821, 1993.

[BEN 73] BENZECRI J.-P., *L'Analyse des Données, Part 1: La Taxinomie, Part 2: L'Analyse des Correspondances*, Dunod, Paris, 1973.

[BEZ 81] BEZDEK J., *Pattern Recognition with Fuzzy Objective Function Algorithms*, Plenum Press, New York, 1981.

[BIE 00] BIERNACKI C., CELEUX G., GOVAERT G., "Assessing a mixture model for clustering with the integrated completed likelihood", *IEEE Transactions on Pattern Analysis and Machine Intelligence*, vol. 22, num. 7, p. 719–725, July 2000.

[BIE 03] BIERNACKI C., CELEUX G., GOVAERT G., "Choosing starting values for the EM algorithm for getting the highest likelihood in multivariate Gaussian mixture models", *Computational Statistics and Data Analysis*, vol. 41, p. 561–575, January 2003.

[BOC 86] BOCK H., "Loglinear models and entropy clustering methods for qualitative data", GAULL W., SCHADER M., Eds., *Classification as a Tool of Research*, p. 19–26, North-Holland, Amsterdam, 1986.

[BOC 89] BOCK H., "Probabilistic aspects in cluster analysis", OPITZ O., Ed., *Conceptual and Numerical Analysis of Data*, p. 12–44, Springer-Verlag, Berlin, 1989.

[BOC 96] BOCK H., "Probability models and hypothesis testing in partitioning cluster analysis", ARABIE P., HUBERT L. J., DE SOETE G., Eds., *Clustering and Classification*, p. 377–453, World Scientific, Singapore, 1996.

[CEL 91] CELEUX G., GOVAERT G., "Clustering criteria for discrete data and latent class models", *Journal of Classification*, vol. 8, num. 2, p. 157–176, 1991.

[CEL 92] CELEUX G., GOVAERT G., "A classification EM algorithm for clustering and two stochastic versions", *Computational Statistics and Data Analysis*, vol. 14, num. 3, p. 315–332, 1992.

[CEL 93] CELEUX G., GOVAERT G., "Comparison of the mixture and the classification maximum likelihood in cluster analysis", *J. Statist. Comput. Simul.*, vol. 47, p. 127–146, 1993.

[CEL 95] CELEUX G., GOVAERT G., "Gaussian parsimonious clustering models", *Pattern Recognition*, vol. 28, num. 5, p. 781–793, 1995.

[CRE 93] CRESSIE N. A. C., *Statistics for Spatial Data*, Wiley, New York, 1993.

[DEG 83] DEGENS P. O., "Hierarchical cluster methods as maximum likelihood estimators", FELSENSTEIN J., Ed., *Numerical Taxonomy*, Heidelberg, Springer-Verlag, p. 249–253, 1983.

[DEM 77] DEMPSTER A. P., LAIRD N. M., RUBIN D. B., "Maximum likelihood from incomplete data via the EM algorithm (with discussion)", *Journal of the Royal Statistical Society*, vol. B 39, p. 1–38, 1977.

[DID 79] DIDAY E., *Optimisation et Classification Automatique*, INRIA, Rocquencourt, 1979.

[EVE 93] EVERITT B. S., *Cluster Analysis*, Arnold, London, 3rd edition, 1993.

[FRI 67] FRIEDMAN H. P., RUBIN J., "On some invariant criteria for grouping data", *Journal of American Statistical Association*, vol. 62, p. 1159–1178, 1967.

[GOO 74] GOODMAN L. A., "Exploratory latent structure models using both identifiable and unidentifiable models", *Biometrika*, vol. 61, p. 215–231, 1974.

[GOV 89] GOVAERT G., "Clustering model and metric with continuous data", DIDAY Y., Ed., *Data Analysis, Learning Symbolic and Numeric Knowledge*, New York, Nova Science Publishers, Inc., p. 95–102, 1989.

[GOV 90a] GOVAERT G., "Classification binaire et modèles", *Revue de Statistique Appliquée*, vol. XXXVIII, num. 1, p. 67–81, 1990.

[GOV 90b] GOVAERT G., Modèle de classification et distance dans le cas discret, Technical report, UTC, Compiègne, France, 1990.

[GOV 96] GOVAERT G., NADIF M., "Comparison of the mixture and the classification maximum likelihood in cluster analysis when data are binary", *Computational Statistics and Data Analysis*, vol. 23, p. 65–81, 1996.

[GOV 03] GOVAERT G., NADIF M., "Clustering with block mixture models", *Pattern Recognition*, vol. 36, p. 463–473, 2003.

[GOV 05] GOVAERT G., NADIF M., "An EM algorithm for the block mixture model", *IEEE Transactions on Pattern Analysis and Machine Intelligence*, vol. 27, p. 643–647, April 2005.

[GOV 06] GOVAERT G., NADIF M., "Fuzzy clustering to estimate the parameters of block mixture models", *Soft Computing*, vol. 10, p. 415–422, March 2006.

[GOV 08] GOVAERT G., NADIF M., "Block clustering with Bernoulli mixture models: comparison of different approaches", *Computational Statistics and Data Analysis*, vol. 52, p. 3233–3245, 2008.

[GOW 74] GOWER J., "Maximal predictive classification", *Biometrics*, vol. 30, p. 643–654, 1974.

[HAR 75] HARTIGAN J. A., *Clustering Algorithms*, Wiley, New York, 1975.

[HAT 86] HATHAWAY R. J., "Another interpretation of the EM algorithm for mixture distributions", *Statistics & Probability Letters*, vol. 4, p. 53–56, 1986.

[LAZ 68] LAZARFIELD P. F., HENRY N. W., *Latent Structure Analysis*, Houghton Mifflin Company, Boston, 1968.

[LER 81] LERMAN I. C., *Classification Automatique et Analyse Ordinale des Données*, Dunod, Paris, 1981.

[LIN 73] LING R. F., "A probability theory of cluster analysis", *Journal of the American Statistical Association*, vol. 68, num. 63, p. 159–164, 1973.

[MAR 75] MARRIOTT F. H. C., "Separating mixtures of normal distributions", *Biometrics*, vol. 31, p. 767–769, 1975.

[MCL 82] MCLACHLAN G. J., "The classification and mixture maximum likelihood approaches to cluster analysis", KRISHNAIAH P. R., KANAL L. N., Eds., *Handbook of Statistics*, vol. 2, Amsterdam, North-Holland, p. 199–208, 1982.

[MCL 88] MCLACHLAN G., BASFORD K., *Mixture Models, Inference and Applications to Clustering*, Marcel Dekker, New York, 1988.

[MCL 97] MCLACHLAN G. J., KRISHMAN K., *The EM Algorithm*, Wiley, New York, 1997.

[MCL 00] MCLACHLAN G. J., PEEL D., *Finite Mixture Models*, Wiley, New York, 2000.

[NEA 98] NEAL R. M., HINTON G. E., "A view of the EM algorithm that justifies incremental, sparse, and other variants", JORDAN M. I., Ed., *Learning in Graphical Models*, p. 355–358, Kluwer Academic Press, 1998.

[NEY 58] NEYMAN J., SCOTT E. L., "Statistical approach to problems of cosmology", *Journal of the Royal Statistical Society, Series B*, vol. 20, num. 1, p. 1–43, 1958.

[RIP 81] RIPLEY B. D., *Spatial Statistics*, Wiley, 1981.

[SCH 76] SCHROEDER A., "Analyse d'un mélange de distributions de probabilité de même type", *Revue de Statistique Appliquée*, vol. 24, num. 1, p. 39–62, 1976.

[SCH 78] SCHWARZ G., "Estimating the number of components in a finite mixture model", *Annals of Statistics*, vol. 6, p. 461–464, 1978.

[SCO 71] SCOTT A. J., SYMONS M. J., "Clustering methods based on likelihood ratio criteria", *Biometrics*, vol. 27, p. 387–397, 1971.

[STO 51] STOUFFER S. A., TOBY J., "Role conflict and personality", *American Journal of Sociology*, vol. 56, p. 395–506, 1951.

[SYM 81] SYMONS M. J., "Clustering criteria and multivariate normal mixture", *Biometrics*, vol. 37, p. 35–43, March 1981.

[TIT 85] TITTERINGTON D., SMITH A. F. M., MAKOV U. E., *Statistical Analysis of Finite Mixture Distributions*, Wiley, New York, 1985.

[VAN 98] VAN CUTSEM B., YCART B., "Indexed dendrograms on random dissimilarities", *Journal of Classification*, vol. 15, p. 93–127, 1998.

[WON 83] WONG M. A., LANE T., "A kth nearest neighbour clustering procedure", *Journal of the Royal Statistical Society (series B)*, vol. 45, num. 3, p. 362–368, 1983.

Chapter 9

Spatial Data Clustering

This chapter aims to provide an overview of different techniques proposed in the literature for clustering spatially located data. We first present the context, using some specific examples as illustrations (section 9.1). We then briefly describe a certain number of non-probabilistic approaches that use spatial data to achieve a classification (section 9.2). Next we present Markov random field models, which have led to significant developments in image segmentation (section 9.3), and we devote part of this coverage of Markov modeling to the problem of parameter estimation (section 9.4). We then illustrate how the Markov approach may be implemented, with reference to a numerical ecology application with binary multivariate data (section 9.5).

9.1. Introduction

9.1.1. *The spatial data clustering problem*

This chapter looks at the clustering of observations originating within distinct sites belonging to a geographic neighborhood. How can a clustering procedure take account of the spatial information contained in this data? Most methods of spatial data reconstruction assume some kind of regularity, i.e. that the data to be reconstructed move slowly through the geographic space. In the applications presented here we make a similar assumption regarding clustering, which can be seen as a piece of data to be reconstructed from observations. A spatial clustering method usually has a dual aim:

Chapter written by Christophe AMBROISE and Mo DANG.

– the partition obtained should be as homogenous as possible, i.e. with maximum resemblance in terms of the variables describing them, between data within the same group;

– the assumption of spatial regularity implies that data which are contiguous within the geographic space will have a greater likelihood of belonging to the same group.

Figure 9.1 shows how much spatial information can impact clustering. If the clustering were only to take account of the observed values, the upper left-hand observation would be placed in the 'medium' group. However, the value of this observation is also quite close to those of the 'high' group, by which it is largely surrounded. Given the assumption of spatial regularity, it would seem appropriate to place this site in the high-value group, even if this means that groups become slightly less homogenous. We therefore see that the two aims mentioned above cannot be satisfied simultaneously and optimally, and that some kind of relative weighting is required to mediate between the opposing criteria of spatial regularity and class cohesion.

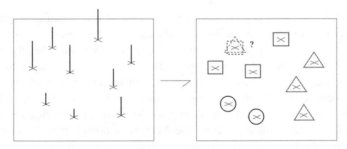

Figure 9.1. *Clustering of spatial data: (a) univariate observations, represented within the geographic space by a vertical line of length x_i and (b) partition into three classes. With no geographic information, the site located top-left would be classified as $\triangle$. With an a priori assumption of spatial regularity, it would be classified as $\square$*

In the situations investigated here, we generally have no prior information about the classes to be obtained and are therefore firmly within the context of *clustering* or *unsupervised learning* (as opposed to *classification* or *unsupervised learning*, where new observations are allocated to pre-existing groups with the classifying rule usually derived from labeled observations).

9.1.2. *Examples of applications*

EXAMPLE 9.1. (SEGMENTATION OF A BIOLOGICAL IMAGE) Figure 9.2 shows a cell culture from a biology laboratory at Compiègne University of Technology. This kind of grayscale image can be seen as a set of 512×512 observations – pixel intensity levels – located in a regular spatial grid. With a color image, each pixel would be described by intensity levels within different spectral bands, i.e. there would be multivariate observations at each site. Biologists at the laboratory would like to be able to measure the surface areas occupied by the different types of cell. The eye easily spots three zones of increasing luminosity spreading from the center towards the edge: the initial cells, the cloud of newly-formed cells and the nutritional substrate. This kind of segmentation of the image into regions is equivalent to partitioning the image such that pixels with similar intensity levels are placed in the same class. An automated segmentation into three classes would provide the biologists with a straightforward measure of the size of the cloud, i.e. the number of pixels belonging to the intermediate class. □

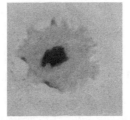

Figure 9.2. 512×512 *image of cell culture*

The segmentation of digital images is an important application of spatial data clustering. A wealth of publications clearly indicates the interest that this field has aroused since the 1980s, due not only to the number of problems to which it gives rise but also to the diversity of its potential applications. Examples are:

– segmentation can reveal different types of plant or crop cover in teledetection images;

– the segmentation of aerial photographs can be used to update maps;

– in mobile robotics, the segmentation of images from sensors is used to localize objects in the robot's environment;

– in medical imaging, the segmentation of tissues can help to evaluate the spread of a pathology.

In most of these applications, image segmentation algorithms attempt to reproduce the same segmentation procedures that would be used by a human expert. Even though substantial progress has been made, machine performance still tends to lag far behind the performance of human visual mechanisms when segmenting grayscale images (see, for example, the preface of [COC 95]). However, when a site (a pixel in the case of an image) is described not by one but by several variables, it is more difficult for the human eye to make a partition than it is with univariate grayscale images. Clustering algorithms can therefore provide invaluable assistance in the analysis and interpretation of multivariate data. This is illustrated by the two following examples.

EXAMPLE 9.2. (PHYSICAL AND CHEMICAL MAP) We wish to map a piece of ground in order to identify areas with the same physical and chemical profile. We have pH, humidity and acidity data collected from 165 boreholes at three different depths. These boreholes, on average 25 m apart, are arranged in an irregular grid pattern. Figure 9.3 shows the spatial distribution of humidity and KC1 pH readings for the horizons (i.e. depths) 1 and 2. A classification of the different areas would provide a multicriteria map which could be used by researchers, land managers and entrepreneurs [BUR 97].

□

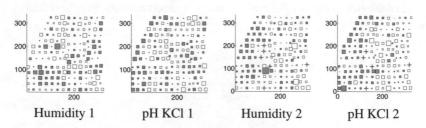

Figure 9.3. *Carreau Blémur data (example 9.2). Relative humidity and KC1 pH at horizons 1 and 2. A gray (or white) square indicates a higher than average (or lower than average) value. The area of the square is proportional to the discrepancy in relation to the average*

EXAMPLE 9.3. (REGIONS WITH SIMILAR FAUNA) Indicators regarding the presence or absence of 22 species of butterfly were gathered over 349 square plots covering the French region of Burgundy. Each square has its corresponding vector of 22 binary values indicating the presence or absence of each species. Figure 9.4 shows the spatial distribution of three of these species.

□

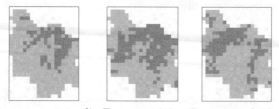

Zygaena purpuralis Zygaena viciae Zygaena ephialtes

Figure 9.4. *Presence/absence of three species of butterfly in the Burgundy region (example 9.3). The presence (or absence) of a species is indicated by a dark (or light) square*

A biologist might wish to classify the squares in order to identify regions with similar habitation profiles, and use the resulting map to form hypotheses (in relation to geology, climate or vegetation) regarding the localization of species in particular places.

9.2. Non-probabilistic approaches

9.2.1. *Using spatial variables*

It is natural to use spatial variables in addition to the other variables which describe sites. Spatial co-ordinates may be weighted in order to control the extent to which spatial information affects the classification or clustering algorithm used. This idea is based on the work of Berry [BER 66], and has also been used in image segmentation by Jain [JAI 91]. Oliver [OLI 89] has criticized this approach for its tendency to separate into different classes sites which are geographically distant but are nevertheless very similar.

9.2.2. *Transformation of variables*

Instead of taking the initial table of individuals/variables as a starting point, it may be desirable to include a preprocessing phase. Preprocessing aims to extract new variables containing spatial information. One possibility is to define a geographic window size and to replace the initial variables by a transformation using the geographic neighbors (i.e. inside the window centered on this individual). A simple weighted average or a median filter are commonly used transformations.

EXAMPLE 9.4. (MEDIAN FILTER) Univariate median filtering involves replacing each observation x_i by the median of the observations x_j inside a window surrounding the pixel i. As for averaging filters, a different weight w_j may be given to each of the pixels inside the window. The weighted median may therefore be defined as

$$\text{med}\{(x_1, w_1), \ldots, (x_n, w_n)\} = \arg\min_a \sum_{i=1}^{n} w_i |x_i - a|.$$

When the weights are integers, the weighted median may be calculated as for a classical median, by replicating w_j times for every observation x_j:

$$\text{med}\{(x_1, w_1), \ldots, (x_n, w_n)\} = \text{med}\{\underbrace{x_1, \ldots, x_1}_{w_1 \text{ times}}, \ldots, \underbrace{x_n, \ldots, x_n}_{w_n \text{ times}}\}.$$

$\square$

9.2.3. *Using a matrix of spatial distances*

If the initial data take the form of a table of distances between individuals, this matrix may be transformed in order to integrate the spatial information. Oliver *et al.* [OLI 89] propose using the geographic distances g_{ij} between nodes i and j to modify the distances or dissimilarities d_{ij} calculated from non-geographic variables.

This means that the chosen clustering method is based on a new dissimilarity matrix $\mathbf{D}^* = \{d_{ij}^*\}$ which mixes geographic and non-geographic information. The usual (unconstrained) clustering algorithms are then used to segment the data.

EXAMPLE 9.5. (k-MEANS AND SPATIAL DISTANCE MATRIX [OLI 89]) Starting from a dissimilarity matrix $\mathbf{D} = \{d_{ij}\}$, the following procedure creates a data partition that avoids excessive geographic fragmentation but nevertheless produces classes which are not totally joined up:

– Modify the dissimilarity matrix

$$d_{ij}^* = d_{ij} \cdot [1 - \exp\left(-g_{ij}/W\right)]$$

where W is an arbitrary coefficient. The greater the value of W, the greater the influence of geographic distances on the new dissimilarity matrix $\mathbf{D}^*$ and the less the fragmentation.

– Transform the matrix $\mathbf{D}^*$ into a table of individuals/variables using factorial analysis.

– Partition this new data table using the k-means algorithm.

□

9.2.4. *Clustering with contiguity constraints*

During the process of agglomerative clustering, clustering may be restricted to entities which are geographically contiguous. Contiguity constraints are perfectly observed here and the classes obtained are connected, meaning that each class forms one single geographic bloc [LEB 78, LEG 87, OPE 77]. This type of clustering will place two spatially distant sites in different classes, no matter how similar the values of their non-geographic variables. In this approach, spatial information is of overriding importance, and the question of modulating the 'amount of spatial information' used in the clustering process simply does not arise.

Producing geographically contiguous classes involves specifying in advance which individuals are neighbors. Defining a contiguity matrix is equivalent to building a non-oriented graph where nodes are elements of the dataset and each edge is a neighborhood relationship.

Clustering with absolute contiguity constraints can be separated into two stages:

1) Defining a contiguity graph: possible methods include Delaunay triangulations, Gabriel graphs and grills.

2) Clustering with constraints: certain classic algorithms may be modified in order to respect the constraints present in the graph.

EXAMPLE 9.6. (CONSTRAINED HIERARCHICAL CLUSTERING [LEB 78])
Hierarchical agglomerative clustering is a simple clustering method, useful for
datasets of a reasonable size (less than 10,000 individuals). It is a straightforward
matter to include spatial constraints:

– *Initialize*: compute the contiguity graph and the distance between each pair of
individuals (every individual being considered as a class).

– *Iterate*: while the number of classes remains greater than one:

   - from among the classes which are neighbors in the graph, join the two that
are the closest with respect to some aggregation criterion; and

   - recompute the distance matrix and the contiguity graph for the new classes.

Seeking the closest classes only from among geographic neighbors means that the
search space is considerably reduced and the procedure accelerated.   □

9.3. Markov random fields as models

Within a statistical framework, several approaches to spatial clustering are
possible. Masson *et al.* [MAS 93] propose the following taxonomy for image
segmentation:

– Global methods treat an image as a random vector. Any hypothesis put forward
concerns the image in its entirety. It is to be noted that a single image will often be the
object of analysis and, in that case, the size of the sample is 1.

– Local methods fall into two types:

   - Contextual: the individuals are small groups of pixels known as contexts. A
context typically comprises a pixel together with its two horizontal and two vertical
neighbors.

   - Blind: the individuals composing the sample are the individual pixels. Here,
no spatial information is taken into account and the methods used are the classical
methods of clustering such as the EM algorithm.

Throughout the rest of this chapter we consider only global methods. For a more
detailed coverage of contextual methods, we refer the reader to [MAS 93]. Blind
(classical) methods are described in Chapter 8.

9.3.1. *Global methods and Bayesian approaches*

Global statistical modeling of images for segmentation assumes the existence of
two random fields. The observed image X corresponds to an initial random field $\mathcal{X} = \{X_s, s \in S\}$ and the segmented image Z corresponds to a second field $\mathcal{Z} = \{Z_s, s \in S\}$ (S being the set of pixels). The random vectors X_s have values in $\mathbb{R}^p$. In the
segmented image, a site's class is represented by an indicator vector z_s belonging to

$\{0,1\}^g$, where g is the number of classes. $z_{sk} = 1$ if site s is in class k, otherwise g is $z_{sk} = 0$. The values in the vector z_s are therefore drawn from a finite set $\Omega = \{\omega_1, \ldots, \omega_g\}$ of g vectors, such that

$$z_s = \omega_k = (0, \ldots, 0, \underbrace{1}_{k\text{th position}}, 0, \ldots, 0)$$

if site s is in class k.

The model takes $\mathcal{X}$ to be a noisy observation of $\mathcal{Z}$. A relation therefore exists between the two fields:

$$\mathcal{X} = R(\mathcal{Z}, N), \tag{9.1}$$

where N is the noise. Segmenting the image presents the following problem: given that one or more realizations of $\mathcal{X}$ are available, how is an estimator $\hat{\mathbf{Z}}$ for $\mathbf{X} = R_{\mathbf{Z}}(\mathbf{Z}, N)$ to be obtained?

This problem may be solved using a Bayesian approach, where an *a priori* distribution $p(\mathbf{Z})$ on the segmented image and a distribution $p(\mathbf{X}|\mathbf{Z})$ on the data are postulated. Note that the probability distribution $p(\mathbf{X}|\mathbf{Z})$ is determined by the relation between fields $\mathcal{X}$ and $\mathcal{Z}$ (equation (9.1)). The *a posteriori* distribution can be expressed by Bayes' theorem:

$$p(\mathbf{Z}|\mathbf{X}) = \frac{p(\mathbf{Z})p(\mathbf{X}|\mathbf{Z})}{p(\mathbf{X})}.$$

The Bayesian strategy is therefore to minimize the *a posteriori* cost:

$$\hat{\mathbf{Z}} = \arg\min_{\mathbf{Z}} \rho(\mathbf{Z}|\mathbf{X})$$

with

$$\rho(\mathbf{Z}|\mathbf{X}) = \mathbb{E}[L(\mathbf{Z}, \mathcal{Z}^*)|\mathbf{X}] = \sum_{\mathbf{Z}^*} L(\mathbf{Z}, \mathbf{Z}^*)p(\mathbf{Z}^*|\mathbf{X}),$$

where $L(\mathbf{Z}, \mathbf{Z}^*)$ is the cost of stating that the segmented image is $\mathbf{Z}$ when it is in fact $\mathbf{Z}^*$. Two cost functions are commonly used in image analysis:

1) $L(\mathbf{Z}, \mathbf{Z}^*) = \mathbb{1}_{\{\mathbf{Z} \neq \mathbf{Z}^*\}}$, the cost '0–1' equal to 0 where the decision is correct and 1 where it is incorrect. In this case the estimator for $\mathbf{Z}$ is the maximum *a posteriori* (MAP):

$$\hat{\mathbf{Z}} = \arg\max_{\mathbf{Z}} p(\mathbf{Z}|\mathbf{X}).$$

2) $L(\mathbf{Z}, \mathbf{Z}^*) = \sum_{s \in S} \mathbb{1}_{\{z_s \neq z_s^*\}}$, which does not consider the segmented image in its entirety, but rather the number of well-classified pixels. In this case, the estimator is that which maximizes the marginal *a posteriori* probabilities (MPM) [MAR 87]:

$$\hat{z}_s = \arg\max_{z_s} p(z_s|\mathbf{X}), \ \forall s \in S.$$

A Bayesian approach to non-supervised segmentation raises the following questions:

1) Which model should be adopted with respect to the observed data ($p(\mathbf{X}|\mathbf{Z})$), and how should *a priori* knowledge about the structure of the segmented image be modeled ($p(\mathbf{Z})$)?

2) How should the parameters be estimated for the selected model?

3) How should an estimator be obtained for the segmented image in order to minimize the chosen cost function?

Global segmentation methods will often decide the choice of model, with the aid of Markov fields used in spatial statistics.

9.3.2. *Markov random fields*

Markov chains are basic stochastic processes that can model dependencies of the random variables $\mathbf{Z}_n$ in relation to a discrete index n.

DEFINITION 9.1.– *A stochastic process $\{\mathbf{Z}_i : i = 1, 2, ...\}$ with values drawn from a finite space is a Markov chain if the realization of $\mathbf{Z}_i$, once all realizations have been completed, depends only on the last value used:*

$$P(\mathbf{Z}_i = z_i | \mathbf{Z}_{i-1} = z_{i-1}, ..., \mathbf{Z}_1 = z_1) = P(\mathbf{Z}_i = z_i | \mathbf{Z}_{i-1} = z_{i-1}). \qquad (9.2)$$

The notion of Markov dependency may be extended in various ways. Markov fields extend it to stochastic processes with indexes belonging to a multidimensional space, rather than simply to a subset of $\mathbb{R}$. There are two types of Markov field:

– Markov fields with continuous indexes, commonly used in theoretical physics; and

– Markov fields with discrete indexes used, among other things, as models for statistics of a spatial nature.

In this chapter we are only concerned with the second category.

9.3.2.1. *Contiguity graph*

When the domain of the index is a subset of $\mathbb{R}^p$ rather than a subset of $\mathbb{R}$, the idea of past and future in relation to an index t no longer applies, and we must revert to a more general concept of neighborhood.

DEFINITION 9.2.– *[GEM 84] Let $S = \{s_1, s_2, ..., s_n\}$ be a set of indexes (in the context of spatial modeling, the index corresponds to the coordinates of a site). $G =$*

$\{G_s, s \in S\}$, a set composed of parts of S, is a system of neighborhood relationships for S if and only if $\forall r, s \in S$,

1) $s \notin G_s$; and
2) $s \in G_r \Leftrightarrow r \in G_s$.

Note that $\{S, G\}$ is a graph.

Another useful concept linked to the idea of a system of neighborhood relationships is that of a clique.

DEFINITION 9.3.– *If G is a system of neighborhood relationships for a set of nodes S, a clique c is a subset of S such that all the elements of c are neighbors according to G.*

The set of indexes (of nodes) within a contiguity graph forms a network. We need to distinguish between regular and irregular-gridded networks. The former are used when modeling the distribution of pixels within an image, or the distribution of a population (comprising, for example, plants or animals) which has been subject to a methodical and uniform sampling. The latter are used to represent the natural distribution of a population.

EXAMPLE 9.7. (CLUSTERING OF RURAL ADMINISTRATIVE DISTRICTS)   The different districts within an administrative region can be characterized by numbers corresponding to their agricultural activity. A district's agricultural activity depends to some degree on its geographic location, and it would seem appropriate to model the spatial distribution of this activity with a Markov field. The coordinates of the districts' geographical centers are used as the set of indexes S, and two districts are considered to be neighbors if they have a common border. This network has an irregular grid. □

In situations where sites are not distributed regularly, Markov modeling requires that relations of contiguity are explicitly defined. One solution is to draw a Voroni diagram and to specify that two sites are contiguous if their respective Voroni polygons have an edge in common.

9.3.2.2. *Markov field distribution*

DEFINITION 9.4.– *Let S be a set of indexes associated with a neighborhood system G and $\mathcal{Z} = \{Z_s, s \in S\}$ a family of random variables with values in Ω. $\mathcal{Z}$ is a Markov field with respect to G if:*

1) $P(\mathcal{Z} = Z) > O, \forall Z \in \Omega$;
2) $P(Z_s = z_s | Z_r, r \neq s) = P(Z_s = z_s | Z_r, r \in G_s), \forall s \in S.$

By this definition, the state of a site depends only on its immediate neighbors but, in practice, the definition cannot be used to define a Markov field without also calling upon the Hammersley–Clifford Theorem (1971) as follows.

THEOREM.– *Let $\mathcal{Z} = \{Z_s, s \in S\}$ be a Markov field on a network S of n nodes associated with a neighborhood system. The probability distribution for the field $\mathcal{Z}$ is a Gibbs distribution:*

$$\pi(\mathbf{S}) = P(\mathcal{Z} = \mathbf{Z}) = \frac{1}{W} e^{-U(\mathbf{Z})},$$

where $W = \sum_{\tilde{\mathbf{Z}} \in \Omega^n} e^{-U(\tilde{\mathbf{Z}})}$ is a normalizing constant and the energy function U has the form:

$$U(\mathbf{Z}) = \sum_{1 \leq i \leq n} V_i(z_i) + \sum_{1 \leq i < j \leq n} \sum V_{i,j}(z_i, z_j) + \cdots + V_{1,2,\cdots,n}(z_1 z_2 \cdots z_n),$$

such that for any $1 \leq i < j \cdots < s \leq n$, the potential function $V_{i,j,\cdots,s}$ may be null if and only if the nodes $i, j, \cdots, s$ form a clique.

EXAMPLE 9.8. (THE ISING MODEL [BIL 92]) The best-known Markov field model come from the topic of statistical mechanics. The Ising model was invented in 1925 to explain certain properties of ferromagnets. The values of the variables Y_s (representing the spin of an atom) are either $+1$ or -1, and correspond to the nodes of a hypercube network S associated with a neighborhood system. In equilibrium, the probability that the system is in a particular configuration $\mathbf{Y}$ is a Gibbs distribution of the energy function:

$$U(\mathbf{Y}) = \frac{1}{T} \left(\alpha \sum_{s \in S} y_s + \beta \sum_{\substack{r,s \in S \\ r \text{ and } s \text{ neighbors}}} y_s y_r \right), \tag{9.3}$$

where α and β are parameters measuring the external magnetic field and the attractive forces and T is the temperature of the system. When $\alpha = 0$ (no external field) and the temperature is high, every configuration becomes equally probable. When, on the other hand, the temperature is low, two configurations are dominant: the configuration where all the spins are $+1$ and the configuration where they are all -1. At low temperatures, the system will remain trapped for substantial lengths of time in one of these two states, which explains the phenomenon of remanent magnetization.   □

EXAMPLE 9.9. (THE STRAUSS MODEL [STR 77]) The Strauss model can be considered as a generalization of the Ising model, in the case where the variables take K discrete values ($K \geq 2$). In the isotropic case (i.e. where there is no particular spatial direction) the Gibbs distribution is defined by the energy function:

$$U(\mathbf{Z}) = -\beta \sum_{\substack{r,s \in S \\ r \text{ and } s \text{ neighbors}}} \mathbb{1}_{\{z_s = z_r\}} = -\beta \sum_{\substack{r,s \in S \\ r \text{ and } s \text{ neighbors}}} z_s \cdot z_r \tag{9.4}$$

where the binary vector z_s denotes the class of node s ($z_{sk} = 1$ if node s belongs to class k). This energy function therefore counts the number of pairs of contiguous nodes which have the same value, and is maximized when the variables of the entire set of nodes are identical. In physics, this model is referred to as the Potts model.   □

9.3.3. *Markov fields for observations and classes*

If the model of the image is a Markov field, then the distributions of the data and the *a priori* and the *a posteriori* probabilities are Gibbs distributions defined:

$$p(\mathbf{X}|\mathbf{Z}) \quad \propto \exp -U^r(\mathbf{X}, \mathbf{Z}, \Phi),$$

$$p(\mathbf{Z}) \quad \propto \exp -U^a(\mathbf{Z}, \beta),$$

$$p(\mathbf{Z}|\mathbf{X}) \quad \propto \exp\{-U^a(\mathbf{Z}, \beta) - U^r(\mathbf{X}, \mathbf{Z}, \Phi)\},$$

where Φ and β are the distribution parameters. The energy U^a is relative to the *a priori* information about the segmented image $\mathbf{Z}$. The closer $\mathbf{Z}$ corresponds to the *a priori* information, the lower the value of U^a. The energy U^r is used to model the relationship between the fields $\mathbf{X}$ and $\mathbf{Z}$.

EXAMPLE 9.10. (CERAMICS AND IMAGE PROCESSING)   As part of the analysis of a ceramic object containing silicon carbide grains, we wish to measure the percentage of silicon carbide present in the material. One possible method is to take a photograph and to determine how much of the material's visible surface area is made up of silicon carbide grains. The ratio of the silicon carbide to the total surface area gives an approximation of the percentage of silicon carbide throughout the material.

The photograph is digitized and the resulting image coded in 256 shades of gray. The first step in the analysis is to segment the image into two classes, i.e. to separate the pixels representing silicon carbide from the rest.

Two contiguous pixels are more likely to belong to the same class than are two pixels chosen at random. This *a priori* information may be modeled using the Strauss model whose energy is given by:

$$U^a(\mathbf{Z}, \beta) = -\beta \sum_{\substack{r,s \in S \\ r \text{ and } s \text{ neighbors}}} z_s \cdot z_r. \tag{9.5}$$

If we consider that the observed image is a degradation of the segmented image such that for each pixel x_s in the class k the noise is Gaussian with mean μ_k and variance σ^2, then the energy relative to the data is:

$$U^r(\mathbf{X}, \mathbf{Z}, \Phi) = \frac{1}{2\sigma^2} \sum_{s \in S} (x_s - \sum_{k=1}^{g} z_{sk}\mu_k)^2 \tag{9.6}$$

where $\Phi = (\boldsymbol{\mu}_1, \ldots, \boldsymbol{\mu}_g, \sigma^2)$ correspond to the parameters of this energy function. $\square$

The information concerning the relations between fields $\mathbf{X}$ and $\mathbf{Z}$ and the *a priori* assumptions about field $\mathbf{Z}$ may be associated with energies. The energy of the *a posteriori* distribution is then the sum of all these energies:

$$U = \sum U_i = U^a + U^r. \tag{9.7}$$

Although the translation of *a priori* knowledge into a Markov-type formulation is fairly straightforward, estimating the model's parameters and segmenting the image present certain difficulties.

To render calculations possible, we usually assume that

$$p_\Phi(\mathbf{X}|\mathbf{Z}) = \prod_{i=1}^n p(\boldsymbol{x}_i|\phi_{\boldsymbol{z}_i}) = \prod_{i=1}^n \prod_{k=1}^g p(\boldsymbol{x}_i|\phi_k)^{z_{ik}}, \tag{9.8}$$

meaning that the distribution of an observation x_i at a node i depends only on the node's class z_i for the entire vector of parameters $\phi_{\boldsymbol{z}_i}$ for this class. This is equivalent to assuming that the noise is spatially uncorrelated and that observations are independent conditionally on the knowledge of the classes. The spatial dependency between contiguous nodes is then based solely on the Markov distribution of the *unobserved* image $\mathbf{Z}$. For this reason, this type of model is also known as a *hidden Markov field*.

9.3.4. *Supervised segmentation*

In this section, the models' parameters (*a priori* and data distributions) are assumed to be known. Given this context, a Bayesian approach to the segmentation problem seeks to obtain a segmented image that minimizes the conditional risk.

If the cost function $\{0, 1\}$ is used, this is equivalent to looking for the most likely image with respect to the *a posteriori* distribution (MAP estimator). In cases where this *a posteriori* distribution is a Gibbs distribution, it is impossible in practice to obtain the MAP by analytical means.

Similarly, if the chosen cost function considers the number of incorrectly classed pixels, the Bayesian approach is equivalent to looking for the image whose pixels are incorrectly classed with maximum posterior probabilities (MPM estimator), and this image cannot be obtained directly.

In both cases, minimizing the conditional risk (MAP or MPM criterion) requires the use of optimization algorithms. The most commonly used approach is to use

Monte Carlo methods linked to simulated annealing procedures [GEM 84]. Note that the MAP estimator may also be obtained via deterministic algorithms, since it is a question of maximizing the energy of the *a posteriori* Gibbs distribution.

The first problem encountered when using Monte Carlo methods is simulating images with *a posterior* distribution. The basis of all existing methods is to look for an irreducible, aperiodic finite-state Markov chain having a unique stationary limit $p(\mathbf{Z}|\mathbf{X})$ on the set of segmented images Ω^n. In other words, if this Markov chain on Ω^n is denoted as $\{\mathcal{Z}^q : q = 1, 2, \cdots\}$, then we have

$$\lim_{q \to \infty} P(\mathcal{Z}^q = \mathbf{Z}|\mathcal{Z}^0 = \mathbf{Z}^0, \ \mathbf{X}) = p(\mathbf{Z}|\mathbf{X}).$$

9.3.4.1. *The Metropolis algorithm*

Metropolis proposed an algorithm where every iteration is of the form:

1) initial selection of a segmented image $\mathbf{Z}^0$;

2) at iteration q:

- From an image $\mathbf{Z}^q$ a new image $\mathbf{Z}^{q+1}$ is created according to a given transition likelihood. Since this likelihood must be symmetrical, $Q(\mathbf{Z}^q|\mathbf{Z}^{q+1}) = Q(\mathbf{Z}^{q+1}|\mathbf{Z}^q)$.

One strategy might be to choose a pixel i at random from the image $\mathbf{Z}^q$ and then to select a new state k for this pixel. The state is chosen at random from among the g possible states according to a uniform distribution. The new image $\mathbf{Z}^{q+1}$ is identical to $\mathbf{Z}^q$, except (possibly) for the pixel i.

- Computation of the ratio

$$\frac{p(\mathbf{Z}^{q+1}|\mathbf{X})}{p(\mathbf{Z}^q|\mathbf{X})} = r.$$

- If the new image is more likely than the original ($r \geq 1$), then the transition $\mathbf{Z}^q \to \mathbf{Z}^{q+1}$ is performed. If the new image is less likely ($r < 1$), then a number u is chosen according to a uniform distribution between 0 and 1, and the transition $\mathbf{Z}^q \to \mathbf{Z}^{q+1}$ is performed if $u \leq r$.

To state this more succinctly, $\mathbf{Z}^q$ is replaced by $\mathbf{Z}^{q+1}$ with a probability $p = \min(1, r)$.

Computing the ratio r does not involve the normalization constant W, and it can be shown [BES 74] that

$$r = \frac{p(z_i^{q+1}|z_{G_i}^{q+1}, \mathbf{X})}{p(z_i^q|z_{G_i}^q, \mathbf{X})}$$

where G_i is the pixel contiguous with i.

To summarize, the Metropolis algorithm simulates a finite-state Markov chain on Ω^n with a transformation matrix defined by

$$
P_{\mathbf{Z}\mathbf{Z}'} = \begin{cases} Q_{\mathbf{Z}\mathbf{Z}'} \cdot \frac{p(\mathbf{Z}'|\mathbf{X})}{p(\mathbf{Z}|\mathbf{X})} & \text{if} \quad p(\mathbf{Z}'|\mathbf{X}) < p(\mathbf{Z}|\mathbf{X}) \\ Q_{\mathbf{Z}\mathbf{Z}'} & \text{if} \quad p(\mathbf{Z}'|\mathbf{X}) \geq p(\mathbf{Z}|\mathbf{X}) \quad \text{and} \quad \mathbf{Z}' \neq \mathbf{Z} \\ 1 - \sum_{\mathbf{Z}':\, \mathbf{Z}' \neq \mathbf{Z}} P_{\mathbf{Z}\mathbf{Z}'} & \text{if} \quad \mathbf{Z} = \mathbf{Z}'. \end{cases}
$$

The transformation matrix $Q = \{Q_{\mathbf{Z}\mathbf{Z}'}\}$ used will be symmetrical. It can be shown that P will be irreducible if Q is irreducible. Since P is reversible,

$$
p(\mathbf{Z}|\mathbf{X}) P_{\mathbf{Z}\mathbf{Z}'} = p(\mathbf{Z}'|\mathbf{X}) P_{\mathbf{Z}'\mathbf{Z}}.
$$

The limit law for the chain will be that which we seek to simulate.

9.3.4.2. *Gibbs sampler*

Image simulation using the Gibbs sampler was proposed by Geman and Geman [GEM 84]. This method defines a visit order for the pixels, and iterates as follows:

1) initial choice of a segmented image $\mathbf{Z}^0$,

2) at iteration q:
   - a pixel i is chosen following the visit order; and
   - z_i^{q+1} is chosen at random according to a distribution $p(z_i|\mathbf{X}; z_{G_i}^q)$.

This Gibbs sampler produces a series of images $\mathbf{Z}^0, \cdots, \mathbf{Z}^q$. For high values of q, we consider that $\mathbf{Z}^q$ is a realization of $p(\mathbf{Z}|\mathbf{X})$. We also have the property:

$$
\lim_{q \to \infty} \frac{1}{q}[f(\mathbf{Z}^1) + \cdots + f(\mathbf{Z}^q)] = \mathbb{E}(f(\mathbf{Z})),
$$

where f is some measurable function and $\mathbf{Z}$ is a random vector according to a distribution $p(\mathbf{Z}|\mathbf{X})$.

9.3.4.3. *Simulated annealing and the MAP estimator*

Even although the sampling methods described above provide simulations according to the desired distribution, they do not directly reveal the segmented image that minimizes the MAP or the MPM criterion.

In the case of MAP, a simulated annealing procedure that makes the series of images $\mathbf{Z}^q$ tend towards MAP can be integrated into the above algorithms [GEM 84].

The idea is to introduce a parameter of temperature T into the distribution $p(\mathbf{Z}|\mathbf{X})$, which can then be written

$$
p_T(\mathbf{Z}|\mathbf{X}) = \frac{\exp(\frac{1}{T} \cdot U_{\Phi,\beta}(\mathbf{Z}, \mathbf{X}))}{W(T)}.
$$

The temperature descends towards zero at each step, which causes $p_T(\mathbf{Z}|\mathbf{X})$ to converge towards a uniform distribution on the global maxima of $p(\mathbf{Z}|\mathbf{X})$. The convergence of the algorithm is clearly demonstrated when the temperature falls slowly. The drawback of a slow fall in temperature is that a very large number of iterations are required before a satisfactory MAP estimator can be obtained.

9.3.4.4. *The ICM algorithm*

To overcome this drawback of a slow fall in temperature, Besag [BES 86] proposed a deterministic algorithm for simulated annealing, starting with a null temperature. Each iteration of the algorithm (known as the iterative conditional mode or ICM algorithm) modifies the class by one pixel as follows:

$$z_i^{q+1} = \arg\max_{z_i} p(z_i|\mathbf{X}; z_{G_i}^q).$$

The advantage of the ICM algorithm is that it converges in less than 10 visits of the entire image, and $p(z^q|\mathbf{X})$ increases at each iteration. The main drawback is that it is heavily impacted by the initial conditions.

9.3.4.5. *The MPM estimator*

If we are seeking to obtain the MPM image then we need to estimate the *a posteriori* marginal probabilities $P(Z_{ik} = 1|\mathbf{X})$ of belonging to the classes. These probabilities may be estimated from a certain number of (image) realizations according to the *a posteriori* distribution [MAR 93]. The marginal probabilities $P(Z_{ik} = 1|\mathbf{X})$ are in fact equal to the expectations $\mathbb{E}(Z_{ik}|\mathbf{X})$, so may be estimated using m_{ik}, the empirically observed rate of occurrence of class k at node i. We use q images, simulated according to the *a posteriori* distribution, to measure this rate:

$$P(\widehat{Z_{ik} = 1}|\mathbf{X}) = \mathbb{E}(\widehat{Z_{ik}|\mathbf{X}}) = m_{ik} = \frac{1}{q}(z_{ik}^1 + \ldots + z_{ik}^q).$$

We therefore obtain a segmentation which is reasonably good in relation to the MPM criterion, by classing each pixel as follows:

$$z_{ik} = \begin{cases} 1 \text{ if } k = \arg\max_\ell m_{i\ell}; \\ \\ 0 \text{ otherwise.} \end{cases}$$

The Metropolis algorithm, a Gibbs sampler, may be used to provide the realizations of the *a posteriori* distribution.

The marginal probabilities $P(Z_{ik} = 1|\mathbf{X}) = \mathbb{E}(Z_{ik}|\mathbf{X})$ may also be estimated deterministically via an approximation of the mean field [WU 95, ZHA 92]. If we use this approximation, the expectations $\mathbb{E}(Z_{ik}|\mathbf{X})$ satisfy a system of fixed-point equations. The expectations may therefore be obtained by reiterating the fixed-point equations until convergence, which generally implies a few dozen iterations.

9.4. Estimating the parameters for a Markov field

The clustering methods described above require prior knowledge of the parameters $\theta = (\beta, \Phi)$, with $\Phi = (\Phi_1, \ldots, \Phi_g)$, of the *a priori* distribution and the distribution of the observations. In practice, these parameters must be estimated from the available information. We first describe some techniques for estimating these parameters when both the observations X and the classification Z are known. This situation corresponds to supervised clustering, i.e. a classifying rule is constructed on the basis of observations whose initial class is known. We then look at the problem of automatic or unsupervised clustering, where only unlabeled observations are available. As with mixture models, we shall remark that estimation procedures in the unsupervised case are often derived from procedures designed to estimate parameters from labeled observations.

In this section, we also look at the hidden Markov field model most commonly used in image segmentation: the partition's *a priori* distribution $p(Z|\beta)$ is the Strauss model, defined by energy equation (9.4), and the distribution of observations conditional on the partition is given by the product (9.8).

9.4.1. *Supervised estimation*

Here we describe techniques for estimating parameters in the case where observations X and a partition Z are both available.

EXAMPLE 9.11. (LABELING OF SATELLITE IMAGES BY EXPERTS) The European Union gives subsidies to farmers according to the type of crops they declare. Satellite imaging may be used to check the veracity of these declarations. On the basis of images X_m and a manual segmentation Z_m, we wish to obtain the parameters of a hidden Markov field model where each class corresponds to a particular type of crop. The images covering the area in question can then be segmented using a Bayes classifier based on the parameters thus identified. $\qquad\square$

One common estimation strategy is to maximize the likelihood of the parameters θ, $L_c(X, Z|\theta)$, which can be written

$$\underbrace{\sum_{k=1}^{g}\sum_{i=1}^{n} z_{ik}\log f(x_i|\Phi_k)}_{\log p(X|Z,\Phi)} \underbrace{- \sum_{\gamma\in\mathcal{C}} V_\gamma(z_\gamma;\beta) \underbrace{- \log \sum_{\tilde{z}} e^{-\sum_{\gamma\in\mathcal{C}} V_\gamma(\tilde{z}_\gamma;\beta)}}_{W(\beta)}}_{\log p(Z|\beta)}$$

where V_γ denotes the potential functions on the cliques γ. As with mixture models, maximizing the likelihood is very straightforward for the $\log p(X|Z, \Phi)$ part that concerns the parameters $\boldsymbol{\Phi}$ of the distribution of observations. It is simply a matter of maximizing the likelihood of the parameters Φ_k on the basis of the observations $\{\boldsymbol{x}_i : z_{ik} = 1\}$ in each class k.

However, it is less straightforward to maximize the $\log p(\mathbf{Z}|\beta)$ part for the parameters β of the *a priori* distribution. The normalizing constant $W(\beta)$ contains a sum over all possible partitions, and this cannot be factorized due to the interactions between neighboring sites. The log-likelihood of the $\log p(\mathbf{Z}|\beta)$ classification would therefore appear impossible to compute, and a direct maximization is not usually feasible. Having remarked that the gradient of the log-likelihood is an expectation linked to the distribution $p(\mathbf{Z}|\beta)$, Younes [YOU 88] proposed a stochastic gradient technique. At each iteration $(q+1)$, the gradient is approached more closely through a simulation of the Markov distribution $p(\mathbf{Z}|\beta^{(q)})$ and the parameters are modified in the direction of the approximate gradient. An alternative approach is to maximize deterministically an approximate expression of the likelihood. This is the approach used in coding sets and pseudo-likelihood, described below.

9.4.1.1. *Coding sets*

Taking advantage of the local nature of the dependencies in a Markov field, Besag [BES 74] proposed maximizing the likelihood over subsets of mutually independent sites. He termed these subsets *coding sets*. Figure 9.5 illustrates the coding sets that are implied by different neighborhood systems on a regular rectangular grid. Each coding set is composed of sites that are not linked by any neighborhood relationship (see Figure 9.5). It is now no longer the likelihood $p(\mathbf{Z}|\beta)$, but rather the likelihood of the classification on a coding set that we wish to maximize, knowing the class of the other sites. This quantity can be factorized to a product of conditional probabilities that we then maximize using a simple gradient ascent.

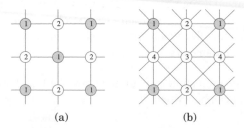

(a)   (b)

Figure 9.5. *Coding sets. (a) Graph of the 4 nearest neighbors: 2 coding sets S_1 and S_2 may be defined, indicated by the numbers 1 or 2. The colored sites highlight one of these sets. (b) Graph of the 8 nearest neighbors: 4 coding sets S_1, S_2, S_3 and S_4 may be defined*

The main drawback of this technique is that a substantial amount of information is lost if the estimation is based only on a coding set. Some authors suggest making as many estimations $\beta^{[1]}, \ldots, \beta^{[q]}$ as there are coding sets $S_1, \ldots, S_q$, and then combining these estimations e.g. by taking their empirical mean [LI 95, p. 137]. This kind of combination has no theoretical foundation. Another weak point is that it is difficult to define coding sets when the contiguity graph has an irregular shape.

9.4.1.2. *Pseudo-likelihood*

To overcome the requirement to combine estimations derived from different coding sets, one intuitive strategy is to extend the previous likelihood, initially restricted to one coding set, by using the product of *all* the conditional probabilities $p(z_i|z_{G_i}, \beta)$ for $i = 1, \ldots, n$. This yields a function of the partition $\mathbf{Z}$ and of the parameters β known as *pseudo-likelihood* [BES 86]:

$$\mathcal{P}(\mathbf{Z}|\beta) = \prod_{i=1}^{n} p(z_i|z_{G_i}, \beta). \tag{9.9}$$

It has been shown that the estimator of the maximum of pseudo-likelihood is asymptotically consistent [GEM 87]. In practice, pseudo-likelihood is very often used as an approximation for equation (9.4) likelihood when estimating the parameters β of a Markov field. This is largely because pseudo-likelihood, just like the likelihood restricted to one coding set, is a product of functions $p(z_i|z_{G_i})$ that are easy to compute and derive. Since pseudo-likelihood is a concave function of the parameters [GEM 87], gradient ascent techniques are frequently used to obtain the parameters that maximize $\mathcal{P}(\mathbf{Z}|\beta)$.

9.4.2. *Unsupervised estimation with EM*

The previous paragraph showed how the maximum likelihood principle makes it possible to estimate the parameters $\theta = (\beta, \Phi)$ of a hidden Markov field when a realization of the observations $\mathbf{X}$ and the partition $\mathbf{Z}$ is available. This knowledge of labeled observations puts us in a discriminatory or supervised clustering situation. In automatic or unsupervised clustering, however, we only have the observations $\mathbf{X}$, meaning that we are addressing a problem of *incomplete data*, as in the case of a mixture model. We are therefore naturally inclined to apply the EM algorithm principle (see Chapter 8) when estimating the parameters of a hidden Markov field. After the parameters have been set to an arbitrary value $\theta^{(0)}$, each new iteration $(q+1)$ modifies the parameters in order to maximize the expectation

$$Q(\theta, \theta^{(q+1)}) = \underbrace{\mathbb{E}\left(\log p(\mathbf{X}|\mathbf{Z}, \Phi^{(q)})|\mathbf{X}, \theta^{(q)}\right)}_{Q_\Phi(\Phi|\theta^{(q)})} + \underbrace{\mathbb{E}\left(\log p(\mathbf{Z}|\beta)|\mathbf{X}, \theta^{(q)}\right)}_{Q_\theta(\beta|\theta^{(q)})}.$$

Decomposing allows the modifications to be considered separately, i.e.

$$\Phi^{(q+1)} = \arg\max_{\Phi} Q_\Phi(\Phi|\theta^{(q)}) \quad \text{and} \quad \beta^{(q+1)} = \arg\max_{\beta} Q_\beta(\beta|\theta^{(q)}).$$

9.4.2.1. *The parameters* Φ *of the distribution of observations*

Since the x_i for any partition $\mathbf{Z}$ are independent, we have

$$Q_{\Phi}(\Phi|\theta^{(q)}) = \sum_{k=1}^{g} \sum_{i=1}^{n} \underbrace{\mathbb{E}\left(Z_{ik}|\mathbf{X}, \theta^{(q)}\right)}_{t_{ik(q+1)}} \log f(x_i|\Phi_k). \qquad (9.10)$$

It is clear that Φ maximizing equation (9.10) is directly possible when the expectations $t_{ik}^{(q+1)}$ are known. Equation (9.10) effectively puts us in the same situation as in mixture models, i.e. the likelihood of the parameters of each distribution $f(\cdot|\Phi_k)$ is maximized, with the individuals x_i given different weights according to the expectations $t_{ik}^{(q+1)}$. For example, in the case of Gaussian $f_{\mathcal{N}}(\cdot|\mu_k, \Sigma_k)$ distributions, the formulae

$$\mu_k^{q+1} = \frac{1}{\sum t_{ik}^{q+1}} \sum_i t_{ik}^{q+1} \mathbf{X}_i$$

and $\Sigma_k^{q+1} = W_k^{-1}$ with

$$W_k = \sum_i t_{ik}^{q+1}(\mathbf{X}_i - \mu_k^{q+1})(\mathbf{X}_i - \mu_k^{q+1})'$$

are obtained for re-estimating the means and the covariance matrices.

The expectations t_{ik}, it should be remembered, are equal to the *a posteriori* probabilities of observations belonging to classes, and they are also used when applying the MPM classifying rule.

One of the methods described above may therefore also be applied here in order to obtain an approximate value for the *a posteriori* probabilities at Expectation step (step E).

Chalmond's Gibbsian EM (GEM) [CHA 89] algorithm uses the first solution:

– a certain number of partitions $\tilde{\mathbf{Z}}^{(1)}, \ldots, \tilde{\mathbf{Z}}^{(m_0+M)}$ are simulated according to the *a posteriori* distribution $p(\mathbf{Z}|\mathbf{X}, \theta^{(q)})$ by means of a Gibbs sampler; and

– the empirically observed rates provide an estimation of the probabilities of class-membership:

$$\hat{t}_{ik}^{(t+1)} = \frac{1}{M} \sum_{t=m_0+1}^{m_0+M} \tilde{z}_{ik}^{(q)}.$$

The procedure proposed by Zhang [ZHA 92] uses the second solution based on an approximation of the mean field: starting with a given value $\hat{\mathbf{t}}^{(0)}$ the membership probabilities are obtained by iterating the system of fixed point equations resulting from the principle of the mean field. In the case of the Potts model, the iteration $(t+1)$ involves computing, for $i = 1, \ldots, n$ and $k = 1, \ldots, g$:

$$
\hat{t}_{ik}^{(t+1)} = \frac{f(\boldsymbol{x}_i|\boldsymbol{\Phi}_k^{(q)})e^{\beta^{(q)}\sum_{j\in G_i} v_{ij}\hat{t}_{jk}^{(t)}}}{\sum_{h=1}^{g} f(\boldsymbol{x}_i|\boldsymbol{\Phi}_h^{(q)})e^{\beta^{(q)}\sum_{j\in G_i} v_{ij}\hat{t}_{jh}^{(t)}}}. \tag{9.11}
$$

When this procedure converges, the approximate probabilities for class-membership are obtained:

$$
\hat{t}_{ik}^{(t+1)} = \hat{t}_{ik}^{[\text{conv}]}.
$$

9.4.2.2. β parameters of the a priori distribution

In section 9.4.1 we saw that, when maximizing $p(\mathbf{Z}|\beta)$, certain obstacles created by the normalizing constant of the *a priori* distribution are encountered. This normalizing constant does not become any simpler in the expectation $Q_\beta(\beta|\boldsymbol{\theta}^{(q)})$.

The solution proposed by Chalmond [CHA 89] is to replace, within $Q_\beta(\beta|\boldsymbol{\theta}^{(q)})$, the likelihood $p(\mathbf{Z}|\beta)$ by the pseudo-likelihood $\mathcal{P}(\mathbf{Z}|\beta)$ (equation (9.9)). Moreover, Chalmond remarks that for an image and a regular neighborhood system, the classes $\boldsymbol{z}_{G_i}$ of a pixel i's neighbors can only take one of a limited number v of possible configurations. Rather than estimate the parameters β of the *a priori* distribution, we may therefore estimate instead the probabilities π_{ku} of observing class k, knowing that the neighbors form the configuration of classes u, where $k = 1, \ldots, g$, $u = 1, \ldots, v$. These probabilities π_{ku} are estimated as the rate of occurrence of the configuration ku in the images simulated at step E according to the *a posteriori* distribution $p(\mathbf{Z}|\mathbf{X}, \boldsymbol{\theta}^{(q)})$. One limitation of this procedure is that it requires a regular neighborhood system in order to be able to use the same type of neighborhood over the entire geographical region under consideration.

Zhang [ZHA 92], on the other hand, suggests approximating the likelihood $p(\mathbf{Z}|\beta)$ with a mean-field approximation of the pseudo-likelihood. This approximation can be interpreted as a pseudo-likelihood in which the labels $\boldsymbol{z}_{G_i}$ of the current site's neighbors are replaced by their expectation $\mathbb{E}(\boldsymbol{Z}_{G_i}|\mathbf{X}, \boldsymbol{\theta}^{(q)})$. This yields an approximation of the expectation $Q_\beta(\beta|\boldsymbol{\theta}^{(q)})$ that can be maximized using a simple gradient ascent.

9.4.2.3. *Initializing EM for a hidden Markov random field*

It should be recalled that the EM algorithm usually only reaches a local likelihood maximum that depends on the initial parameters $\boldsymbol{\theta}^{(0)}$ selected. Unlike mixture models, here we cannot compare different solutions $\hat{\theta}$ in respect of their likelihood. In the expression of the likelihood for a hidden Markov field

$$L(\boldsymbol{\theta}) = \log p(\mathbf{X}|\boldsymbol{\theta}) = \log \sum_{\mathbf{Z}} p(\mathbf{X}|\mathbf{Z}, \boldsymbol{\Phi}) p(\mathbf{Z}|\beta),$$

the sum of all the possible partitions cannot be factorized, which makes it impossible in practice to compute the likelihood. The usual strategy of running the EM algorithm several times from randomly chosen initial positions and retaining the solution with the highest likelihood cannot, therefore, be applied here. Instead, proposed solutions are generally derived from 'reasonable' parameter values.

Regarding the parameters of the *a priori* distribution, these are usually set under the assumption of a null spatial interaction and equiprobable classes [CHA 89, ZHA 92]:

$$V_{ij}^{[2]}(\cdot, \cdot; \beta^{(0)}) = 0 \quad \text{and} \quad V_i^{[1]}(\cdot; \beta^{(0)}) = 0.$$

For the Potts model, this means setting $\beta^{(0)} = 0$. In the parameter configuration proposed by Chalmond [CHA 89], it means taking $P_{kj}^{(0)} = 1/g, \forall k, j$.

A strategy frequently used in setting the parameters of the distribution of observations is (1) to define a partition for the observations using the mobile-centers method and (2) to estimate the initial parameters for the distributions $\boldsymbol{\Phi}_k^{(0)}$ by maximizing the likelihood within each class thus defined (e.g. [ZHA 92]).

9.4.3. *Classification likelihood and inertia with spatial smoothing*

We saw above that when the parameters of the hidden Markov field are known, the ICM algorithm is often chosen, because of its speed, to approximate MAP clustering. When the parameters are not known, an unsupervised variant of ICM may be used. This variant can be interpreted as an algorithm for maximizing the classification likelihood of a hidden Markov field. The classification likelihood is written:

$$C_{mark}(\mathbf{Z}, \boldsymbol{\theta}) \triangleq \log p(\mathbf{X}, \mathbf{Z}|\boldsymbol{\theta}) \tag{9.12}$$

$$= \sum_{i=1}^{n} \sum_{k=1}^{g} z_{ik} \log f(\boldsymbol{x}_i|\boldsymbol{\Phi}_k) - U^{\text{prio}}(\mathbf{Z}; \beta) - \log W(\beta).$$

As with mixture models, an iterative procedure can be defined for maximizing alternately this criterion for the classification and the parameters. We start with initial parameters $\boldsymbol{\theta}^{(0)}$, then perform the following steps at each new iteration $(q + 1)$:

– Allocation step:

$$\mathbf{Z}^{(q+1)} = \arg\max_{\mathbf{Z}} C_{mark}(\mathbf{Z}, \boldsymbol{\theta}^{(q)}) = \arg\max_{\mathbf{Z}} p(\mathbf{Z}|\mathbf{X}, \boldsymbol{\theta}^{(q)}).$$

This is the identification of the MAP classification depending on parameters of the previous iteration. If the MAP is approximated locally by the ICM algorithm, the unsupervised variant of the algorithm proposed by Besag [BES 86] is obtained. If the global MAP is approximated by simulated annealing, it is the *adaptive simulated annealing* algorithm proposed by Lakshmanan and Derin [LAK 89] that is obtained.

– Representation step:

$$\boldsymbol{\theta}^{(q+1)} = \arg\max_{\boldsymbol{\theta}} p(\mathbf{X}, \mathbf{Z}^{(q+1)}|\boldsymbol{\theta}).$$

This involves seeking the parameters $\boldsymbol{\theta}$ that maximize the likelihood of the complete data. It is therefore possible to use one of the methods described in section 9.4.1 for estimating the parameters of a hidden Markov field when the observations $\mathbf{X}$ and a partition (in this case $\mathbf{Z}^{(q+1)}$) are available. Besag [BES 86], for example, proposes that the parameters of the *a priori* distribution are estimated in order to maximize the pseudo-likelihood $\mathcal{P}(\mathbf{Z}^{(q+1)}|\beta)$.

9.4.3.1. *Associated inertia criterion*

There exists an interesting link between the criterion of classification likelihood $C_{mark}(\mathbf{Z}, \boldsymbol{\theta})$ and a criterion of intraclass inertia with spatial smoothing. If a Gaussian distribution $\mathcal{N}(\boldsymbol{\mu}_k, \boldsymbol{\Sigma}_k)$ for each class is used as the law for observations $p(\mathbf{X}|\mathbf{Z}, \Phi)$, and a Potts field with a specific β parameter as the *a priori* law, then the classification likelihood can be written:

$$C_{mark}(\mathbf{Z}, \boldsymbol{\theta}) = -\underbrace{\sum_{k=1}^{g} \sum_{i:z_{ik}=1} d_{\boldsymbol{\Sigma}_k}(\boldsymbol{x}_i, \boldsymbol{\mu}_k)}_{\text{intraclass inertia}} + \beta \underbrace{\sum_{i,j \text{ neighbors}} \mathbb{1}_{\{z_i=z_j\}}}_{\text{spatial regularity}} + \text{Cst}(\beta)$$

where $d_{\boldsymbol{\Sigma}_k}(\boldsymbol{x}_i, \boldsymbol{\mu}_k)$ is a Mahanalobis distance between the observation $\boldsymbol{x}_i$ and the center of its class $\boldsymbol{\mu}_k$. Maximizing $C_{mark}(\mathbf{Z}, \boldsymbol{\theta})$ therefore means seeking a partition $\mathbf{Z}$ that is not only homogenous within the observation space $\boldsymbol{x}_i$ (low intraclass inertia), but also smooth within the geographic space (high spatial regularity). The relative importance of these two aspects is governed by the parameter β.

9.4.3.2. *Fuzzy classification likelihood*

We have seen, in the case of mixture models, that extending the classification likelihood to fuzzy partitions gives rise to a criterion, the alternate optimization of which is equivalent to the EM algorithm (see Chapter 8, section 8.3.6). A similar equivalence has been identified in the case of hidden Markov fields [DAN 98, p. 107–114]. In the case of the Potts model, an extension of the classification likelihood to the cover fuzzy partitions can be written:

$$F_{potts}(\mathbf{C}, \boldsymbol{\theta}) = C_R(\mathbf{C}, \boldsymbol{\Phi}) + \beta\, G(\mathbf{C}) - \log W(\beta) + E(\mathbf{C}) \tag{9.13}$$

where $\mathbf{C} = (c_{ik})_{i=1,n}^{k=1,g}$ denotes a fuzzy classification matrix ($c_{ik} \in [0; 1]$ and $\sum_{k=1}^{g} c_{ik} = 1$) and $E(\mathbf{C}) = -\sum_{i=1}^{n} \sum_{k=1}^{g} c_{ik} \log c_{ik}$ represents the entropy. Maximizing $F_{potts}(\mathbf{C}, \boldsymbol{\theta})$ iteratively gives rise to a family of algorithms termed neighborhood EM (NEM) [AMB 97]. Starting from an initial value $\boldsymbol{\theta}^{(0)}$ for the parameters, each iteration $(q + 1)$ is divided into two steps:

– *Fuzzy allocation*: identification of the fuzzy classification matrix satisfying

$$\mathbf{C}^{(q+1)} = \arg\max_{\mathbf{c}} F_{potts}(\mathbf{C}, \boldsymbol{\theta}^{(q)}).$$

– *Representation*: the identification of the parameters $\boldsymbol{\theta}^{(q+1)}$ maximizing $F_{potts}(\mathbf{C}^{(q+1)}, \boldsymbol{\theta})$ is divided into two parts:

$$\boldsymbol{\Phi}^{(q+1)} = \arg\max_{\boldsymbol{\Phi}} \sum_{k=1}^{g} \sum_{i=1}^{n} c_{ik}^{(q+1)} f(\boldsymbol{x}_i | \boldsymbol{\Phi}_k) \tag{9.14}$$

and

$$\beta^{(q+1)} = \arg\max_{\beta} \beta\, G(\mathbf{C}^{(q+1)}) - \log W(\beta). \tag{9.15}$$

Dang [DAN 98, p. 107–114] shows that these two steps are equivalent, given certain conditions, to the E and M steps of the EM algorithm proposed by Zhang [ZHA 92]. Celeux *et al.* [CEL 03] placed this type of approach in a highly generalized theoretical framework and showed that a large number of algorithms use this identical principle.

9.4.4. *Other methods of unsupervised estimation*

Younes [YOU 89] proposes using a stochastic gradient procedure adapted to this situation of unlabeled observation. In the case under discussion, the gradient of the log-likelihood is considered as the sum of two expectations. At each iteration $(q + 1)$, the first expectation may be approximated by simulating the partition's *a posteriori* distribution $p(\mathbf{Z}|\mathbf{X}, \boldsymbol{\theta}^{(q)})$ and the second expectation by simulating the combined distribution of the observations and the partition, i.e. $p(\mathbf{X}, \mathbf{Z}|\boldsymbol{\theta}^{(q)})$.

Pieczynski [PIE 94] proposes a generalized estimation procedure, known as *iterative conditional expectation* (ICE), in the presence of hidden data. The basic assumption behind this procedure is that a technique is available for estimating the parameters from the complete data, i.e. that we are capable of defining an estimator $\hat{\theta}(\mathbf{X}, \mathbf{Z})$. We have seen that this is indeed the case for a hidden Markov field model. The ICE procedure involves approximating the estimation $\hat{\theta}(\mathbf{X}, \mathbf{Z})$ on the basis of the observed values $\mathbf{X}$ alone. The best approximation, with respect to the mean quadratic error, is the expectation that can be written as

$$\theta \approx \mathbb{E}\left(\hat{\theta}(\mathbf{X}, \mathbf{Z})|\mathbf{X}, \theta\right).$$

Since the parameters θ appear on both sides of the approximation, they may be computed iteratively by adjusting the parameters from their value $\theta^{(q)}$ at the previous iteration by

$$\theta^{(q+1)} = \mathbb{E}\left(\hat{\theta}(\mathbf{X}, \mathbf{Z})|\mathbf{X}, \theta^{(q)}\right).$$

In the case of hidden Markov fields, the expectation $\mathbb{E}\left(\hat{\theta}(\mathbf{X}, \mathbf{Z})|\mathbf{X}, \theta^{(q)}\right)$ cannot be calculated explicitly, but it may be approximated by MCMC sampling of the *a posteriori* distribution $p(\mathbf{Z}|\mathbf{X}, \theta^{(q)})$.

Finally, we should mention estimation by Bayesian inference, which is an approach that has recently been applied to hidden Markov fields [BAR 00]. The estimation of the number of classes can be included in a coherent fashion with this approach. However, Bayesian inference does introduce an additional layer of complexity, given that each parameter requires its *a priori* law to be modeled and its *a posteriori* law to be simulated by MCMC sampling.

9.5. Application to numerical ecology

We shall now describe how the techniques presented above can be applied to example 9.3 referred to in our introduction. This application is the outcome of work performed jointly with G. Govaert, P. Aubry and C. Dutreix. We are particularly grateful to them for permission to publish certain results to illustrate the ideas we are attempting to convey here.

9.5.1. *The problem*

Between 1976 and 1995, biologists working in the field listed the presence or absence of $d = 22$ species of butterfly of the *Zygaenidae* family over $n = 349$ UTM squares (10 km × 10 km) covering the French region of Burgundy [DUT 95]. Figure 9.4 in our introduction shows 3 of the 22 maps indicating the spatial distribution of each species. These observations are presented as a binary matrix $\mathbf{X} = (x_i^j)$, where $x_i^j = 1$ if species j is present within plot i, $i = 1, \ldots, n; j = 1, \ldots, d$.

The aim of our analysis is to partition the UTM squares into a small number of homogenous classes based on the presence or absence of the different species of *Zygaenidae*. Each class should be characterized by a typical combination of species present. It would also seem reasonable to suppose that the different classes are subject to some spatial displacement over time. Dynamic processes within a species relating to growth, reproduction, mortality and migration tend to be propagated spatially. Moreover, these processes are linked to climatic structures and to the spatial distribution of the types of plants on which the different species feed – the distribution of plants being linked to the geochemical properties of the soil.

9.5.2. The model: Potts field and Bernoulli distributions

9.5.2.1. *A priori law* $p(\mathbf{Z}|\beta)$

The assumption of spatial regularity can be modeled using a Potts (or Strauss) *a priori* law. We use a contiguity graph for the 8 closest neighbors, with the weight of each neighbor diminishing exponentially with the distance from the site being considered.

9.5.2.2. *Observations law* $p(\mathbf{X}|\mathbf{Z}, \mathbf{\Phi})$

We continue to assume that the observations x_i are mutually independent for a given classification $\mathbf{Z}$, which corresponds to equation (9.8). Regarding the distribution $p(x_i|\phi_k)$ within a given class, we shall assume that the presence or absence of the p species follows p mutually independent Bernoulli distributions $\mathcal{B}(\phi_k^j), j = 1, \ldots, p$. In addition, as [GOV 90] suggests, each Bernoulli distribution $\mathcal{B}(\phi_k^j = \{a_k^j, \varepsilon_k^j\})$ may be characterized by its center $a_k^j \in \{0, 1\}$ (the most probable binary value) and its dispersion $\varepsilon_k^j \in]0; \frac{1}{2}[$ (the probability of generating a value different from that of the center). The observations law is then given by:

$$p(\mathbf{X}|\mathbf{Z}, \mathbf{\Phi}) = \prod_{i=1}^{n} \prod_{k=1}^{g} \left(\prod_{j=1}^{p} (\varepsilon_k^j)^{|x_i^j - a_k^j|} (1 - \varepsilon_k^j)^{1 - |x_i^j - a_k^j|} \right)^{z_{ik}}.$$

Characterizing the distributions in this manner has two advantages [GOV 90]:

– The conditional log-likelihood $\log p(\mathbf{X}|\mathbf{Z}, \mathbf{\Phi})$ can be seen as the opposite of an L_1-norm intraclass inertia; the classification-likelihood criterion can therefore also be seen as a criterion of inertia with spatial smoothing (see section 9.4.3 for the case of Gaussian distributions).

– Assuming that the central value $\mathbf{a}_k = (a_k^1, \ldots, a_k^p)$ is the most characteristic descriptor of a class, we may set equality constraints on dispersions, e.g. $\varepsilon_k^j = \varepsilon_j$ (for a given variable, every class has the same dispersion). Imposing this kind of constraint leads to more *parsimonious* models, i.e. models with fewer parameters to estimate.

9.5.3. *Estimating the parameters*

Since the number of classes g is not known in advance, we attempt to determine this value by minimizing the information criteria AIC [AKA 73], AIC3 [BOZ 81] and BIC [SCH 78]. These suggest $g = 5, 6$ and 8 classes, respectively. A partition into 6 classes appeared to the biologists to be the most appropriate in that, starting from a partition into 5 classes, it allowed a separation into two fairly distinct groups (whereas partitions into 7 or more classes were seen to subdivide the groups artificially).

Estimating the parameters was achieved by applying the EM algorithm to the above model. The algorithm was initialized with random parameter values. From the results obtained using the different initial values, those retained were that which maximized the criterion of fuzzy classification pseudo-likelihood defined by Dang [DAN 98, p. 119–120].

9.5.4. *Resulting clustering*

At the convergence of the EM algorithm, step E provides an estimation of the *a posteriori* probabilities $t_{ik} = P(\mathbf{Z}_i = \omega_k|\mathbf{X})$ that sites $i = 1, \ldots, 349$ belong to classes $k = 1, \ldots, 6$. These probabilities can be used to compute an MPM (marginal posterior mode) partition:

$$\hat{\mathbf{z}}_i = \omega_{\hat{k}} \quad \text{where} \quad \hat{k} = \arg\max_k \left\{ t_{ik} = P(\mathbf{Z}_i = \omega_k|\mathbf{X}) \right\}.$$

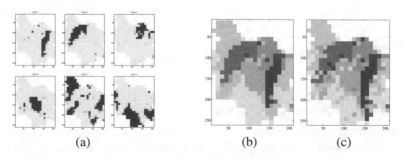

        (a)   (b)   (c)

Figure 9.6. *Clustering of plots of land according to presence/absence indicators. Markov model (with spatial smoothing): (a) probabilities that t_{ik} belong to the 6 classes (the dark squares indicate high probabilities) and (b) MPM classification. (c) Map obtained using a Bernoulli mixture model and therefore without spatial smoothing*

Figures 9.6a and b show the probabilities t_{ik} of belonging to classes, together with the classification $\hat{\mathbf{Z}}$ thus obtained. In comparison to the partition depicted in Figure 9.6c, obtained without spatial smoothing, we remark that the spatial regularization simplified the partition but without modifying its general character.

The partition without spatial smoothing gave well-defined ecological regions, due to the fact that the species in question have relatively continuous spatial distributions in the region being studied. Smoothing then removed certain ambiguities and corrected sampling errors. As for interpreting the resulting partition, the biologists found that the regions thus defined corresponded to the presence or absence of limestone, given that the caterpillars of some *Zygaenidae* develop on plants which essentially grow on lime-rich soils.

9.6. Bibliography

[AKA 73] AKAIKE H., "Information theory and an extension of the maximum likelihood principle", PETROV B., CSAKI F., Eds., *Second International Symposium on Information Theory*, Budapest, Akademiai Kiado, p. 267–281, 1973.

[AMB 97] AMBROISE C., DANG M. V., GOVAERT G., "Clustering of spatial data by the EM algorithm", SOARES A., GÓMEZ-HERNANDEZ J., FROIDEVAUX R., Eds., *geoENV I - Geostatistics for Environmental Applications*, vol. 9 of *Quantitative Geology and Geostatistics*, Dordrecht, Kluwer Academic Publisher, p. 493–504, 1997.

[BAR 00] BARKER S. A., RAYNER P. J. W., "Unsupervised image segmentation using Markov random field models", *Pattern Recognition*, vol. 33, p. 587–602, 2000.

[BER 66] BERRY B., Essay on commodity flows and the spatial structure of the Indian economy, Research paper num. 111, University of Chicago, Department of Geography, 1966.

[BES 74] BESAG J., "Spatial interaction and the statistical analysis of lattice systems", *Journal of the Royal Statistical Society*, vol. 35, p. 192–236, 1974.

[BES 86] BESAG J., "Spatial analysis of dirty pictures", *Journal of the Royal Statistical Society*, vol. 48, p. 259–302, 1986.

[BIL 92] BILLOIRE A., "Simulations de Monte Carlo en physique théorique", BOULEAU N., TALAY D., Eds., *Probabilités Numériques*, p. 145–162, INRIA, Paris, 1992.

[BOZ 81] BOZDOGAN H., Multi-Sample Cluster Analysis and Approaches to Validity Studies in Clustering Individuals, PhD thesis, Department of Mathematics, University of Illinois at Chicago, IL 60680, 1981.

[BUR 97] BURROUGH P., VAN GAANS P., HOOTMANS R., "Continuous classification in soil survey: spatial correlation, confusion and boundaries", *Geoderma*, vol. 77, p. 115–135, 1997.

[CEL 03] CELEUX G., FORBES F., PEYRARD N., "EM procedures using mean field-like approximations for Markov model-based image segmentation", *Pattern Recognition*, vol. 36, p. 131–144, 2003.

[CHA 89] CHALMOND B., "An iterative Gibbsian technique for reconstruction of m-ary images", *Pattern Recognition*, vol. 22, num. 6, p. 747–761, 1989.

[COC 95] COCQUEREZ J. P., PHILIPP S., *Analyse d'Images: Filtrage et Segmentation*, Masson, Paris, 1995.

[DAN 98] DANG M. V., Classification de Données Spatiales: Modèles Probabilistes et Critères de Partitionnement, PhD thesis, University of Technology, Compiègne, 1998, 250 pages.

[DUT 95] DUTREIX C., Eléments d'écologie des peuplements des Zygénides en région Bourgogne (Lepidoptera Zygaenidae), PhD thesis, University of Bourgogne, 1995.

[GEM 84] GEMAN S., GEMAN D., "Stochastic relaxation, Gibbs distributions, and the Baysian restoration of images", *IEEE Transactions on Pattern Analysis and Machine Intelligence*, vol. PAMI-6, p. 721–741, 1984.

[GEM 87] GEMAN S., GRAFFIGNE C., "Markov random field image models and their applications to computer vision", GLEASON A. M., Ed., *Proceedings of the International Congress of Mathematicians: Berkeley, August 3–11*, p. 1496–1517, 1987.

[GOV 90] GOVAERT G., "Classification binaire et modèles", *Revue de Statistique Appliquée*, vol. 38, num. 1, p. 67–81, 1990.

[JAI 91] JAIN A., FARROKHNIA F., "Unsupervised texture segmentation using Gabor filters", *Pattern Recognition*, vol. 24, num. 12, p. 1167–1186, 1991.

[LAK 89] LAKSHMANAN S., DERIN H., "Simultaneous parameter estimation and segmentation of Gibbs random fields", *IEEE Transactions on Pattern Analysis and Machine Intelligence*, vol. 11, p. 799–813, 1989.

[LEB 78] LEBART L., "Programme d'agrégation avec contraintes (C.A.H. contiguïté)", *Cahier de l'Analyse des Données*, vol. 3, p. 275–287, 1978.

[LEG 87] LEGENDRE P., "Constrained clustering", *Developments in Numerical Ecology*, vol. G 14, p. 289–307, Springer, 1987.

[LI 95] LI S., *Markov Random Field Modeling in Computer Vision*, Springer-Verlag, New York, 1995.

[MAR 87] MARROQUIN J., MITTER S., POGGIO T., "Probabilistic solution of illed posed problems in computational vision", *Journal of the American Statistical Association*, vol. 82, p. 76–89, 1987.

[MAR 93] MARROQUIN J., GIROSI F., Least squares quantization in PCM's, A.I. Memo num. 1390, Massachusetts Institute of Technology Artificial Intelligence Laboratory, 1993.

[MAS 93] MASSON P., PIECZINSKY W., "SEM algorithm and unsupervised statistical segmentation of satellite images", *IEEE Transactions on Geoscience and Remote Sensing*, vol. 31, num. 3, p. 618–633, 1993.

[OLI 89] OLIVER M., WEBSTER R., "A geostatistical basis for spatial weighting in multivariate classification", *Mathematical Geology*, vol. 21, p. 15–35, 1989.

[OPE 77] OPENSHAW S., "A geographical solution to scale and aggregation problems in region-building, partitioning and spatial modelling", *Transactions of the Institute of British Geographers*, vol. 2, p. 459–472, 1977.

[PIE 94] PIECZYNSKI W., "Champs de Markov cachés et estimation conditionnelle itérative", *Traitement du Signal*, vol. 11, num. 2, p. 141–153, 1994.

[SCH 78] SCHWARZ G., "Estimating the number of components in a finite mixture model", *Annals of Statistics*, vol. 6, p. 461–464, 1978.

[STR 77] STRAUSS D., "Clustering on coloured lattices", *Journal of Applied Probability*, vol. 14, p. 135–143, 1977.

[WU 95] WU C., DOERSCHUK P., "Cluster expansions for the deterministic computation of Bayesian estimators based on Markov random fields", *IEEE Transactions on Pattern Analysis and Machine Intelligence*, vol. 17, num. 3, p. 275–293, March 1995.

[YOU 88] YOUNES L., "Estimation and annealing for Gibbsian fields", *Annales de l'Institut Henri Poincaré*, vol. 24, num. 2, p. 269–294, 1988.

[YOU 89] YOUNES L., "Parametric inference for imperfectly observed Gibbsian fields", *Probability Theory and Related Fields*, vol. 82, p. 625–645, 1989.

[ZHA 92] ZHANG J., "The mean field theory in EM procedures for Markov random fields", *IEEE Transactions on Signal Processing*, vol. 40, num. 10, p. 2570–2583, October 1992.

List of Authors

Christophe AMBROISE
Laboratoire statistique et génome
CNRS
Evry Val d'Essonne University
France

Gérard D'AUBIGNY
Équipe méthodologie statistique et sciences sociales
Pierre Mendès France University
Grenoble
France

Philippe BESSE
Laboratoire de statistique et probabilités
CNRS
Paul Sabatier University
Toulouse
France

Hervé CARDOT
Institut de mathématiques de Bourgogne
CNRS
University of Bourgogne
Dijon
France

Henri CAUSSINUS
Laboratoire de statistique et probabilités
CNRS
Paul Sabatier University
Toulouse
France

Gilles CELEUX
Projet Select
INRIA Saclay Île-de-France
Orsay
France

Mo DANG
Laboratoire Heudiasyc
CNRS
Compiègne University of Technology
France

Gérard GOVAERT
Laboratoire Heudiasyc
CNRS
Compiègne University of Technology
France

Mohamed NADIF
Laboratoire d'informatique
Paris Descartes University
France

Ndèye NIANG
Conservatoire national des arts et métiers
Paris
France

Jérôme PAGÈS
Laboratoire de mathématiques appliquées
Agrocampus Ouest
Rennes
France

Anne RUIZ-GAZEN
Laboratoire de statistique et probabilités
CNRS
Paul Sabatier University
Toulouse
France

Gilbert SAPORTA
Conservatoire national des arts et métiers
Paris
France

Index